Multinary Alloys Based on III-V Semiconductors

Multinary Alloys Based on III-V Semiconductors

Vasyl Tomashyk

CRC Press is an imprint of the
Taylor & Francis Group, an **informa** business

CRC Press
Taylor & Francis Group
6000 Broken Sound Parkway NW, Suite 300
Boca Raton, FL 33487-2742

First issued in paperback 2020

ISBN-13: 978-1-4987-7833-6 (hbk)
ISBN-13: 978-0-367-78063-0 (pbk)

Visit the Taylor & Francis Web site at
http://www.taylorandfrancis.com

and the CRC Press Web site at
http://www.crcpress.com

This book is dedicated to the memory of my best friend, Dr Sci. Petro Feychuk (1948–2008).

Contents

List of Symbols and Acronyms

Bu^n *n*-butyl, C_4H_9
Bu^t *t*-butyl, C_4H_9
***c*-** cubic
DMF dimethylformamide
DSC differential scanning calorimetry
DTA differential thermal analysis
EDX energy-dispersive X-ray spectroscopy
EPMA electron probe microanalysis
Et ethyl, C_2H_5
G gas
HRTEM high resolution transmission electron microscopy
L liquid
LPE liquid phase epitaxy
MBE molecular beam epitaxy
M mol
Me methyl, CH_3
MOCVD metal-organic chemical vapor deposition
MOVPE metal-organic vapor-phase epitaxy
ppm parts per million
Pr^i *i*-propyl, C_3H_7
S solid
T temperature (in K)
TG thermogravimetry
THF tetrahydrofurane, C_4H_8O
(X) solid solution based on X
XRD X-ray diffraction
XPS X-ray photoelectron spectroscopy

Preface

A significant volume of semiconductor devices and circuits employs III-V semiconductors, the most commonly used crystal material, after Si and Ge, for integrated circuits. For electronic applications, these semiconductors offer the basic advantage of higher electron mobility, which translates into higher operating speeds. In addition, devices made with III-V semiconductor compounds provide lower-voltage operation for specific functions, radiation hardness (especially important for satellites and space vehicles), and semi-insulating substrates (avoiding the presence of parasitic capacitance in switching devices). Among these semiconductors, nitrides, which have recently been intensively studied, occupy a special place. Group III metal nitrides have been widely explored materials for electronic applications. Formation of quaternary and multinary alloys in the wurtzite or sphalerite phase offers a promising perspective of tuning their direct band gaps in a very wide region that is crucial for optoelectronics. In contrast to II–VI semiconductors, B(Al,Ga,In)N alloys represent a special class of materials with unique properties due to several remarkable properties that makes them attractive and suitable for reliable applications. The intrinsic group of superhard materials includes cubic boron nitride and some quaternary and multinary compounds, which possess an innate hardness. Multinary silicon aluminum oxynitrides are technical important materials for high-temperature engineering applications. Sialons are well known to exhibit useful physical properties, such as high strength, good wear resistance, high decomposition temperature, good oxidation resistance, excellent thermal shock properties, and resistance to corrosive environments. They have traditionally been investigated with an aim towards applications such as cutting tools, burners, welding nozzles, heat exchanger, and engine applications. AlN is also a ceramic material with mechanical and chemical properties suitable for use in structural applications at elevated temperatures.

Though ternary and quaternary phase equilibria in the systems based on boron, aluminum, gallium, and indium pnictides were collected and published in the handbooks *Ternary Alloys Based on III-V Semiconductors* by V. Tomashyk (Taylor & Francis, 2017) and *Quaternary Alloys Based on III-V Semiconductors* by V. Tomashyk (Taylor & Francis, 2018), data pertaining to the phase equilibria in multinary systems based on these semiconductor compounds are preferentially dispersed in the scientific literature. This reference book is intended to illustrate an up-to-date experimental and theoretical information about phase relations based on III-V semiconductor systems with five and more components. This book also practically contains all multinary compounds, including almost 250 minerals, in which compositions there are B, Al, Ga, or In and N, P, As, or Sb as such materials could be used as precursors at obtaining nanosized III-V semiconductors. For example, borophosphates have been synthesized and researched for more than 20 years. They have received much attention because of their variable structural chemistry and potential concerning applications in the fields of catalysis and optics, sorption and separation, photonic technologies, ion exchange, and so on. Metallic glasses which include boron and phosphorus in their compositions are also presented in this book.

This book includes the data about 498 multinary systems based on III-V semiconductors, including literature data from 897 papers; these data are illustrated in 21 figures. The information is divided into nine chapters: the first seven chapters are devoted to the multinary systems based on BN, BP, BAs, AlN, AlP, AlAs, and AlSb compounds (BSb compound does not exist). The last chapters include data on the system based on gallium and indium pnictides, respectively. The chapters are structured so that first, the Group III element numbers in the periodic system are presented in increasing order, that is, from B to In compounds, and then the pnictogen numbers are given in increasing order, that is, from nitrides to antimonides. The same principle is used for further description of the systems in every chapter, that is, in increasing order initially from the second, then from the third (fourth, fifth, and so on) components number in the periodic system (the first component is B, Al, Ga, or In and the last component is N, P, As, or Sb).

The missing and new finding concerning the ternary and quaternary systems based on the III-V semiconductors are given in Appendices A and B, respectively.

This book will be helpful for researchers in industrial and national laboratories and universities, and graduate students majoring in materials science and engineering, and solid-state chemistry. It will also be suitable for phase relation researchers, inorganic chemists, and semiconductor and solid-state physicists.

Author

Professor Vasyl Tomashyk is the executive director and head of the department of V.Ye. Lashkaryov Institute for Semiconductor Physics of the National Academy of the Sciences of Ukraine and graduated from Chernivtsi State University in Ukraine in 1972 (master of chemistry). He is a doctor of chemical sciences (1992), professor (1999), and author of about 640 publications in scientific journals and conference proceedings and ten books (five of them published by CRC Press), which are devoted to physical–chemical analysis, the chemistry of semiconductors, and chemical treatment of semiconductor surfaces.

Tomashyk is a specialist at a high international level in the field of solid-state and semiconductor chemistry, including physical–chemical analysis and the technology of semiconductor materials. He was the head of research topics within the International program "Copernicus." He is a member of Materials Science International Team (Stuttgart, Germany, since 1999) that prepares a series of prestigious reference books under the title *Ternary alloys* and *Binary alloys*, and has published 35 chapters in the Landolt–Börnstein New Series. Tomashyk works actively with young researchers and graduate students, and under his supervision, 20 PhD theses were prepared. For many years, he was also a professor at Ivan Franko Zhytomyr State University in Ukraine.

CHAPTER 1

Systems Based on BN

1.1 BORON–HYDROGEN–LITHIUM–CARBON–NITROGEN

The Li[**BH_3(CN)**] quinary compound is formed in the B–H–Li–C–N system (Wittig and Raff 1951). It was obtained by the interaction of Li[BH_4] with excess of HCN at 100°C. This compound precipitates from ether solution in well-formed crystals with the addition of 2 M of dioxane.

1.2 BORON–HYDROGEN–SODIUM–CARBON–NITROGEN

Some compounds are formed in the B–H–Na–C–N quinary system. **$NaBH_2(CN)_2$** was obtained as follows (Spielvogel et al. 1984). To a solution of $C_6H_5NH_2{\cdot}BH_2CN$ (0.15 mM) in 400 mL of dry tetrahydrofuran in a 1 L flask was added NaCN (0.15 mM), followed by an additional 100 mL of dry tetrahydrofuran under N_2. The mixture was stirred at room temperature for 0.5 h and then heated at reflux for 72 h, at which time the mixture had turned to a slightly grayish color. The mixture was cooled and filtered under N_2. The solvent was removed under vacuum to give a semisolid material. To this crude $NaBH_2(CN)_2$ was added CH_2Cl_2 until it was barely cloudy, and the solution was kept in the refrigerator to give extremely hygroscopic white needles of the title compound.

$NaBH_3CN$ is very hygroscopic and has an experimental density of 1.999 ± 0.005 g cm^{-3} (Wade et al. 1970). It was prepared by the next way. To the three-necked round-bottom flask was added tetrahydrofuran (1 L) and 98.5% $NaBH_4$ (2.09 M). The flask was then purged with dry N_2. HCN (16.7%) in tetrahydrofuran (2.33 M of 98% HCN) was placed in the dropping funnel. The system was repurged briefly. The HCN solution was added slowly to the rapidly stirred slurry of $NaBH_4$ at room temperature. Evolution of H_2 occurred slowly as soon as addition began. The flask was maintained at about 25°C in a water bath, since the reaction is slightly exothermic. The mixture was poststirred for 1 h and then gradually heated to reflux until H_2 evolution ceased.

When cooled to room temperature, the reaction mixture was purged with N_2 and briefly evacuated. The small amount of undissolved solid was removed by filtration. The clear, faintly yellow filtrate was dried on a rotary vacuum evaporator. Approximately half of the solvent was removed before heat was applied; drying was finished at 60°C under vacuum. White solid $NaBH_3CN$ was recovered.

In other preparations, the reaction mixture was never taken above room temperature (Wade et al. 1970). Subsequent examination showed that these unrefluxed solutions contained two anionic species. The normal cyanotrihydroborate ($NaBH_3CN$) predominated, but sodium isocyanotrihydroborate (**$NaBH_3NC$**) was also present. The isomer mixture is isolated by vacuum stripping the tetrahydrofuran at room temperature.

$NaBH_3CNBH_3$ was synthesized when the dried NaCN (3.5 g) was added to tetrahydrofuran (25 mL) (Wade et al. 1970). To the stirred slurry of NaCN in tetrahydrofuran, 1 M (BH_3·tetrahydrofuran) was added. The reaction flask was maintained at 25°C with a water bath. After all the (BH_3·tetrahydrofuran) was added, the reaction mixture was stirred for 5 h, during which time the NaCN dissolved. The solution was then filtered to remove suspended particles, and the solvent removed in vacuum at 50°C. A white hygroscopic solid was isolated, which was identified as $NaBH_3CNBH_3$.

$Na_2[B(CN)_3]·3NH_3$ crystallizes in the orthorhombic structure with the lattice parameters $a = 799.41 \pm 0.06$, $b = 1{,}062.56 \pm 0.09$, $c = 1{,}153.36 \pm 0.08$ pm at 110 K and a calculated density of 1.261 g cm^{-3} (Bernhardt et al. 2011a, 2011b). Yellow crystals of this compound were obtained from ammonia at 0°C.

1.3 BORON–HYDROGEN–SODIUM–PHOSPHORUS–OXYGEN–MANGANESE–NITROGEN

The **$Na_5(NH_4)Mn_3[B_9P_6O_{33}(OH)_3]·1.5H_2O$** multinary compound, which crystallizes in the hexagonal structure with the lattice parameters $a = 1{,}193.31 \pm 0.02$ and $c = 1{,}212.90 \pm 0.04$ pm and a calculated density of 2.635 g cm^{-3}, is formed in the B–H–Na–P–O–Mn–N system (Lin et al. 2008b). Transparent, colorless single crystals of the title compound were synthesized hydrothermally from a mixture of H_3BO_3 (32.2 mM), $Mn(CH_3COO)_2·4H_2O$ (3 mM), $(NH_4)_2HPO_4$ (6.4 mM), NaF (5 mM), and water (133.4 mM). The educt mixture was transferred into a Teflon-lined stainless steel autoclave (25 mL) and kept at 240°C for 5 days. The autoclave was cooled down to ambient temperature by removing out of the oven. The reaction products were washed with hot distilled water (60°C) until the boric acid was completely removed. Finally, the solids were dried in air at 60°C.

1.4 BORON–HYDROGEN–POTASSIUM–NEPTUNIUM–OXYGEN–NITROGEN

The **$K_2[(NpO_2)_3B_{10}O_{16}(OH)_2(NO_3)_2]$** multinary compound, which crystallizes in the monoclinic structure with the lattice parameters $a = 659.9 \pm 0.5$, $b = 1{,}602.6 \pm 1.2$, $c = 1{,}105.3 \pm 0.9$ pm, and $\beta = 90.922 \pm 0.013°$ at 100 K, is formed in the B–H–K–Np–O–N system (Wang et al. 2010). It was prepared using the next procedure. $NpCl_5$ (10 mg in 30 μL of water), $Th(NO_3)_4·5H_2O$ (22.7 mg), KNO_3 (9.8 mg), and H_3BO_3 (47.4 mg) were mixed together in a Teflon-lined autoclave. The mixture was heated at 220°C for 3 days, then cooled at 1°C h^{-1} to 160°C, and quenched. The product consisted of a colorless glass of thorium borate coated with pale pink crystals of the title compound.

1.5 BORON–HYDROGEN–POTASSIUM–CARBON–NITROGEN

The **$K[BH(CN)_3]$** quinary compound is formed in the B–H–K–C–N system (Landmann et al. 2015). It could be obtained by the following three methods.

1. A three-necked round-bottom flask equipped with a mechanical stirrer and a connection tube to an Ar line was charged with potassium (607 mM) and liquid NH_3 (250 mL) at –78°C. $K[BF(CN)_3]$ (279 mM) was carefully added in several portions for 12 h. In the course of the addition, the reaction mixture turned green. The mixture was stirred for an additional 1 h at –78°C. Then the mixture was warmed to room temperature within 5 h and NH_3 was allowed to evaporate. The residue was dried in a vacuum and subsequently washed with tetrahydrofuran (3 × 100 mL) in an Ar atmosphere. The solid remainder was suspended in tetrahydrofuran (500 mL) and a mixture of water (11 mL) in tetrahydrofuran (50 mL) was slowly added, resulting in the decolorization of the reaction mixture. The mixture was dried with K_2CO_3 and the solid material was filtered off. The volume of the solution was reduced to 50 mL and colorless title compound precipitated upon addition of CH_2Cl_2.
2. $K[BF(CN)_3]$ (170 mM) was placed in a three-necked round-bottom flask equipped with a mechanical stirrer, and it was dissolved in tetrahydrofuran (500 mL). Potassium (63.9 mM) was added in small pieces and subsequently a solution of naphthalene (15.6 mM) in tetrahydrofuran (10 mL) was added in a single portion. The suspension was stirred for 6 h at room temperature. Excess potassium was separated from the green suspension and washed with tetrahydrofuran (2 × 50 mL). A mixture of water (333 mM) and tetrahydrofuran (15 mL) was slowly added until the suspension became colorless. The suspension was filtered off, washed with tetrahydrofuran (2 × 20 mL), and dried over K_2CO_3. The volume of the combined organic layers was reduced to 50 mL and colorless $K[BH(CN)_3]$ was precipitated upon addition of CH_2Cl_2.
3. $Na[BF(CN)_3]$ (28.6 mM) was dissolved in liquid NH_3 (40 mL) at –78°C. Sodium (57.4 mM) was carefully added to the reaction mixture. Subsequently, the mixture was allowed to warm to room temperature, which resulted in the evaporation of NH_3. The residue was dissolved in a mixture of H_2O (50 mL) and tetrahydrofuran (200 mL) at 0°C and K_2CO_3 (~5 g) was added until two layers had formed. The aqueous layer was separated and saturated with K_2CO_3 (~50 g). The viscous solution was extracted with tetrahydrofuran (3 × 50 mL). The combined organic layers were dried with K_2CO_3 and concentrated to a volume of 10 mL. A colorless $K[BH(CN)_3]$ was precipitated upon addition of CH_2Cl_2.

1.6 BORON–HYDROGEN–SILVER–CARBON–NITROGEN

Some compounds are formed in the B–H–Ag–C–N quinary system (Györi et al. 1983). **$AgBH(NC)_3$** was synthesized as follows. To a solution of AgCN (300 mM) in Me_2S (120 mL), a 2.00 M solution of $Me_2S{\cdot}BHBr_2$ (50.0 mL, 100.0 mM) was added during 25 min with stirring and cooling with cold water. After a further 30 min stirring, the mixture was filtered and the yellow-colored lower oily layer was separated and added dropwise to diglyme (200 mL) at 80°C–85°C for 2 h in a fast N_2 stream to remove Me_2S. The powdery product was filtered off, washed with ether (2 × 40 mL), dried, and suspended in water (80 mL). To this suspension was added a solution of KCN (479.9 mM) in water (50 mL). The suspension was stirred for 15 min, filtered, dried in N_2 stream, and kept under reduced pressure (1 Pa) for 1.5 h at 110°C to give the title compound.

$AgBH_2(NC)_2$ was obtained as a solution of AgCN (136.38 mL) in Me_2S (20 mL) was added to a 2.80 M solution of $Me_2S{\cdot}BH_2Br$ (67.2 mM) in Me_2S. The crystals that separated were filtered off, washed with Me_2S (2 × 4 mL), and dried. The filtrate was concentrated to 25 mL, and the solid that precipitated was filtered off, washed with Me_2S (2 × 4 mL), and dried. The filtrate was added dropwise to 300 mL of refluxing ether. The separated material was filtered off and dried. The combined solid product fractions were suspended in water (25 mL) and then stirred with an aqueous solution (15 mL) of KCN (150.0 mM) with stirring. After 15 min, the mixture was filtered, washed with water (5 × 10 mL), and dried under reduced pressure in a N_2 stream to a constant weight, yielding $AgBH_2(NC)_2$.

For preparing **$AgBH_n(NC)_{4-n}$** (n = 1, 2), to an aqueous solution (15 mL) of $LiBH_2(CN)_2{\cdot}C_4H_8O_2$ (5.0 mM) or $NaBH(CN)_3$ (5.0 mM) was added a 1.0 M aqueous $AgNO_3$ solution (0.5 mL). After 0.5 h stirring, the mixture was treated with carbon, then filtered, and 1.0 M $AgNO_3$ solution (4.5 mL) was added to the filtrate. The precipitated Ag salts were filtered off, washed with water (3 × 3 mL), dried, and purified, by extraction with MeCN.

All experiments were carried out in dry oxygen-free solvents under dry N_2 using the Schlenk technique.

1.7 BORON–HYDROGEN–BERYLLIUM–PHOSPHORUS–OXYGEN–NITROGEN

The **$NH_4Be[BP_2O_8]{\cdot}0.33H_2O$** multinary compound, which crystallizes in the cubic structure with the lattice parameters a = 1,246.42 ± 0.04 pm, is formed in the B–H–Be–P–O–N system (Zhang et al. 2003). In a typical reaction, a mixture of $Be(OH)_2$, H_3BO_3, 85% H_3PO_4, $(NH_4)H_2PO_4$, and distilled water were mixed with stirring for 1 h, until the components were dissolved completely, and the pH of the mixture was 2.0. After stirring at room temperature for 2 h, the clear solution composed of $1Be(OH)_2/2H_3BO_3/8H_3PO_4/1(NH_4)H_2PO_4/70H_2O$ was transferred to a Teflon-coated steel autoclave and heated at 180°C for 4 days. The crystalline products were recovered by filtration and washed with distilled water. Transparent, nearly globular crystals of the title compound were obtained.

1.8 BORON–HYDROGEN–MAGNESIUM–PHOSPHORUS–OXYGEN–NITROGEN

The **$NH_4Mg(H_2O)_2[BP_2O_8]{\cdot}H_2O$** multinary compound, which crystallizes in the hexagonal structure with the lattice parameters a = 952.9 and c = 1,573.6 pm, is formed in the B–H–Mg–P–O–N system (Baykal et al. 2000). It was synthesized under hydrothermal conditions. The hydrothermal synthesis of this compound was started from a mixture of $MgCl_2{\cdot}6H_2O$, H_3BO_3, and $(NH_4)_2HPO_4$ (molar ratio 3:1.5:5.5). Required amount of H_2O and concentrated HNO_3 was added to dissolve this solid mixture at 90°C, and then the total volume was reduced to 10 mL by evaporation of water. The highly viscous solution (pH < 1) was filled into Teflon-coated autoclave (degree of filling 60%) and treated at 160°C for 3 days.

1.9 BORON–HYDROGEN–ZINC–PHOSPHORUS–OXYGEN–NITROGEN

The **$NH_4Zn[BP_2O_8]$** and **$(NH_4)_{16}Zn_{16}[B_8P_{24}O_{96}]$** multinary compounds are formed in the B–H–Zn–P–O–N system. The first of them crystallizes both in the monoclinic [a = 1,284.8 ± 0.3, b = 1,289.6 ± 0.3, c = 851.9 ± 0.3 pm, and β = 91.024 ± 0.003° (Kniep et al. 2000)] and in the triclinic [a = 743.7 ± 0.1, b = 761.2 ± 0.1, and c = 785.0 ± 0.1 pm and α = 119.05 ± 0.02°, β = 101.59 ± 0.01°, and γ = 103.43 ± 0.01° and a calculated density of 2.687 g cm^{-3} (Kniep et al. 1999b, 1999c)] structure.

Monoclinic modification of the title compound was prepared by solvothermal reaction (with ethylene glycol) from a mixture of ZnO (0.700 g), B_2O_3 (0.298 g), $(NH_4)_2HPO_4$ (2.264 g) (molar ratio 2:1:4), and 5 mL of 85% H_3PO_4 (pH = 1.5) in a Teflon autoclave at 165°C (Kniep et al. 2000). Triclinic modification was obtained under mild hydrothermal conditions in a Teflon autoclave (20 mL). For this, ZnO (2.33 g), B_2O_3 (0.99 g), and $(NH_4)_2HPO_4$ (7.55 g) (molar ratio 2:1:4) were heated at about 80°C in demineralized water (10 mL) and treated under stirring with 85% H_3PO_4 (6 mL), until the components had dissolved completely. The clear solutions were concentrated to about 15–20 mL to give a highly viscous gel. The pH value was between 1 and 1.5. The highly viscous gel was transferred to the Teflon autoclave (degree of filling of about 70%) and stored for 3–4 days at about 170°C. The crystalline reaction product was filtered off in vacuum, washed with demineralized water, and dried at 60°C. Single crystals grow to a length of about 0.8 mm under these conditions.

$(NH_4)_{16}Zn_{16}[B_8P_{24}O_{96}]$ crystallizes in the cubic structure with the lattice parameters a = 1,317.3 ± 0.6 pm and a calculated density of 2.688 g cm^{-3} (Yang et al. 2005). This compound was synthesized by using boric acid as a flux at the temperature slightly above the melting point of boric acid (170°C). Typically, H_3BO_3 (1.0 g), $Zn(CH_3COO)_2{\cdot}2H_2O$ (0.315 g), and $(NH_4)_2HPO_4$ (0.423 g) were added into a 15 mL Teflon-lined stainless steel autoclave directly. The vessel was heated under autogenous pressure at 240°C for 5 days, and then cooled to room temperature. The final product, containing large single crystals in the form of colorless cubes, was washed with hot water (50°C) until the residual boric acid was completely removed, and then dried in air. After the calcination of this compound at 800°C for 3 h, the final products are the mixture of $Zn_2P_2O_7$ and BPO_4.

1.10 BORON–HYDROGEN–ZINC–PHOSPHORUS–OXYGEN–COBALT–NITROGEN

The **$(NH_4)_{16}Zn_{13.4}Co_{2.6}[B_8P_{24}O_{96}]$** and **$NH_4(Zn_{1-x}Co_x)[BP_2O_8]{\cdot}(0 \leq x \leq 0.14)$** multinary phases are formed in the B–H–Zn–P–O–Co–N system. The first of them crystallizes in the cubic structure with the lattice parameters a = 1,315.58 ± 0.15 pm and a calculated density of 2.686 g cm^{-3} (Yang et al. 2005). It was prepared by using boric acid as a flux at the temperature slightly above the melting point of boric acid (170°C). Typically, $Zn(CH_3COO)_2{\cdot}2H_2O$, $Co(NO_3)_2{\cdot}6H_2O$, $(NH_4)_2HPO_4$, and H_3BO_3 (molar ratio 0.8:0.2:2.2:11.2) were added into a 15 mL Teflon-lined stainless steel autoclave directly. The vessel was heated under autogenous pressure at 240°C for 5 days, and then cooled to room temperature. The final product, containing large blue cubic-like crystals, was washed with hot water (50°C) until the residual boric acid was completely removed, and then dried in air. After the calcination of this compound at 800°C for 3 h, the final products are the mixture of $Zn_2P_2O_7$, $Co_2P_2O_7$, and BPO_4.

$NH_4(Zn_{1-x}Co_x)[BP_2O_8]$ ($0 \leq x \leq 0.14$) crystallizes in the triclinic structure with the lattice parameters $a = 743.19 \pm 0.11$, $b = 759.97 \pm 0.05$, and $c = 784.02 \pm 0.02$ pm and $\alpha = 118.9940 \pm 0.0009°$, $\beta = 101.6597 \pm 0.0009°$, and $\gamma = 103.4308 \pm 0.0014°$ and a calculated density of 2.688 g cm^{-3} at $x = 0.12$ (Schäfer et al. 2000). To obtain these phases, reactions were carried out with gels of molar composition yCoO/(1 – y)ZnO/0.5B_2O_3/2$(NH_4)_2HPO_4$ with $y = 0$, 0.2, 0.4, 0.6, 0.8, and 1. In a typical reaction ($y = 0.2$), ZnO (1.860 g), CoO (0.428 g), $(NH_4)_2HPO_4$ (7.546 g), and B_2O_3 (0.995 g) were mixed with 15 mL distilled water. A 10 mL of 85% H_3PO_4 was slowly added with stirring and then the total volume of the mixture was reduced to approximately 16 mL by heating at 100°C. The resulting gel was then held at 165°C for 3 days in a Tefon autoclave (20 mL; degree of filling 80%; pH = 1, 5). After filtration, the solid product was dried at 60°C for 1 day. The title compound was obtained as a pale blue powder. It is stable up to 500°C.

1.11 BORON–HYDROGEN–CADMIUM–PHOSPHORUS–OXYGEN–NITROGEN

The **$NH_4Cd(H_2O)_2(BP_2O_8)·0.72H_2O$** multinary compound, which crystallizes in the hexagonal structure with the lattice parameters $a = 969.78 \pm 0.16$, $c = 1{,}602.6 \pm 0.4$ pm and a calculated density of 2.886 g cm^{-3}, is formed in the B–H–Cd–P–O–N system (Ge et al. 2005). The title compound was synthesized under mild hydrothermal conditions in a Teflon-lined stainless steel autoclave. In a typical synthesis, the starting materials $CdCl_2·2.5H_2O$ (0.459 g), $(NH_4)HB_4O_7·3H_2O$ (2.840 g), $(NH_4)_2HPO_4$ (2.640 g), 85% H_3PO_4 (3.5 mL), and H_2O (10 mL) were used to prepare a mixture. The mixture was heated under stirring until the components were dissolved into a clear solution. A proper amount of aqueous ammonia and H_3PO_4 were added to adjust the pH at 2.0–2.5. Then, the reactants were transferred to an autoclave (degree of filling 70%) and stored at 150°C for 6 days and slowly cooled at a rate of 1°C min^{-1} to –30°C. The so-obtained crystalline products were washed in hot distilled water (about 60°C) and dried in air. The pH value of 2.0–2.5 is rather important in obtaining pure compounds.

1.12 BORON–HYDROGEN–ALUMINUM–PHOSPHORUS–OXYGEN–NITROGEN

The **$NH_4Al[BP_2O_8(OH)]$** multinary compound, which crystallizes in the monoclinic structure with the lattice parameters $a = 923.4 \pm 0.2$, $b = 837.0 \pm 0.1$, $c = 941.3 \pm 0.2$ pm, and $\beta = 103.670 \pm 0.007°$, is formed in the B–H–Al–P–O–N system (Mi et al. 2002b). The colorless prisms of this compound were synthesized under mild hydrothermal conditions. The reaction was carried out with a mixture of $NH_4H_2PO_4$ (2.300 g), $Al(H_2PO_4)_3$ (3.179 g), H_3BO_3 (0.618 g), and 85% H_3PO_4 (3 mL) in aqueous solution (molar ratio of NH_4/Al/B/P = 2:1:1:7.5). The mixture was filled in 20 mL Teflon autoclave (degree of filling about 50%) The autoclave was placed in an oven with subsequent heating at 140°C for 7 days.

1.13 BORON–HYDROGEN–GALLIUM–PHOSPHORUS–OXYGEN–NITROGEN

The **$NH_4Ga[BP_2O_8(OH)]$** multinary compound, which crystallizes in the monoclinic structure with the lattice parameters $a = 928.1 \pm 0.2$, $b = 829.2 \pm 0.2$, $c = 956.4 \pm 0.2$ pm, and $\beta = 102.65 \pm 0.03°$, is formed in the B–H–Ga–P–O–N system (Li et al. 2002b). It was synthesized under mild hydrothermal conditions. The reaction was carried out with a mixture of Ga (0.07 g) dissolved in 18% HCl with H_3BO_3 (0.309 g) and $NH_4H_2PO_4$ (1.150 g) (molar ratio of NH_4/Ga/B/P = 12:1:4:12) in aqueous solution. The mixture was sealed in glass tube (after adding 1 mL of H_2O to achieve a degree of filling of 30%) and subsequently heated at 130°C for 14 days.

1.14 BORON–HYDROGEN–INDIUM–PHOSPHORUS–OXYGEN–NITROGEN

The **$NH_4In[BP_2O_8(OH)]$** multinary compound, which crystallizes in the triclinic structure with the lattice parameters $a = 529.80 \pm 0.04$, $b = 848.80 \pm 0.04$, and $c = 839.01 \pm 0.05$ pm and $\alpha = 93.077 \pm 0.005°$, $\beta = 93.331 \pm 0.006°$, and $\gamma = 80.634 \pm 0.005°$, is formed in the B–H–In–P–O–N system (Mi et al. 2002e). It was prepared under mild hydrothermal conditions. The reaction was carried out with a mixture of In (0.057 g) dissolved in 18% HCl (1 mL), $(NH_4)_2HPO_4{\cdot}12H_2O$ (2.088 g), and H_3BO_3 (0.310 g) (molar ratio of In/B/P = 1:10:12) in aqueous solution. The mixture was sealed in glass tube (after adding 1 mL of H_2O to achieve a degree of filling of 30%) and subsequently heated at 140°C for 15 days.

1.15 BORON–HYDROGEN–CARBON–OXYGEN–NITROGEN

The **$[(MeCO_2)_2BNH_2]$** quinary compound, which melts at 144.5°C–146°C, is formed in the B–H–C–O–N system (Brennan et al. 1960). $B_3Cl_3N_3H_3$ reacts with anhydrous acetic acid to form this compound. Monoclinic ($a = 2,281$, $b = 1,041$, $c = 1,760$ pm, $\beta = 92°45'$ and an experimental density of 1.36 g cm^{-3}) and triclinic crystalline forms were obtained by recrystallization of the material from acetic anhydride and glacial acetic acid, respectively.

1.16 BORON–HYDROGEN–CARBON–SULFUR–NITROGEN

The **H_3NBH_2SCN** and **$B_3(SCN)_3N_3H_3$** quinary compounds are formed in the B–H–C–S–N system. H_3NBH_2SCN melts at 119°C with decomposition (Aftandilian et al. 1961). To obtain it, NH_4SCN (0.52 M) was dissolved in dimethoxyethane (200 mL). The solution was cooled to −80°C and B_2H_6 (263 mM) was added. On warming, H_2 (588 mM) was evolved. The solvent was then removed at reduced pressure to leave a nonvolatile residue that crystallized on standing to give a solid. This residue was taken up in 400 mL of dry ether and filtered; addition of petroleum ether to the filtrate gave the title compound.

$B_3(SCN)_3N_3H_3$ is nearly white, faintly yellow nonsublimable powder melting with decomposition in vacuum at 147°C–150°C (Brennan et al. 1960). Upon exposure to the atmosphere, it turns yellow.

1.17 BORON–HYDROGEN–TITANIUM–PHOSPHORUS–OXYGEN–NITROGEN

The $\mathbf{NH_4Ti_2[B(PO_4)_4]{\cdot}H_2O}$ multinary compound, which crystallizes in the tetragonal structure with the lattice parameters $a = 631.11 \pm 0.02$ and $c = 1{,}636.5 \pm 0.1$ pm, is formed in the B–H–Ti–P–O–N system (Zhou et al. 2011b). The synthesis of the title compound was performed under hydrothermal conditions. First, Ti (1.966 g) was dissolved in concentrated HCl (100 mL) and kept in a sealed bottle for several days to get a clear blue solution, which is named afterwards Ti*. Then, the mixture of Ti* (5 mL), H_3BO_3 (1.236 g), 85% H_3PO_4 (0.5 mL), and $NH_4H_2PO_4$ (0.861 g) (NH_4/Ti/B/P molar ratio 3.75:1:10:8.75) was transferred to a Teflon autoclave. The resulting pH values were lower than 1. The synthesis was conducted at 180°C for 3 days. After that, the autoclave was directly taken out of the oven. The product was filtered and washed with deionized water for several times and dried at 50°C for 8 h.

1.18 BORON–HYDROGEN–PHOSPHORUS–VANADIUM–OXYGEN–NITROGEN

The $\mathbf{NH_4V[BP_2O_8(OH)]}$ and $\mathbf{(NH_4)_5V_3[BP_3O_{19}]{\cdot}H_2O}$ multinary compounds are formed in the B–H–P–V–O–N system. The first of them crystallizes in the monoclinic structure with the lattice parameters $a = 942.5 \pm 0.2$, $b = 826.9 \pm 0.2$, $c = 969.7 \pm 0.2$ pm, and $\beta = 102.26 \pm 0.02°$ and a calculated density of 2.579 ± 0.001 g cm^{-3} (Kritikos et al. 2001). It which was prepared from a mixture of NH_4VO_3 (1.00 mM), H_3PO_4 (0.40 mL, 6.88 mM), $C_6H_{12}N_2$ (1,4-diazabicyclo[2,2,2]octane) (1.07 mM), H_3BO_3 (2.10 mM), and H_2O (2.00 mL, 111 mM) which was added to a 5 mL Teflon-lined stainless steel autoclave and heated under autogenous pressure (180°C–190°C, 3 days to 8 weeks). The initial pH of the solution was approximately 2.5 and, after completion of the reaction, the pH value decreased to 2.0. The title compound appeared as rod-shaped, pale green crystals. The yield increased with higher temperature, lower pH (adjustment with concentrated HCl), and prolonged reaction time. However, none of the reaction conditions tested gave a completely single phase of this compound. The crystalline borophosphate phase was mechanically separated from other phases using an optical microscope.

An alternative synthesis (Kritikos et al. 2001), with comparable yield, was obtained by using MeOH as reducing agent; NH_4VO_3 (1.00 mM), H_3PO_4 (0.50 mL, 7.31 mM), H_3BO_3 (2.10 mM), MeOH (0.2 mL, 4.93 mM), and H_2O (2.00 mL, 111 mM) were added to a 5 mL Teflon-lined stainless steel autoclave and heated under autogenous pressure (180°C–190°C, 3–7 days). Green, rod-shaped crystals were obtained.

$(NH_4)_5V_3[BP_3O_{19}]{\cdot}H_2O$ crystallizes in the triclinic structure with the lattice parameters $a = 898.61 \pm 0.07$, $b = 917.66 \pm 0.08$, and $c = 1{,}365.6 \pm 0.1$ pm and $\alpha = 82.763 \pm 0.001°$, $\beta = 89.169 \pm 0.002°$, and $\gamma = 68.154 \pm 0.001°$ and a calculated density of 2.066 g cm^{-3} (Bontchev et al. 2000). This compound forms easily at room temperature. In a typical reaction, VO_2 (6 mM), 85% H_3PO_4 (0.36 mL, 6 mM), H_3BO_3 (6 mM), and 30% NH_4OH (1.05 mL, 10 mM) were mixed with 10 mL of H_2O. Thirty percent H_2O_2 (1.0 mL, 30 mM) was added as an oxidizing agent dropwise with constant stirring. The solution was left at room temperature, and after 3 days, the solid residue consisted of bright orange-yellow crystals of the title compound up to several millimeters in size together with small amounts of H_3BO_3 and $VOPO_4{\cdot}2H_2O$ was obtained. The same product was obtained when V_2O_3 was used as vanadium source.

$(NH_4)_5V_3[BP_3O_{19}]{\cdot}H_2O$ could also be obtained by oxidation of an aqueous solution of NH_4-vanadium borophosphate electrochemically or by using aqueous hydrogen peroxide (Bontchev et al. 2000). In a typical electrolysis reaction, 0.01 M solution of NH_4-vanadium borophosphate (40 mL, pH = 7.34) was electrolyzed at 1.5 V and 0.1 mA for 24 h. During this time, the solution changed color from blue to green and then to yellow and the pH dropped to 3.56. Bright orange-yellow crystals of the title compound formed after 3 days together with small amounts of H_3BO_3 and $VOPO_4{\cdot}2H_2O$. Similarly, oxidation of a solution of NH_4-vanadium borophosphate (0.10 g) in 10 mL of H_2O with an aqueous solution of 30% H_2O_2 (0.5 mL) also gave this compound. The blue solution rapidly changed its color to yellow, and the pH changed from 7.28 to 3.32. Crystals of the title compound formed on evaporation.

1.19 BORON–HYDROGEN–PHOSPHORUS–VANADIUM–OXYGEN–IRON–NITROGEN

The $\mathbf{NH_4(Fe_{0.53}V_{0.47})[BP_2O_8(OH)]}$ multinary phase, which crystallizes in the monoclinic structure with the lattice parameters $a = 940.4 \pm 0.4$, $b = 831.6 \pm 0.5$, $c = 970.6 \pm 0.3$ pm, and $\beta = 102.28 \pm 0.04°$ and a calculated density of 2.591 ± 0.001 g cm^{-3}, is formed in the B–H–P–V–O–Fe–N system (Kritikos et al. 2001). To prepare this compound, a mixture of NH_4VO_3 (0.710 mM), $FeCl_3$ (0.635 mM), H_3PO_4 (0.40 mL, 6.88 mM), MeOH (0.20 mL, 4.93 mM), $(NH_4)_2B_4O_7{\cdot}4H_2O$ (1.90 mM), and H_2O (2.00 mL, 111 mM) was added to a 5 mL Teflon-lined stainless steel autoclave. The pH was adjusted to 0.2 with concentrated HCl. The steel autoclave was heated under autogenous pressure at 180°C for 6–8 days. Yellow brown, rod-shaped, crystals of the title compound were obtained.

1.20 BORON–HYDROGEN–PHOSPHORUS–OXYGEN–NITROGEN

Three quinary compounds are formed in the B–H–P–O–N system. $\mathbf{NH_4[B_3PO_6(OH)_3]{\cdot}0.5H_2O}$ crystallizes in the triclinic structure with the lattice parameters $a = 436.65 \pm 0.02$, $b = 936.80 \pm 0.04$, and $c = 1{,}082.67 \pm 0.08$ pm and $\alpha = 81.532 \pm 0.009°$, $\beta = 85.369 \pm 0.009°$, and $\gamma = 83.641 \pm 0.008°$ and a calculated density of 1.808 g cm^{-3} (Liu and Zhao 2007). The title compound was prepared under solvothermal conditions. $(NH_4)_2B_4O_7$ (1.04 g), $NH_4H_2PO_4$ (0.9 g), and 5 mL of glycol were placed in a Teflon-lined stainless steel autoclave and heated to 130°C for 5 days, followed by cooling to room temperature. Colorless rod-shaped crystals were obtained.

$\mathbf{(NH_4)_2[B_3PO_7(OH)_2]}$ crystallizes in the monoclinic structure with the lattice parameters $a = 442.6 \pm 0.3$, $b = 1{,}277.2 \pm 0.5$, $c = 1{,}608.2 \pm 0.5$ pm, and $\beta = 100.60 \pm 0.01°$ (Hauf and Kniep 1996a). This compound was obtained by hydrothermal treatment of concentrated solutions of $NH_4H_2PO_4$, $(NH_4)_2HPO_4$, and B_2O_3 at 150°C.

$\mathbf{(NH_4)_4(H_2B_2P_4O_{16})}$ crystallizes in the tetragonal structure with the lattice parameters $a = 723.92 \pm 0.10$, $c = 1{,}487.0 \pm 0.3$ pm (Xing et al. 2010). It was prepared ionothermally from a mixture of ionic liquid (1-ethyl-3-methylimidazolium bromide), H_3BO_3, and $(NH_4)_2HPO_4$

(molar ratio 10:1:2) at 200°C for 5 days in a Teflon-lined stainless steel autoclave. The autoclave was subsequently allowed to cool to room temperature.

The stage-1 intercalation compound, which is characterized by the interplanar spacing of 690 pm, can be formed by simple drying of *h*-BN in H_3PO_4 (Kovtyukhova et al. 2013). To obtain it, 70–200 mg of *h*-BN was added to 0.5–1 mL of 85% H_3PO_4 in a glass vial, stirred with a glass stick and allowed to settle. The excess acid (required to ensure complete wetting of the *h*-BN powder) was then decanted away. A drop of the remaining thick *h*-BN/H_3PO_4 suspension was cast on a glass or Si wafer and dried in air at 120°C–250°C.

1.21 BORON–HYDROGEN–PHOSPHORUS–OXYGEN–FLUORINE–NITROGEN

The $\mathbf{NH_4BPO_4F}$ multinary compound, which crystallizes in the cubic structure with the lattice parameter $a = 759.13 \pm 0.09$ and a calculated density of 2.168 g cm^{-3}, is formed in the B–H–P–O–F–N system (Li et al. 2004). It was synthesized under mild conditions in a water-free flux of H_3BO_3 and $NH_4H_2PO_4$, both with low melting points of 185°C and 180°C, respectively. The reaction made of H_3BO_3, $NH_4H_2PO_4$, and NaF (molar ratio 3:3:2) was heated in a Teflon-lined stainless steel autoclave at 240°C. The product was washed with hot deionized water, and colorless truncated-cube-like crystals of NH_4BPO_4F were recovered as the only solid phase. This compound is stable in air up to 396°C.

1.22 BORON–HYDROGEN–PHOSPHORUS–OXYGEN–MANGANESE–NITROGEN

Some multinary compounds are formed in the B–H–P–O–Mn–N system. $\mathbf{(NH_4)_{0.5}Mn_{1.25}(H_2O)_2[BP_2O_8]\cdot 0.5H_2O}$ crystallizes in the hexagonal structure with the lattice parameters $a = 951.04$ and $c = 1{,}571.08$ pm (Birsöz et al. 2007). The title compound was synthesized under mild hydrothermal conditions starting with the precursor mixture as follows: $MnCl_2\cdot 2H_2O$ (4.876 g), H_3BO_3 (0.744 g), $(NH_4)_2HPO_4$ (2.384 g), and HCl (15 mL) were mixed in distilled water (15 mL) and treated under stirring until components dissolved completely. The clear solution was heated without boiling to reduce the volume to about 12 mL and then transferred to Teflon autoclaves (degree of filling 65%) and heated at 180°C for 7 days. The pale pink single phase was filtered off in a vacuum, washed with distilled water, and dried at 65°C.

$\mathbf{NH_4Mn(H_2O)_2BP_2O_8\cdot H_2O}$ also crystallizes in the hexagonal structure with the lattice parameters $a = 965.2 \pm 0.2$, $c = 1{,}579.2 \pm 0.5$ pm, and a calculated density of 2.547 g cm^{-3} (Shi et al. 2006). To obtain it, a mixture of $MnCl_2\cdot 4H_2O$, H_3BO_3, 85% H_3PO_4, NH_2CSNH_2, and H_2O (molar ratio 0.5:3:1:1:14) was loaded in a thick wall Pyrex tube, which was sealed under vacuum and heated at 110°C for 6 days. The product was filtered, washed with deionized water, and dried at room temperature.

Given the proximity of the chemical composition and parameters of the hexagonal cell, it is possible that these two compounds are in fact one compound.

$\mathbf{(NH_4)_4[Mn_9B_2(OH)_2(HPO_4)_4(PO_4)_6]}$ crystallizes in the monoclinic structure with the lattice parameters $a = 3{,}260.3 \pm 0.7$, $b = 1{,}061.7 \pm 0.2$, $c = 1{,}071.8 \pm 0.2$ pm, and $\beta = 108.26 \pm 0.03°$ and a calculated density of 2.971 g cm^{-3} at 173 ± 2 K (Yang et al. 2006b). It was prepared by a hydrothermal method. Typically, a reaction mixture of $Mn(CH_3COO)_2 \cdot 4H_2O$ (1 mM), H_3BO_3 (16.2 mM), $(NH_4)_2HPO_4$ (3.2 mM), and H_2O (55.6 mM) was loaded into a 15 mL Teflon-lined stainless steel autoclave directly and then heated under autogenous pressure at 200°C for 5 days. The final product containing light pink stick-like crystals was washed with hot water (50°C) until the residual boric acid was completely removed, and then it was dried in air.

$(NH_4)_6[Mn_3B_6P_9O_{36}(OH)_3] \cdot 4H_2O$ also crystallizes in the monoclinic structure with the lattice parameters a = 2,375.6 ± 0.4, b = 1,430.2 ± 0.2, c = 1,232.43 ± 0.19 pm, and β = 92.207 ± 0.002° and a calculated density of 2.089 g cm^{-3} (Yang et al. 2011a). The title compound was obtained by the boric acid flux method. Typically, a mixture of H_3BO_3 (16.2 mM), $MnCl_2 \cdot 4H_2O$ (1.0 mM), 85% H_3PO_4 (0.27 mL, 4.0 mM), and acetamide (8.5 mM) was directly added into a 15 mL Teflon-lined stainless steel autoclave and heated at 200°C for 7 days. The final product, as colorless transparent rod-like single crystals, was washed with hot water (50°C) until the residual H_3BO_3 was completely removed and then dried in air.

$(NH_4)_7Mn_4(H_2O)[B_2P_4O_{15}(OH)_2]_2[H_2PO_4][HPO_4]$ crystallizes in the orthorhombic structure with the lattice parameters a = 1,710.2 ± 0.2, b = 1,071.95 ± 0.15, and c = 2,228.3 ± 0.3 pm and a calculated density of 2.270 g cm^{-3} (Yang et al. 2012). It was synthesized by using a flux-like method. Typically, 1-alkyl-3-methylimidazolium bis(oxalato)borate (0.326 g), $Mn(CH_3COO)_2 \cdot 4H_2O$ (0.245 g), $NH_4H_2PO_4$ (0.403 g), and 85% H_3PO_4 (0.15 mL) were charged into a 15 mL Teflon-lined stainless steel autoclave directly. The vessel was heated under autogenous pressure at 160°C for 5 days and then cooled to room temperature. The final product containing large rectangular single crystals with light pink color was washed with distilled water for several times and then dried in air.

1.23 BORON–HYDROGEN–PHOSPHORUS–OXYGEN–IRON–NITROGEN

The **$(NH_4)_{0.4}Fe^{II}{}_{0.55}Fe^{III}{}_{0.5}(H_2O)_2[BP_2O_8] \cdot 0.6H_2O$** and **$NH_4Fe[BP_2O_8(OH)]$** multinary compounds are formed in the B–H–P–O–Fe–N system. The first of them has two polymorphic modifications, which crystallize in the hexagonal structure with the same X-ray diffraction (XRD) pattern. Its first modification has the next lattice parameters: a = 948.3 ± 0.4, c = 1,569.7 ± 0.5 pm, and a calculated density of 2.514 g cm^{-3} (Huang et al. 2001b). The second compound crystallizes in the monoclinic structure with the lattice parameters a = 937.0 ± 0.1, b = 830.9 ± 0.5, c = 968.0 ± 0.1 pm, and β = 102.05 ± 0.01° and a calculated density of 2.628 g cm^{-3} (Huang et al. 2001b) [a = 939.3 ± 0.3, b = 828.5 ± 0.2, c = 968.9 ± 0.2 pm, and β = 102.07 ± 0.04° and a calculated density of 2.618 ± 0.001 g cm^{-3} (Kritikos et al. 2001)]. The title compounds were synthesized under mild hydrothermal conditions (Huang et al. 2001b). $FeCl_2 \cdot 4H_2O$ (28.5 mM), $(NH_4)_2HPO_4$ (57 mM), and H_3BO_3 (28.5 mM) were mixed together in 10 mL of deionized water with stirring. Eighty-five percent of H_3PO_4 (7.753 g) was added after the mixture was homogenized. The resulting gray gel was transferred to a Teflon autoclave (20 mL; degree of filling of ~70%; pH = 1.0–1.5) and held at 180°C for 3 days. The reaction products were separated by vacuum filtration, washed with deionized water, and dried for 1 day at 60°C. Three kinds of crystals were separated from this reaction: black crystals (purple-colored in transmitted light) with hexagonal bipyramidal and trapezoedric shape (the first modification of the first compound), pale gray translucent crystals with the same shape (the second modification of the first compound), and pale pink translucent rod-shaped crystals of ($NH_4Fe[BP_2O_8(OH)]$). The relative amount of the reaction products can be controlled by the experimental conditions. These are stirring time (preparation of the reaction gels), reaction temperature, reaction time, pH value, and degree of filling of the autoclave. Among these, reaction temperature is the most important factor. At 100°C–140°C, the small pale gray crystals of the second modification of the first compound constitute the main part of the product, while at 180°C, larger black crystals of the first modification of the first compound were found as the majority phase, and at 220°C, pale pink crystals of $NH_4Fe[BP_2O_8(OH)]$ were found together with black opaque aggregates of $NH_4Fe_2(PO_4)_2$. The oxidation states of iron in the first compound has a remarkable flexibility, depending on the preparation conditions, as confirmed by product color (ranging from black to pale gray) and measurement of magnetic susceptibilities. This indicates, together with small variations in the chemical composition, the presence of a homogeneity range.

$NH_4Fe[BP_2O_8(OH)]$ could also be synthesized by the next way (Kritikos et al. 2001). A mixture of $FeCl_3$ (1.00 mM), H_3PO_4 (0.40 mL, 6.88 mM), $(NH_4)_2B_4O_7{\cdot}4H_2O$ (1.00 mmol), NH_4Cl (2.00 mM), and H_2O (2.50 mL, 139 mM) was added to a 5 mL Teflon-lined stainless steel autoclave. The pH value was adjusted to 0.5 with concentrated HCl. The steel autoclave was heated under autogenous pressure in 180°C for 5 days. Very pale pink, rod-shaped crystals appeared.

1.24 BORON–HYDROGEN–PHOSPHORUS–OXYGEN–COBALT–NITROGEN

Three multinary compounds are formed in the B–H–P–O–Co–N system.

$\mathbf{[(NH_4)_xCo_{(3-x)/2}](H_2O)_2[BP_2O_8]{\cdot}(1-x)H_2O\ (x \approx 0.5)}$ crystallizes in the hexagonal structure with the lattice parameters $a = 949.14 \pm 0.02$, $c = 1{,}558.25 \pm 0.04$ pm, and a calculated density of 2.634 g cm^{-3} at $x = 0.5$ (Schäfer et al. 2002). Beside the purple title compound, another blue colored phase was obtained. Chemical composition and X-ray powder patterns reveal very close relationships in direction of possible polymorphous modifications. The second phase exhibits a superstructure of the first phase with a change in coordination numbers around Co from 6 and 5 to 6 and 4. These two phases were prepared under mild hydrothermal conditions at 170°C in Teflon digestion bombs (20 mL, degree of filling of ~50%–60%). In detail, the first modification was synthesized from a mixture of $CoCl_2$ (14 mM), B_2O_3 (7 mM), and $(NH_4)_2HPO_4$ (28 mM). The educts were suspended in H_2O and 1.5 mL of 17% HCl was added to reach a pH value of 1. The resulting blue gel was held at the desired temperature for 3 days leading to hexagonal bypiramidal as well as trapezoedric-shaped crystals. The second modification was prepared in a similar way, but in absence of Cl^- using CoO and 85% H_3PO_4 instead of $CoCl_2$ and HCl, respectively.

$\mathbf{(NH_4)_7Co_4(H_2O)[B_2P_4O_{15}(OH)_2]_2[H_2PO_4][HPO_4]}$ crystallizes in the orthorhombic structure with the lattice parameters $a = 1{,}692.06 \pm 0.07$, $b = 1{,}055.92 \pm 0.05$, and $c = 2{,}199.97 \pm 0.10$ pm (Yang et al. 2010b). It was obtained using 1-ethyl-3-methylimidazolium bis(oxalato)borate (ionic liquid) as a boron source together with $Co(CH_3COO)_2{\cdot}4H_2O$, $NH_4H_2PO_4$, and 85% H_3PO_4. Typically, a Teflon-lined stainless steel autoclave (15 mL) was charged with ionic liquid (0.300 g), $Co(CH_3COO)_2{\cdot}4H_2O$ (0.250 g), $NH_4H_2PO_4$ (0.406 g), and 85% H_3PO_4 (0.250 g). After 6 days of crystallization, the product was cooled, filtered, washed thoroughly with H_2O several times, and then dried in air. The pink prism crystals were obtained.

$(NH_4)_8Co_2[B_4P_8O_{30}(OH)_4]$ crystallizes in the triclinic structure with the lattice parameters $a = 969.28 \pm 0.03$, $b = 987.47 \pm 0.03$, $c = 1{,}001.25 \pm 0.02$ pm and $\alpha = 62.057 \pm 0.002°$, $\beta = 82.456 \pm 0.002°$, $\gamma = 76.095 \pm 0.02°$, and a calculated density of 2.152 g cm^{-3} (Su et al. 2011). The title compound was prepared using the ionic liquid 1-ethyl-3-methylimidazolium bromide as a solvent (as ionic liquid, 1-methyl-3-butylimidazolium bromide can also be used). The mixture of $Co(NO_3)_2{\cdot}6H_2O$ (0.29 g), $(NH_4)_2HPO_4$ (0.4 g), H_3BO_3 (0.124 g), and ionic liquid (1.0 g) (molar ratio 1:3:2:5.24) was heated in 15 mL Teflon-lined autoclave under autogenous pressure at 200°C for several days followed by cooling to room temperature. The product was filtered off, washed with deionized water, rinsed with ethanol, purified ultrasonically, and dried in vacuum desiccators at ambient temperature. Single-phased pink tetragonal-stick crystals were obtained.

1.25 BORON–HYDROGEN–SULFUR–OXYGEN–NITROGEN

The stage-1 intercalation compound, which is characterized by the interplanar spacing of 740 pm, can be formed by simple drying of *h*-BN in 96% H_2SO_4 or 20% solution of SO_3 in H_2SO_4 (Kovtyukhova et al. 2013). To obtain such compound, 70–200 mg of *h*-BN was added to 0.5–1 mL of H_2SO_4 or 20% solution of SO_3 in H_2SO_4 in a glass vial, stirred with a glass stick, and allowed to settle. The excess acid (required to ensure complete wetting of the *h*-BN powder) was then decanted away. A drop of the remaining thick *h*-BN/H_2SO_4 suspension was cast on a glass or Si wafer and dried in air at 70°C–200°C.

1.26 BORON–HYDROGEN–CHLORINE–OXYGEN–NITROGEN

The stage-1 intercalation compound, which is characterized by the interplanar spacing of 660 pm, can be formed by simple drying of *h*-BN in $HClO_4$ (Kovtyukhova et al. 2013). To obtain it, 70–200 mg of *h*-BN was added to 0.5–1 mL of 85% $HClO_4$ in a glass vial, stirred with a glass stick and allowed to settle. The excess acid (required to ensure complete wetting of the *h*-BN powder) was then decanted away. A drop of the remaining thick *h*-BN/$HClO_4$ suspension was cast on a glass or Si wafer and dried in air at 90°C–120°C.

1.27 BORON–LITHIUM–CARBON–FLUORINE–NITROGEN

The **$Li[BF(CN)_3]$** and **$Li[BF_2(CN)_2]$** quinary compounds are formed in the B–Li–C–F–N system (Bernhardt et al. 2003a). The reaction of $Li[BF_4]$ with Me_3SiCN leads selectively, depending on the reaction time and temperature, to the mixture of these cyanofluoroborates.

1.28 BORON–SODIUM–CARBON–FLUORINE–NITROGEN

The **$Na[BF(CN)_3]$** quinary compound, which melts at 256°C and decomposes at 273°C to yield a mixture of $Na[B(CN)_4]$ and $Na[BF_4]$, is formed in the B–Na–C–F–N system (Sprenger et al. 2015). To prepare this compound, $Na[BF(CN)_3]{\cdot}Me_3SiCN$ was heated to 80°C in a vacuum overnight.

1.29 BORON–POTASSIUM–CARBON–FLUORINE–NITROGEN

The **K[BF(CN)$_3$]**, **K[BF$_2$(CN)$_2$]**, and **K[BF$_3$(CN)]** quinary compounds are formed in the B–K–C–F–N system. K[BF(CN)$_3$] melts at 215°C ± 5°C (Bernhardt et al. 2003a, 2003b) and crystallizes in the triclinic structure with the lattice parameters $a = 651.9 \pm 0.1$, $b = 731.9 \pm 0.1$, $c = 763.3 \pm 0.2$ pm and $\alpha = 68.02 \pm 0.03°$, $\beta = 74.70 \pm 0.03°$, $\gamma = 89.09 \pm 0.03°$, and a calculated density of 1.505 g cm^{-3} (Bernhardt et al. 2003a). It was synthesized using various methods.

1. All volatile substances were removed from a solution of Li[BF(CN)$_3$]·Me$_3$SiCN (19 mM) in H_2O (20 mL) within 2 h under vacuum (Bernhardt et al. 2003a). The residue was dissolved in 30% H_2O_2 (20 mL, 176 mM) and heated at about 50°C for 1 h. Subsequently, 37% HCl (5 mL, 60 mM) and Pr$^n{}_3$N (5 mL, 26 mM) were added. The mixture was extracted twice with CH_2Cl_2 (50 mL), the organic phase was dried with $MgSO_4$, and stirred for 1 h with a solution of KOH (53 mM) in a minimum amount of H_2O. The organic phase was separated off, and the residue was washed with CH_2Cl_2 (50 mL). The title compound was extracted from this residue with MeCN or tetrahydrofuran (100 mL), the organic phase was dried with K_2CO_3 and concentrated on a rotary evaporator. The resulting K[BF(CN)$_3$] was washed with CH_2Cl_2 (50 mL) and freed in vacuum of the solvent.
2. The mixture of K[BF$_4$] (48 mL) with Me$_3$SiCN (50 mL, 375 mM) in a 100-mL round-bottom flask was stirred for one month under N_2 using reflux, and subsequently, all volatile products were removed under vacuum (Bernhardt et al. 2003a). The residue was mixed with 30% H_2O_2 (50 mL, 441 mM) and heated at about 50°C for 1 h. Subsequently, 37% HCl (15 mL, 180 mM) and Pr$^n{}_3$N (10 mL, 52 mM) were added. The mixture was extracted twice with CH_2Cl_2 (100 mL), the organic phase was dried with $MgSO_4$ and stirred for 1 h with a solution of KOH (8 g) in a minimum amount of H_2O. The organic phase was separated off, and the residue washed with CH_2Cl_2 (100 mL). From the residue the title compound was extracted with tetrahydrofuran (300 mL). The organic phase was dried with K_2CO_3 and concentrated on a rotary evaporator. The resulting K[BF(CN)$_3$] was washed with CH_2Cl_2 (100 mL) and freed in vacuum of the solvent.
3. A 2-L round-bottom flask equipped with a valve with a stem and fitted with a gas-tight magnetic stirrer head was charged with Na[BF$_4$] (1.09 M), Me$_3$SiCN (1.3 L, 9.75 M), and Me$_3$SiCl (0.1 L, 0.79 M) (Sprenger et al. 2015). The reaction mixture was stirred for 1 h at 50°C. Subsequently, the overpressure was released, the vessel was closed, and the reaction mixture was heated to 100°C (temperature of the heat-on attachment of the magnetic stirrer) for 6 h. Upon cooling to room temperature, crystalline Na[BF(CN)$_3$]·Me$_3$SiCN formed, and this was filtered off. The solvate molecules were removed in a vacuum at 80°C overnight. The crude product was dissolved in H_2O (50 mL) and the solution was gradually treated with an aqueous solution of 35% H_2O_2 (50 mL) and K_2CO_3 (150 g) until the red solution decolorized. Excess H_2O_2 was destroyed by the addition of $K_2S_2O_3$. The aqueous layer was extracted with tetrahydrofuran (10 × 80 mL). The combined organic layers were dried with K_2CO_3, filtered,

and the solution was concentrated to a volume of ~50 mL. CH_2Cl_2 (200 mL) was added, which resulted in the formation of a colorless precipitate that was collected by filtration and dried in a vacuum.

The second fraction of K[BF(CN)$_3$] was isolated from the filtrate, which was obtained after removal of the crystalline Na[BF(CN)$_3$]·Me_3SiCN. The trimethylsilyl cyanide was removed under reduced pressure, and the brownish semisolid remainder was purified and transferred to the potassium salt.

4. A glass tube (50 mL) equipped with a valve with a stem and fitted with a magnetic stirring bar was charged with Na[BF$_4$] (22.77 mM), Me_3SiCN (30.0 mL, 224.97 mM), and Me_3SiCl (2.50 mL, 19.79 mM) (Sprenger et al. 2015). The reaction mixture was stirred at 50°C (oil bath temperature) for 1 h and then the reaction vessel was vented. The vessel was closed again and the mixture was heated to 100°C (oil bath temperature) for further 2 h. The mixture was kept at room temperature, all volatiles were removed under reduced pressure, and the light red, solid residue was dissolved in a minimum amount of H_2O (~5 mL). The aqueous solution was treated with an aqueous solution of 35% H_2O_2 (1–2 mL) and K_2CO_3 (2–3 g) until the red solution became colorless. All volatiles were removed using a rotary evaporator. The remaining white solid was extracted with acetone (2 × 100 mL), the combined organic layers were dried with K_2CO_3, filtered, and concentrated to a volume of ~15 mL. CH_2Cl_2 (20 mL) was added, and a colorless precipitate formed was collected by filtration and dried in a vacuum.
5. A 2-L three-necked round-bottom flask equipped with a dropping funnel with a valve with a stem, an inside thermometer, and a reflux condenser that was cooled to 25°C–35°C was fitted with a magnetic stirring bar and charged with Na[BF$_4$] (0.91 M), Me_3SiCN (1.2 L, 8.99 M), and Me_3SiCl (100.0 mL, 0.79 M) (Sprenger et al. 2015). The heat-on attachment of the magnetic stirrer was heated to 100°C. The mixture was stirred for 5 h, and the temperature inside the vessel reached a maximum of 87°C. Most of the Me_3SiF that was formed in the course of the reaction passed through the reflux condenser (25°C–35°C). No significant loss of Me_3SiCl and Me_3SiCN was observed. The Me_3SiF was collected at 0°C. After cooling of the reaction mixture to room temperature, all volatiles were distilled off. The reddish brown residue was dried at 50°C under vacuum overnight. The brownish black solid remainder was dissolved in aqueous solution of 35% H_2O_2 (100.0 mL), and K_2CO_3 (150 g) was added in portions. Excess H_2O_2 was quenched with $K_2S_2O_3$, and the aqueous layer was subsequently extracted with tetrahydrofuran (5 × 80 and 5 × 50 mL). The combined organic layers were dried with K_2CO_3, filtered, and the solution was concentrated to approximately 40–50 mL. The addition of CH_2Cl_2 (150 mL) resulted in the precipitation of a colorless solid that was collected by filtration and dried in a vacuum.
6. A glass finger (50 mL) equipped with a valve with a stem and fitted with a magnetic stirring bar was charged with K[BF$_2$(CN)$_2$] (35.73 mM), Me_3SiCN (24.0 mL, 179.97 mM), and Me_3SiCl (1.60 mL, 12.66 mM) (Sprenger et al. 2015). The reaction mixture was heated to 100°C for 3 h. All volatiles were removed under reduced pressure. The solid

residue was dissolved in an aqueous solution of 35% H_2O_2 (20.0 mL), and solid K_2CO_3 (10 g) was added in small portions while stirring. After 1 h, all volatiles were removed using a rotary evaporator. The white solid remainder was extracted with tetrahydrofuran (3 × 15 mL). The combined organic layers were dried with K_2CO_3, filtered, and concentrated to 3–4 mL. $CHCl_3$ (30 mL) was added, and immediately, a colorless precipitate formed was collected by filtration and dried in a vacuum.

$K[BF_2(CN)_2]$ melts at 185°C ± 5°C (Bernhardt et al. 2003a, 2003b) and crystallizes in the tetragonal structure with the lattice parameters a = 1,316.59 ± 0.08 and c = 3,819.76 ± 0.26 pm, and a calculated density of 1.685 g cm^{-3} (Bernhardt and Willner 2009) [a = 1,315.96 ± 0.03 and c = 3,841.83 ± 0.08 pm, and a calculated density of 1.677 g cm^{-3} (Bernhardt et al. 2003a)]. Some methods were used to prepare this compound.

1. $Li[BF_2(CN)_2 \cdot 2Me_3SiCN$ (20 mM) was dissolved in H_2O (20 mL), and after 2 h, all volatile substances were removed under vacuum (Bernhardt et al. 2003a). The residue was dissolved in 30% H_2O_2 (20 mL, 176 mM), and the obtained solution was heated at ~50°C for 1 h. Subsequently, this solution was reacted with 37% HCl (5 mL, 60 mM) and $Pr^n{}_3N$ (5 mL, 26 mM). The mixture was then extracted with CH_2Cl_2 (50 mL), and the organic phase was dried with $MgSO_4$ and stirred vigorously for 1 h with a solution of KOH (53 mM) in a minimum amount of H_2O. The organic phase was separated off, and the residue was washed with CH_2Cl_2 (50 mL). The title compound was extracted from the residue with MeCN (100 mL). The organic phase was dried with K_2CO_3 and concentrated on a rotary evaporator. The resulting $K[BF_2(CN)_2]$ was washed with CH_2Cl_2 and freed in vacuum of the solvent.

2. A round-bottom flask (250 mL) fitted with a magnetic stirring bar and a dropping funnel (50 mL) was charged with KCN (205.8 mM) and acetonitrile (25 mL) in an Ar atmosphere (Sprenger et al. 2015). The suspension was cooled to 0°C and a solution of $BF_3 \cdot MeCN$ in acetonitrile (15.2%–16.8% BF_3, 100 mL, ~205 mM) was added dropwise within 4 h. The yellow suspension was filtered at 0°C. All volatiles were removed under reduced pressure, the yellow solid was taken up into an aqueous solution of 35% H_2O_2 (20.0 mL) at room temperature, and the solution was stirred for additional 4 h. All volatiles were removed with a rotary evaporator and the solid residue was extracted with MeCN (2 × 100 mL). The combined organic phases were dried with K_2CO_3, filtered, and the solution was concentrated to a volume of 5 mL. Upon addition of CH_2Cl_2 (50 mL), light yellow $K[BF_2(CN)_2]$ precipitated was filtered off and dried in a vacuum.

3. To 63 mM of finely ground KCN in 50 mL flask with a valve, $BF_3 \cdot Et_2O$ (41 mM) and MeCN (30 mL) were added (Bernhardt et al. 2003b). The reaction mixture was stirred at room temperature for 3 h, then all volatiles were removed in vacuum, the residue was dissolved in MeCN (50 mL) and freed from KCN and $K[BF_4]$ by filtration. After removal of the acetonitrile in vacuum, a mixture of solid was obtained. By recrystallization from H_2O, pure colorless $K[BF_2(CN)_2]$ was prepared.

4. KCN (1.0 M) and MeCN (200 mL) were placed in a 500-mL round-bottom flask with dropping funnel (Bernhardt et al. 2003b). At room temperature, $BF_3{\cdot}Et_2O$ (400 mM) was added dropwise with stirring. After further stirring (1.5 h) at room temperature, the solution was filtered and the filter residue was washed with MeCN (~300 mL). The combined acetonitrile phases were concentrated on a rotary evaporator. The title compound was obtained as a crude product. It was dissolved in H_2O (200 mL) and concentrated HCl (30 mL) and $Pr^n{}_3N$ were added (35 mL). The obtained mixture was extracted with CH_2Cl_2 (200 mL). The dichloromethane phase was dried with $MgSO_4$ and dissolved with KOH (25 g) in a little amount of H_2O with vigorous stirring. The viscous aqueous phase was separated and washed with CH_2Cl_2. The product was extracted with MeCN (300 mL), dried with K_2CO_3, and concentrated on a rotary evaporator. The white $K[BF_2(CN)_2]$ was washed with CH_2Cl_2 and dried in vacuum.

$K[BF_3(CN)]$ melts at 170°C ± 5°C (Bernhardt et al. 2003b) and crystallizes in the orthorhombic structure with the lattice parameters $a = 1{,}334.86 \pm 0.15$, $b = 652.39 \pm 0.07$, and $c = 1{,}000.85 \pm 0.11$ pm and a calculated density of 2.026 g cm^{-3} (Bernhardt and Willner 2009).

The synthesis of this compound was carried out by the interaction of $BF_3{\cdot}Et_2O$ and powdered KCN in MeCN at −40°C (Bernhardt et al. 2003b). To a finely ground KCN (1.0 M) in 500 mL flask with a valve, $BF_3{\cdot}Et_2O$ (50 mL, 0.4 M) and MeCN (200 mL) were added. The mixture was slowly warmed to room temperature with vigorous stirring. After 1.5 h at room temperature, all volatiles were removed in vacuum, and the residue was extracted first with tetrahydrofuran (200 mL) and then with MeCN (500 mL). The residue of the extraction consisted of KCN and a little amount of $K[BF_4]$. Tetrahydrofuran of the first fraction was removed and the mixture of $K[BF_2(CN)_2]$ and $K[BF_3(CN)]$ (molar ratio ~1:1) was obtained. The removal of MeCN from the second fraction leads to the formation of only $K[BF_3(CN)]$.

1.30 BORON–COPPER–LANTHANUM–NICKEL–NITROGEN

The $\mathbf{La_3(Ni_{1-x}Cu_x)_2B_2N_{3-\delta}}$ ($x \leq 0.1$) solid solutions are formed in the B–Cu–La–Ni–N quinary system (Michor et al. 1998). Polycrystalline samples of these solid solutions were arc melted from La ingots, Cu and Ni powders or ingots, and *h*-BN powder.

1.31 BORON–CALCIUM–CARBON–FLUORINE–NITROGEN

The $\mathbf{Ca_{15}(CBN)_6(C_2)_2F_2}$ quinary compound, which crystallizes in the cubic structure with the lattice parameter $a = 1{,}653.6 \pm 0.4$ pm and a calculated density of 2.67 g cm^{-3} (Reckeweg et al. 2010) [$a = 1{,}656.8 \pm 0.1$ pm and a calculated density of 2.5884 g cm^{-3} (Wörle et al. 1997)], is formed in the B–Ca–C–F–N system.

Transparent dark red crystals of the title compound were obtained as follows (Reckeweg et al. 2010). Ca (9 mM), CaF_2 (90.51 mM), *h*-BN (3.02 mM), and graphite (5.83 mM) were arc-welded into a Ta container, which was sealed in an evacuated silica tube. This tube was placed upright in a box furnace and heated to 1,030°C within 12 h. After two days' reaction time, the furnace was switched off and allowed to cool to room temperature. All manipulations were performed in a glove box under Ar atmosphere.

This compound was also obtained in sealed Nb ampoules from a stoichiometric mixture of *h*-BN, graphite, and CaO with twofold excess of Ca metal (Wörle et al. 1997). The Nb ampoules were filled in a glove box under Ar atmosphere with finely ground and well-mixed components. After sealing under Ar atmosphere by arc welding, the ampoules were placed in an Ar-filled quartz tube and heated to 1,000°C for 20 h. $Ca_{15}(CBN)_6(C_2)_2F_2$ was obtained in the form of dark single crystals with metallic luster. As a powdered sample it shows light brown color.

The title compound is air-sensitive and decomposes in few minutes when exposed to moisture air (Wörle et al. 1997; Reckeweg et al. 2010). Above 1,119°C, it decomposes in vacuum completely.

1.32 BORON–CALCIUM–CARBON–CHLORINE–NITROGEN

The **Ca_3Cl_2CBN** quinary compound, which crystallizes in the orthorhombic structure with the lattice parameters $a = 1{,}386.7 \pm 0.9$, $b = 384.7 \pm 0.3$, and $c = 1{,}124.7 \pm 0.6$ pm and a calculated density of 2.53 g cm^{-3}, is formed in the B–Ca–C–Cl–N system (Meyer 1991). It was obtained from the reaction of Ca and $CaCl_2$ with $CaCN_2$, B, and C or with BN and C in sealed Ta containers at 900°C.

1.33 BORON–CALCIUM–CARBON–BROMINE–NITROGEN

The **Ca_3Br_2CBN** quinary compound, which crystallizes in the orthorhombic structure with the lattice parameters $a = 1{,}444.3 \pm 0.2$, $b = 390.64 \pm 0.06$, and $c = 1{,}139.2 \pm 0.2$ pm, is formed in the B–Ca–C–Br–N system (Womelsdorf and Meyer 1994). It was obtained from the reaction of Ca and $CaBr_2$ with BN and C in sealed Ta containers at 950°C.

1.34 BORON–STRONTIUM–ALUMINUM–EUROPIUM–SILICON–NITROGEN

The **$Sr_{2-y}Eu_yB_{2-2x}Si_{2+3x}Al_{2-x}N_{8+x}$** multinary phase, which crystallizes in the hexagonal structure, is formed in the B–Sr–Al–Eu–Si–N system. The lattice parameter *a* shifts from 477.58 pm for $x = 0$ to 479.16 pm for $x = 0.1$ and 479.78 pm for $x = 0.2$, and the parameter *c* shifts from 976.92 pm for $x = 0$ to 978.96 pm for $x = 0.1$ and 980.52 pm for $x = 0.2$ (Ten Kate et al. 2016) [$a = 479.88 \pm 0.01$ and $c = 978.02 \pm 0.02$ pm and a calculated density 3.675 g cm^{-1} for $Sr_{1.90}Eu_{0.10}B_{1.76}Si_{2.36}Al_{1.88}N_{8.12}$ (Funahashi et al. 2014)]. Eu-doped $Sr_{2-y}Eu_yB_{2-2x}Si_{2+3x}Al_{2-x}N_{8+x}$ powders were prepared via a solid-state reaction synthesis. α-Si_3N_4 or β-Si_3N_4, Sr_3N_2, BN, AlN, and EuN powders were thoroughly mixed with mortar and pestle (Funahashi et al. 2014; Ten Kate et al. 2016). Starting materials were mixed according to the appropriate molar ratio in the title compound, except for Sr_3N_2, for which a 5% excess was used in all samples. The powder mixture was then pressed into pellets of ~1 g, and the pellets were put in a molybdenum crucible. A pellet made of carbon black was also added to the crucible to create a reducing atmosphere in the sample container. Mixing of the starting materials, pressing into pellets, and transferring into the crucible were all performed inside a N_2-filled glove box. The crucibles were heated in a nitrogen gas pressure sintering furnace at 1,670°C–1,800°C and 1 MPa for 2 h. After they were heated, the samples were crunched, and the powder was analyzed without any further treatment.

1.35 BORON–STRONTIUM–CARBON–BROMINE–NITROGEN

The **Sr_3Br_2CBN** quinary compound, which crystallizes in the orthorhombic structure with the lattice parameters $a = 1{,}448.4 \pm 0.2$, $b = 405.46 \pm 0.05$, and $c = 1{,}170.1 \pm 0.1$ pm and a calculated density of 3.58 g cm^{-3}, is formed in the B–Sr–C–Br–N system (Womelsdorf and Meyer 1994). It was obtained from the reaction of Ca and $CaBr_2$ with BN and C in sealed Ta containers at 950°C.

1.36 BORON–BARIUM–ALUMINUM–SILICON–NITROGEN

The **$Ba_5B_2Al_4Si_{32}N_{52}$** quinary compound, which crystallizes in the triclinic ($a = 978.79 \pm 0.11$, $b = 979.20 \pm 0.11$, and $c = 1{,}272.26 \pm 0.15$ pm and $\alpha = 96.074 \pm 0.004°$, $\beta = 112.330 \pm 0.003°$, and $\gamma = 94.080 \pm 0.004°$ and a calculated density of 3.645 g cm^{-3}) is formed in the B–Ba–Al–Si–N system (Yoshimura et al. 2017). To obtain this compound, powders of barium nitride labeled "Ba_3N_2", strontium nitride labeled "Sr_3N_2" EuN, AlN, and α-Si_3N_4 were used as starting materials. The Ba_3N_2 and Sr_3N_2 powders were mixtures of Ba_3N_2, Ba_2N, BaN_2, and BaH_2, and Sr_2N, SrN_2, SrN, and SrH, respectively. The powders of Ba_3N_2 (135.8 mg), Sr_3N_2 (15.8 mg), EuN (3.7 mg), AlN (136.7 mg), and α-Si_3N_4 (207.9 mg) were weighed and mixed with an alumina pestle and mortar in an Ar-filled glove box, and charged in a BN crucible with a lid. The mixture was heated in a graphite resistance furnace from room temperature to 800°C below 8×10^{-3} Pa of N_2. From 800°C to 2,000°C, the sample was heated at 0.85 MPa of N_2 at a heating rate 20°C min^{-1}. After heating at 2,000°C and 0.85 MPa of N_2 for 4 h, the sample was heated to 2,050°C at the same heating rate. This temperature was held for 4 h, and then the sample was cooled down to 1,200°C at a rate of 20°C min^{-1} and finally to room temperature in the furnace by shutting off the electric power to the heater, maintaining the N_2 pressure of 0.85 MPa. Some colorless transparent platelet single crystals of the title compound were grown on the surface of the powder sample and on the inner side and bottom wall of the BN crucible.

1.37 BORON–BARIUM–PRASEODYMIUM–SILICON–OXYGEN–NITROGEN

The **$Ba_4Pr_7[Si_{12}N_{23}O][BN_3]$** multinary compound, which crystallizes in the hexagonal structure with the lattice parameters $a = 1{,}225.7 \pm 0.1$ and $c = 544.83 \pm 0.09$ pm and a calculated density of 5.299 g cm^{-3}, is formed in the B–Ba–Pr–Si–O–N system (Orth et al. 2001). It was prepared as follows. In a characteristic reaction batch Ba (0.99 mM), Pr (1.00 mM), $Si(NH)_2$ (1.67 mM), $BaCO_3$ (0.10 mM), and 37.9 mg poly(boron amide imide) were thoroughly mixed in a glove box under a purified argon atmosphere. Under nitrogen (100 kPa), the mixture was then transferred into a tungsten crucible positioned in the center of a quartz glass reactor of the radio-frequency furnace. Within 1 h, the crucible was heated to 1,000°C and then within 4 h to 1,650°C. The reaction mixture was kept at this temperature for 4 h and finally it was cooled to 1,200°C within 9 h and subsequently quenched to room temperature. Dark green crystals of the title compound were obtained. The crystals were embedded in a glass melt of very high mechanical hardness. The crystals had to be mechanically separated from this glass phase by the use of severe mechanical force. Usually *h*-BN was observed as a further by-product, and it was separated in an inert solvent (cyclohexane) by the application of ultrasound.

1.38 BORON–BARIUM–NEODYMIUM–SILICON–OXYGEN–NITROGEN

The $\mathbf{Ba_4Nd_7[Si_{12}N_{23}O][BN_3]}$ multinary compound, which crystallizes in the hexagonal structure with the lattice parameters $a = 1,222.6 \pm 0.1$ and $c = 544.6 \pm 0.1$ pm and a calculated density of 5.383 g cm^{-3}, is formed in the B–Ba–Nd–Si–O–N system (Orth et al. 2001). It was synthesized in the same manner as $Ba_4Pr_7[Si_{12}N_{23}O][BN_3]$ and has been obtained as large orange-brown crystals.

1.39 BORON–BARIUM–SAMARIUM–SILICON–OXYGEN–NITROGEN

The $\mathbf{Ba_4Sm_7[Si_{12}N_{23}O][BN_3]}$ multinary compound, which crystallizes in the hexagonal structure with the lattice parameters $a = 1,215.97 \pm 0.05$ and $c = 542.80 \pm 0.05$ pm and a calculated density of 5.562 g cm^{-3}, is formed in the B–Ba–Sm–Si–O–N system (Orth et al. 2001). It was synthesized in the same manner as $Ba_4Pr_7[Si_{12}N_{23}O][BN_3]$ and has been obtained as large dark red crystals. This compound exhibits a thermal stability up to 1,400°C and it is resistant against hydrolysis. At higher temperatures, a rapid decomposition and the formation of gaseous products and an X-ray amorphous solid residue was observed.

1.40 BORON–ALUMINUM–CARBON–SILICON–TITANIUM–NITROGEN

$\mathbf{AlN\text{–}SiC\text{–}TiB_2}$. Annealing treatments were carried out at different temperatures (600°C–1,200°C) for various holding times (2–10 h) to evaluate the microstructural and mechanical properties changes of the AlN–SiC–TiB_2 ceramic composites prepared by self-propagating high temperature synthesis and hot isostatic pressing (Zhou et al. 2013). The experimental results show that the annealing treatment is beneficial to improve the mechanical properties. At the same time, a solid solution enriched in SiC is depleted and fine TiB_2 grains are precipitated. Due to the precipitation strengthening by the SiC-rich solid solution and the grain boundary strengthening by fine TiB_2 particles, the improvement of the mechanical properties was obvious with higher annealing temperatures and longer holding times. The ceramic composites were investigated through XRD, scanning electron microscopy (SEM), and transmission electron microscopy (TEM).

1.41 BORON–LANTHANUM–CERIUM–NICKEL–NITROGEN

The $\mathbf{La_{3-x}Ce_xNi_2B_2N_{3-\delta}}$ ($x \leq 0.3$) solid solutions, which crystallize in the tetragonal structure, are formed in the B–La–Ce–Ni–N quinary system (Ali et al. 2011a, 2011b). Powder XRD indicated that the *a* lattice parameter decreases with increasing Ce fraction; however, the *c* lattice parameter shows a nonmonotonic variation with a maximum at about $x = 0.2$.

Polycrystalline samples of these solid solutions were prepared by inductive levitation melting in a three-step process (Ali et al. 2011a, 2011b). In the first step, stoichiometric amounts of Ni and B were melted several times in Ar atmosphere. In the second step, La and Ce metals were premelted in vacuum and then melted with NiB forming $(La,Ce)_3Ni_2B_2$ precursor alloys. In a third step, these alloys were repeatedly melted in Ar/N_2 atmosphere such that the nitrogen stoichiometry is slowly increased to reach a composition close to $\delta \approx 0.3$. The $La_{3-x}Ce_xNi_2B_2N_{3-\delta}$ alloys were flipped after each melting cycle to improve their homogeneity. The samples were finally annealed in a high-vacuum furnace at 1,100°C for 1 week.

1.42 BORON–LANTHANUM–PRASEODYMIUM–NICKEL–NITROGEN

The $\mathbf{La_{3-x}Pr_xNi_2B_2N_{3-\delta}}$ solid solutions, which crystallize in the tetragonal structure, are formed in the B–La–Pr–Ni–N quinary system (Ali et al. 2011b). Polycrystalline samples of these solid solutions were prepared by inductive levitation melting of $(La,Pr)_3Ni_2B_2$ precursor alloys in Ar/N_2 atmosphere such that the nitrogen stoichiometry is slowly increased. The samples were finally annealed in a vacuum furnace at 1,100°C for 1 week. After this initial heat treatment, smaller pieces were sealed in quartz ampoules for additional heat treatments with final quenching of the ampoules in water.

1.43 BORON–LANTHANUM–NEODYMIUM–NICKEL–NITROGEN

The $\mathbf{La_{3-x}Nd_xNi_2B_2N_{3-\delta}}$ solid solutions, which crystallize in the tetragonal structure, are formed in the B–La–Nd–Ni–N quinary system (Ali et al. 2011b). Polycrystalline samples of these solid solutions were prepared by inductive levitation melting of $(La,Nd)_3Ni_2B_2$ precursor alloys in Ar/N_2 atmosphere such that the nitrogen stoichiometry is slowly increased. The samples were finally annealed in a vacuum furnace at 1,100°C for 1 week. After this initial heat treatment, smaller pieces were sealed in quartz ampoules for additional heat treatments with final quenching of the ampoules in water.

1.44 BORON–LANTHANUM–COBALT–NICKEL–NITROGEN

The $\mathbf{La_3(Ni_{1-x}Co_x)_2B_2N_{3-\delta}}$ ($x \leq 0.3$) solid solutions are formed in the B–La–Co–Ni–N quinary system, which crystallize in the tetragonal structure with the lattice parameters $a = 372.30 \pm 0.01$ and $c = 2{,}042.89 \pm 0.08$ pm for $x = 0.3$ and $\delta = 0$ (Michor et al. 1998). Polycrystalline samples of these solid solutions were arc melted from La ingots, Co and Ni powders or ingots, and *h*-BN powder.

1.45 BORON–PHOSPHORUS–CHLORINE–BROMINE–NITROGEN

The $\mathbf{[Br_2BNPCl_3]_2}$ quinary compound, which crystallizes in the orthorhombic structure with the lattice parameters $a = 1{,}334.93 \pm 0.08$, $b = 1{,}189.55 \pm 0.07$, and $c = 1{,}026.42 \pm 0.06$ pm and a calculated density of 2.624 g cm^{-3}, is formed in the B–P–Cl–Br–N system (Jäschke and Jansen 2002). It was prepared by the following procedure. To a –78°C solution of BBr_3 (0.075 M) in CH_2Cl_2 (250 mL) was slowly added a solution of $Me_3SiNPCl_3$ (0.075 M) in CH_2Cl_2 (250 mL). The heating of the reaction mixture was gradually increased in such a way that the reaction solution was held 2 h at –50°C and further stirred at –35°C for 2 h before leaving finally warm to room temperature. The resulting white solid was filtered off with exclusion of air, washed twice with little amounts of CH_2Cl_2, and dried in vacuum. From the filtrate after removal of the solvent in a vacuum, further solid was obtained. For purification, the crude product was recrystallized from CH_2Cl_2. At 5°C large, colorless, distorted octahedral crystals of the title compound were isolated by filtration.

1.46 BORON–ARSENIC–OXYGEN–SULFUR–FLUORINE–NITROGEN

The **$(BN)_{\geq 9}AsF_5SO_3F$** multinary compound, which crystallizes in the hexagonal structure with the lattice parameters $a = 254 \pm 1$ and $c = 825 \pm 2$ pm, is formed in the B–As–O–S–F–N system (Shen et al. 1999a). Intercalation of $(BN)_{\sim 3}SO_3F$ with AsF_5 at ~20°C leads to a formation of a dark blue title compound, which is stable at least up to 100°C. The same product was obtained at the intercalation of $(BN)_{\sim 3}SO_3F$ with $AsF_5 + F_2$ mixture.

1.47 BORON–OXYGEN–SULFUR–FLUORINE–NITROGEN

The **$(BN)_{\sim 3}SO_3F$** quinary compound, which crystallizes in the hexagonal structure with the lattice parameters $a = 250 \pm 1$ and $c = 802 \pm 1$ pm, is formed in the B–O–S–F–N system (Shen et al. 1999a). *h*-BN was intercalated at ~20°C by $S_2O_6F_2$ to give a deep blue solid of the title compound. According to the data of Bartlett et al. (1978), the composition of this compound is $(BN)_4SO_3F$.

CHAPTER 2

Systems Based on BP

2.1 BORON–HYDROGEN–LITHIUM–COPPER–OXYGEN–PHOSPHORUS

The **$LiCu(H_2O)[BP_2O_8]{\cdot}2H_2O$** and **$LiCu_2BP_2O_8(OH)_2$** multinary compounds are formed in the B–H–Li–Cu–O–P system. The first of them crystallizes in the hexagonal structure with the lattice parameters $a = 952.0 \pm 0.4$ and $c = 1{,}543.8 \pm 0.6$ pm (Boy and Kniep 2001a). Turquoise, hexagonal bipyramids of the title compound were prepared by hydrothermal treatment of a mixture of $Li_2B_4O_7$, LiH_2PO_4, and $CuCl_2{\cdot}2H_2O$ (molar ratio 1:8:1). A concentrated aqueous solution of pH = 1–1.5 (by adding HCl) was held at 160°C for 2 weeks in a Teflon autoclave (20 mL, degree of filling of 50%).

$LiCu_2BP_2O_8(OH)_2$ crystallizes both in the orthorhombic [$a = 531.95 \pm 0.10$, $b = 793.52 \pm 0.15$, and $c = 1{,}746.4 \pm 0.3$ pm (Yang et al. 2013)] and in the monoclinic [$a = 1{,}509.74 \pm 0.19$, $b = 476.17 \pm 0.06$, $c = 965.85 \pm 0.12$ pm, and $\beta = 91.019 \pm 0.001°$ and a calculated density of 3.528 g cm^{-3} (Zheng and Zhang 2009a)] structure.

Orthorhombic modification of the title compound was prepared by the boric acid flux reaction of a mixture containing H_3BO_3, LiH_2PO_4, and $Cu(CH_3COO)_2{\cdot}H_2O$ (molar ratio 8:2:1) at 200°C for 5 days (Yang et al. 2013). Adding traces of H_2O could improve the degree of crystallinity. In addition, increasing the molar ratio of $LiH_2PO_4/Cu(CH_3COO)_2{\cdot}H_2O$ from 2.0 to 3.5 led to impurity of monoclinic modification.

Blue block crystals of the monoclinic modification were synthesized hydrothermally from a mixture of $Cu(NO_3)_2$, $Li_2B_4O_7$, H_2O, and 85% H_3PO_4 (Zheng and Zhang 2009a). In a typical synthesis, $Cu(NO_3)_2$ (0.725 g) was dissolved in a mixture of H_2O (5 mL), $Li_2B_4O_7$ (1.691 g), and 85% H_3PO_4 (2 mL) with constant stirring. Finally, the mixture was kept in a 30-mL Teflon-lined steel autoclave at 170°C for 6 days. Then, the autoclave was slowly cooled to room temperature.

2.2 BORON–HYDROGEN–LITHIUM–MAGNESIUM–OXYGEN–PHOSPHORUS

The **$LiMg(H_2O)_2[BP_2O_8]{\cdot}H_2O$** multinary compound, which crystallizes in the hexagonal structure with the lattice parameters $a = 941.39 \pm 0.01$ and $c = 1{,}571.13 \pm 0.03$ pm and a calculated density of 2.364 g cm^{-3} at 173 ± 2 K (Lin et al. 2008c), is formed in the

B–H–Li–Mg–O–P system. Transparent, colorless single crystals of the title compound were hydrothermally synthesized. A mixture of $MgCl_2{\cdot}6H_2O$ (534.4 mg), $LiOH{\cdot}H_2O$ (5.042 g), H_3BO_3 (1.568 g), and 85% H_3PO_4 (10 mL) in an approximate molar ratio Mg/Li/B/P = 1:46:10:65 was dissolved in 5 mL of distilled water with stirring. The resulting solution (pH = 1.5) was transferred into a Teflon-lined autoclave (30 mL, degree of filling 67%) and held at 190°C for 4 days under autogenous pressure. Then the autoclave was cooled to room temperature by turning off the power. Products were filtered off, washed with distilled water, and dried at room temperature.

2.3 BORON–HYDROGEN–LITHIUM–ZINC–OXYGEN–PHOSPHORUS

The **$LiZn(H_2O)_2[BP_2O_8]{\cdot}H_2O$** multinary compound, which crystallizes in the hexagonal structure with the lattice parameters $a = 944.67 \pm 0.08$ and $c = 1{,}563.97 \pm 0.13$ pm and a calculated density of 2.688 g cm^{-3} (Zhang et al. 2011) [$a = 946.9 \pm 0.2$ and $c = 1{,}566.7 \pm 0.2$ pm (Boy and Kniep 2001b)], is formed in the B–H–Li–Zn–O–P system. The title compound was prepared by hydrothermal treatment of a mixture of $Li_2B_4O_7$ (1.30 g), LiH_2PO_4 (6.40 g), and ZnO (0.63 g) (molar ratio 1:8:1). A concentrated aqueous solution of pH = 1–1.5 (by adding H_3PO_4) was held at 170°C for 2 weeks in a Teflon autoclave (20 mL, degree of filling 50%). Colorless hexagonal bipyramids of $LiZn(H_2O)_2[BP_2O_8]{\cdot}H_2O$ were obtained.

2.4 BORON–HYDROGEN–LITHIUM–CADMIUM–OXYGEN–PHOSPHORUS

The **$LiCd(H_2O)_2[BP_2O_8]{\cdot}H_2O$** multinary compound, which crystallizes in the hexagonal structure with the lattice parameters $a = 971.11 \pm 0.06$ and $c = 1{,}601.2 \pm 0.1$ pm, is formed in the B–H–Li–Cd–O–P system (Ge et al. 2003a). It was prepared under mild hydrothermal conditions. A mixture of $CdCl_2{\cdot}2.5H_2O$ (603 mg), LiOH (5.042 g), H_3BO_3 (1.568 g), and 85% H_3PO_4 (10 mL) was heated at 90°C in deionized water (10 mL) under constant stirring until the components were completely dissolved. The solution (pH = 2.0–2.5) was transferred to a Teflon autoclave (27 mL, degree of filling 70%) and heated at 170°C for 4 days. Colorless crystals of $LiCd(H_2O)_2[BP_2O_8]{\cdot}H_2O$ were obtained.

2.5 BORON–HYDROGEN–LITHIUM–TITANIUM–OXYGEN–PHOSPHORUS

The **$[Li_{0.34}(H_3O)_{0.66}]\{Ti_2[B(PO_4)_4]\}{\cdot}H_2O$** multinary compound, which crystallizes in the tetragonal structure with the lattice parameters $a = 631.84 \pm 0.02$ and $c = 1{,}644.06 \pm 0.08$ pm, is formed in the B–H–Li–Ti–O–P system (Zhou et al. 2011b). The synthesis of the title compound was performed under hydrothermal conditions. First, Ti (1.966 g) was dissolved in concentrated HCl (100 mL) and kept in a sealed bottle for several days to get a clear blue solution, which is named afterwards Ti*. Then, the mixture of Ti* (5 mL), H_3BO_3 (1.236 g), and LiH_2PO_4 (1.039 g) (Li/Ti/B/P molar ratio 5:1:10:5) was transferred to a Teflon autoclave. The resulting pH values were lower than 1. The synthesis was conducted at 180°C for 3 days. After that, the autoclave was directly taken out of the oven. The product was filtered and washed with deionized water several times and dried at 50°C for 8 h.

2.6 BORON–HYDROGEN–LITHIUM–VANADIUM–OXYGEN–PHOSPHORUS

The **$Li_3V_2[BP_3O_{12}(OH)][HPO_4]$** multinary compound, which crystallizes in the monoclinic structure with the lattice parameters $a = 937.83 \pm 0.05$, $b = 823.13 \pm 0.04$, $c = 1,775.04 \pm 0.09$ pm, and $\beta = 119.394 \pm 0.003°$, is formed in the B–H–Li–V–O–P system (Lin et al. 2010). It was obtained from a mixture of H_3BO_3 (33.0 mM), VCl_3 (2.0 mM), LiH_2PO_4 (10 mM), LiCl (14 mM), and H_2O (2 mL) under hydrothermal conditions. The educts were transferred into Teflon-lined stainless steel autoclaves (25 mL) and held at 280°C for 7 days followed by cooling down to ambient temperature. The reaction products were washed with hot water (60°C) until soluble components were completely removed and finally dried in air at 60°C.

2.7 BORON–HYDROGEN–LITHIUM–OXYGEN–PHOSPHORUS

The **$Li[B_3PO_6(OH)_3]$** quinary compound, which crystallizes in the orthorhombic structure with the lattice parameters $a = 1,148.4 \pm 0.5$, $b = 871.9 \pm 0.5$, and $c = 1,387 \pm 1$ pm, is formed in the B–H–Li–O–P system (Hauf and Kniep 1997a). It was prepared by hydrothermal treatment of a mixture of $LiOH{\cdot}2H_2O$, P_2O_5, and B_2O_3 (molar ratio 11:1:3.5; concentration solution in HCl) at 160°C.

2.8 BORON–HYDROGEN–LITHIUM–OXYGEN–MANGANESE–PHOSPHORUS

The **$LiMn(H_2O)_2[BP_2O_8]{\cdot}H_2O$** multinary compound, which crystallizes in the hexagonal structure with the lattice parameters $a = 957.65 \pm 0.04$ and $c = 1,585.7 \pm 0.1$ pm, is formed in the B–H–Li–O–Mn–P system (Zhuang et al. 2008). It was obtained using the presence of boric acid as a flux. A mixture of $MnCO_3$ (0.1149 g), H_3BO_3 (1.484 g), and LiH_2PO_4 (0.6235 g) was ground to a homogeneous powder, which was transferred to a Teflon autoclave (10 mL; degree of filling 10%), where it was heated at 170°C for 4 days.

2.9 BORON–HYDROGEN–LITHIUM–OXYGEN–COBALT–PHOSPHORUS

The **$LiCo(H_2O)_2[BP_2O_8]{\cdot}H_2O$** multinary compound, which crystallizes in the hexagonal structure with the lattice parameters $a = 943.43 \pm 0.03$ and $c = 1,572.80 \pm 0.08$ pm, is formed in the B–H–Li–O–Co–P system (Menezes et al. 2008e). Single crystals of the title compound were prepared hydrothermally from a mixture of $Li_2B_4O_7$ (1.3579 g), $Co(CH_3COO)_2{\cdot}4H_2O$ (1.0004 g), 85% H_3PO_4 (0.3870 g and 2.7771 g) (molar ratio Li/Co/B/P = 1:1:2:6. The mixture was heated by stirring together 10 mL of deionized water at 90°C for 3 h. The reaction mixture was then filled into a Teflon-lined autoclave (10 mL, degree of filling 40%) and heated under autogenous pressure at 170°C for 5 days. The reaction product was separated by filtration, washed with hot water, filtered and washed again with acetone, and finally dried in air at 60°C. Pink hexagonal bipyramids of $LiCo(H_2O)_2[BP_2O_8]{\cdot}H_2O$ were obtained.

2.10 BORON–HYDROGEN–LITHIUM–OXYGEN–NICKEL–PHOSPHORUS

The **$LiNi(H_2O)_2[BP_2O_8]{\cdot}H_2O$** multinary compound, which crystallizes in the hexagonal structure with the lattice parameters $a = 933.59 \pm 0.03$ and $c = 1,574.97 \pm 0.11$ pm, is formed in the B–H–Li–O–Ni–P system (Zheng and Zhang 2009b). Green block-shaped crystals were synthesized hydrothermally from a mixture of $Ni(NO_3)_2$, $Li_2B_4O_7$, water, and 85% H_3PO_4. In a typical synthesis, $Ni(NO_3)_2{\cdot}6H_2O$ (0.87 g) was dissolved in a mixture of water (5 mL), $Li_2B_4O_7$ (1.691 g), and 85% H_3PO_4 (2 mL) under constant stirring. Finally, the mixture was kept in a 30-mL Teflon-lined steel autoclave at 170°C for 6 days. The autoclave was slowly cooled to room temperature.

2.11 BORON–HYDROGEN–SODIUM–POTASSIUM–OXYGEN–CHROMIUM–PHOSPHORUS

The $\mathbf{Na_{10}K_5[NaCr_8B_4P_{12}O_{60}H_8]{\cdot}10H_2O}$ multinary compound, which crystallizes in the orthorhombic structure with the lattice parameters a = 1,046.33 ± 0.09, b = 1,504.91 ± 0.11, and c = 1,872.14 ± 0.15 pm and a calculated density of 2.712 g cm^{-3} (Liu et al. 2010), is formed in the B–H–Na–K–O–Cr–P system. The title compound was synthesized using boric acid as a flux in a closed system. A typical example was to transfer the homogenized mixture of $CrCl_3{\cdot}6H_2O$, $Na_3PO_4{\cdot}12H_2O$, KF, and H_3BO_3 (molar ratio 1:6:3:12) into a 23-mL Teflon-lined stainless steel autoclave, and then kept at 170°C in an oven for 7 days to give green rod crystals.

2.12 BORON–HYDROGEN–SODIUM–COPPER–OXYGEN–PHOSPHORUS

Two multinary compounds are formed in the B–H–Na–Cu–O–P system.

$\mathbf{Na_4Cu_3[B_2P_4O_{15}(OH)_2]{\cdot}2HPO_4}$ crystallizes in the monoclinic structure with the lattice parameters a = 1,747.7 ± 0.4, b = 510.1 ± 0.1, c = 2,243.7 ± 0.4 pm, and β = 102.69 ± 0.08° (Boy et al. 1998b). Turquoise transparent prisms of the title compound were prepared by hydrothermal treatment of a mixture of $CuCl_2{\cdot}2H_2O$, $Na_2B_4O_7{\cdot}10H_2O$, and $Na_2HPO_4{\cdot}2H_2O$ (molar ratio 2:1:7) in 18% HCl at 150°C.

$\mathbf{Na_2Cu[B_3P_2O_{11}(OH)]{\cdot}0.67H_2O}$ (Yang et al. 2006c) or $\mathbf{Na_6Cu_3\{B_6P_6O_{27}(O_2BOH)_3\}{\cdot}2H_2O}$ (Liu et al. 2006) crystallizes in the hexagonal structure with the lattice parameters a = 1,155.4 ± 0.2 and c = 1,231.4 ± 0.3 pm and a calculated density of 2.862 g cm^{-3} (Yang et al. 2006c) [a = 1,154.81 ± 0.06 and c = 1,227.67 ± 0.07 pm (Liu et al. 2006)]. $Na_2[CuB_3P_2O_{11}(OH)]{\cdot}0.67H_2O$ was synthesized under hydrothermal conditions (Yang et al. 2006c). $Cu(CH_3COO)_2{\cdot}H_2O$ (5 mM), 3 mL of 14.6 M H_3PO_4, and $Na_2B_4O_7{\cdot}10H_2O$ (20 mM) were charged into a 50-mL Teflon reactor, and heated at 200°C for 4 days. After cooling to room temperature, the solid products were washed extensively with hot water (80°C) until the soluble components were completely removed. Sky-blue crystals of the title compound were obtained. $Na_6Cu_3\{B_6P_6O_{27}(O_2BOH)_3\}{\cdot}2H_2O$ was synthesized by the reaction of $Cu(NO_3)_2$ and $NH_4H_2PO_4$ with an excess of H_3BO_3 at 170°C in a closed system, where the boric acid acted as both reactant and flux (Liu et al. 2006).

2.13 BORON–HYDROGEN–SODIUM–SILVER–OXYGEN–PHOSPHORUS

The $\mathbf{Na_{1.89}Ag_{0.11}[BP_2O_7(OH)]}$ multinary phase, which crystallizes in the orthorhombic structure with the lattice parameters a = 683.98 ± 0.14, b = 2,086.5 ± 0.4, and c = 1,318.9 ± 0.3 pm and a calculated density of 2.711 g cm^{-3}, is formed in the B–H–Na–Ag–O–P system (Kniep and Engelhardt 1998). It was synthesized under hydrothermal conditions from a mixture of $AgNO_3$, $Na_2HPO_4{\cdot}2H_2O$, and $Na_2B_4O_7{\cdot}10H_2O$. The nitrate solution (pH = 1.5–2.0) was placed in a Teflon autoclave (20 mL; degree of filling 50%–60%) and heated at 165°C–170°C under autogenous pressure from a few days up to 3 weeks. Colorless crystals of the title compound were obtained by filtration; then, they were washed with hot H_2O and dried at 50°C–70°C.

2.14 BORON–HYDROGEN–SODIUM–BERYLLIUM–OXYGEN–PHOSPHORUS

The **$NaBe[BP_2O_8]\cdot 0.33H_2O$** multinary compound, which crystallizes in the cubic structure with the lattice parameters $a = 1{,}241.30 \pm 0.01$ pm, is formed in the B–H–Na–Be–O–P system (Zhang et al. 2003). In a typical reaction, a mixture of $Be(OH)_2$, H_3BO_3, 85% H_3PO_4, NaOH, and distilled water (molar ratio 1:2:7.2:1:70) was mixed with stirring for 1 h, until the components were dissolved completely, and the pH of the mixture was 2.0. After stirring at room temperature for 2 h, the clear solution was transferred to a Teflon-coated steel autoclave and heated at 180°C for 4 days. The crystalline products were recovered by filtration and washed with distilled water.

2.15 BORON–HYDROGEN–SODIUM–MAGNESIUM–OXYGEN–PHOSPHORUS

The **$NaMg(H_2O)_2[BP_2O_8]\cdot H_2O$** and **$Na_2Mg[B_3P_2O_{11}(OH)]\cdot 0.67H_2O$** multinary compounds are formed in the B–H–Na–Mg–O–P system. The first of them crystallizes in the hexagonal structure with the lattice parameters $a = 942.8 \pm 0.5$ and $c = 1{,}582.0 \pm 0.8$ pm and a calculated density of 2.42 g cm^{-3} (Kniep et al. 1997a, 1997b; Baykal et al. 2000). It was synthesized under hydrothermal conditions (Baykal et al. 2000). The hydrothermal synthesis was started from mixtures (totally 8 g) of $MgCl_2\cdot 6H_2O$, $NaBO_3\cdot 4H_2O$, and $(NH_4)_2HPO_4$ in various molar ratios (Baykal et al. 2000). Required amount of H_2O and concentrated HNO_3 was added to dissolve this solid mixture at 90°C, and then the total volume was reduced to 10 mL by evaporation of water. The highly viscous solution ($pH < 1$) was filled into a Teflon-coated autoclave (degree of filling 60%) and treated at 160°C for 3 days.

$NaMg(H_2O)_2[BP_2O_8]\cdot H_2O$ was also synthesized under mild hydrothermal conditions by Kniep et al. (1997a, 1997b). The experiment was carried out at 150°C in sealed glass ampoules (10–12 mL, degree of filling 25%–30%) by mixing $MgHPO_4$ (1 mM), $Na_2B_4O_7$ (2 mM), 18% HCl (0.85 mL), and H_2O (1 mL) within approximately 2 weeks. Solid reaction product (colorless crystals) was obtained after washing with hot water (~50% with respect to the amount of solid starting material).

$Na_2Mg[B_3P_2O_{11}(OH)]\cdot 0.67H_2O$ also crystallizes in the hexagonal structure with the lattice parameters $a = 1{,}177.1 \pm 0.1$ and $c = 1{,}210.0 \pm 0.1$ pm and a calculated density of 2.537 g cm^{-3} (Yang et al. 2006c). It was synthesized under hydrothermal conditions. $Mg(NO_3)_2\cdot 2H_2O$ (5 mM), 3 mL of 14.6 M H_3PO_4, and $Na_2B_4O_7\cdot 10H_2O$ (20 mM) were charged into a 50-mL Teflon reactor and heated at 200°C for 4 days. After cooling to room temperature, the solid products were washed extensively with hot water (80°C) until the soluble components were completely removed. Colorless crystals of the title compound were obtained.

2.16 BORON–HYDROGEN–SODIUM–ZINC–OXYGEN–PHOSPHORUS

Three multinary compounds are formed in the B–H–Na–Zn–O–P system.

$NaZn[BP_2O_8]{\cdot}H_2O$ crystallizes in the hexagonal structure with the lattice parameters $a = 954.04 \pm 0.02$ and $c = 1{,}477.80 \pm 0.03$ pm and (Boy et al. 2001f). It was obtained at a dehydration of **$NaZn(H_2O)_2[BP_2O_8]{\cdot}H_2O$.**

$NaZn(H_2O)_2[BP_2O_8]{\cdot}H_2O$ also crystallizes in the hexagonal structure with the lattice parameters $a = 946.2 \pm 0.2$ and $c = 1{,}583 \pm 5$ pm and a calculated density of 2.736 g cm^{-3} (Boy et al. 2001f) [$a = 945.6 \pm 0.2$ and $c = 1{,}582.8 \pm 0.5$ pm (Kniep et al. 1997a, 1997b)]. Colorless crystals of this compound were prepared under mild hydrothermal conditions in a Teflon autoclave (20 mL) (Kniep et al. 1997a, 1997b; Boy et al. 2001f). For this, $ZnCl_2$ (1.05 g), $Na_2B_4O_7{\cdot}10H_2O$ (1.46 g), and $Na_2HPO_4{\cdot}2H_2O$ (4.83 g) (molar ratio 2:1:7) were heated at 80°C–100°C in demineralized water (10 mL) and treated under stirring with 37% HCl (5 mL), until the components were completely dissolved. The clear solution was concentrated to about 12 mL by heating to give a highly viscous gel (pH = 1), which was then transferred to a Teflon autoclave (degree of filling 50%–55%) and stored at a temperature of 120°C–170°C for up to 2 weeks. The crystalline reaction product was separated from the mother liquor by vacuum filtration, washed with demineralized water, and dried at 60°C.

$Na_2Zn[B_3P_2O_{11}(OH)]{\cdot}0.67H_2O$ also crystallizes in the hexagonal structure with the lattice parameters $a = 1{,}193.6 \pm 0.2$ and $c = 1{,}236.3 \pm 0.3$ pm and a calculated density of 2.683 g cm^{-3} (Yang et al. 2006c). It was synthesized under hydrothermal conditions. $Zn(CH_3COO)_2{\cdot}H_2O$ (5 mM), 3 mL of 14.6 M H_3PO_4, and $Na_2B_4O_7{\cdot}10H_2O$ (20 mM) were charged into a 50-mL Teflon reactor and heated at 200°C for 4 days. After cooling to room temperature, the solid products were washed extensively with hot water (80°C) until the soluble components were completely removed. Colorless crystals of the title compound were obtained.

2.17 BORON–HYDROGEN–SODIUM–CADMIUM–OXYGEN–PHOSPHORUS

The **$NaCd(H_2O)_2[BP_2O_8]{\cdot}0.8H_2O$** multinary compound, which crystallizes in the hexagonal structure with the lattice parameters $a = 971.3 \pm 0.1$ and $c = 1{,}613.6 \pm 0.3$ pm, is formed in the B–H–Na–Cd–O–P system (Ge et al. 2003b). Colorless crystals of this compound were obtained under mild hydrothermal conditions. A mixture of $CdCl_2{\cdot}2.5H_2O$ (0.459 g), $Na_2B_4O_7{\cdot}10H_2O$ (1.900 g), $NaBO_2{\cdot}4H_2O$ (6.436 g), and 5 mL of 86% H_3PO_4 was heated at 90°C in deionized H_2O (10 mL) under constant stirring until the components were completely dissolved. The clear solution (pH = 2) was transferred to a Teflon autoclave (degree of filling 70%) and heated at 170°C for 4 days.

2.18 BORON–HYDROGEN–SODIUM–ALUMINUM–OXYGEN–PHOSPHORUS

The **$NaAl[BP_2O_7(OH)_3]$** multinary compound, which crystallizes in the monoclinic structure with the lattice parameters $a = 1{,}049.7 \pm 0.2$, $b = 799.3 \pm 0.2$, $c = 907.7 \pm 0.2$ pm, and $\beta = 117.26 \pm 0.03°$, is formed in the B–H–Na–Al–O–P system (Koch and Kniep 1999). It was prepared by hydrothermal treatment of a mixture of $Al(H_2PO_4)_3$, $Na_2B_4O_7{\cdot}10H_2$), and 85% H_3PO_4 in a Teflon autoclave (degree of filling 75%) at 165°C for 2 weeks.

2.19 BORON–HYDROGEN–SODIUM–GALLIUM–OXYGEN–PHOSPHORUS

The **NaGa[BP$_2$O$_7$(OH)$_3$]** multinary compound, which crystallizes in the monoclinic structure with the lattice parameters *a* = 1,040.8 ± 0.3, *b* = 809.4 ± 0.2, *c* = 909.9 ± 0.2 pm, and *β* = 116.64 ± 0.02°, is formed in the B–H–Na–Ga–O–P system (Huang et al. 2001a). It was synthesized under mild hydrothermal conditions. The reaction was carried out with a mixture of Ga (0.139 g) dissolved in 1 mL of 18% HCl with $Na_2HPO_4 \cdot 12H_2O$ (1.075 g) and $Na_2B_4O_7 \cdot 10H_2O$ (0.4767 g) (molar ratio of Ga/P/B = 2:3:6) in aqueous solution. The mixture was sealed in glass tubes (after adding 1 mL H_2O to achieve a degree of filling 30%) with subsequent heating at 135°C for 60 days.

2.20 BORON–HYDROGEN–SODIUM–INDIUM–OXYGEN–PHOSPHORUS

The **NaIn[BP$_2$O$_7$(OH)$_3$]** and **NaIn[BP$_2$O$_8$(OH)]** multinary compounds are formed in the B–H–Na–In–O–P system. The first of them crystallizes in the monoclinic structure with the lattice parameters *a* = 1,036.8 ± 0.2, *b* = 852.0 ± 0.1, *c* = 941.5 ± 0.2 pm, and *β* = 115.951 ± 0.005° (Huang et al. 2002a). It was synthesized under mild hydrothermal conditions. The reaction was carried out with a mixture of In (0.230 g) dissolved in 1 mL of 18% HCl with $Na_2HPO_4 \cdot 12H_2O$ (1.074 g) and $Na_2B_4O_7 \cdot 10H_2O$ (1.525 g) (molar ratio of In/P/B = 2:3:16) in aqueous solution. The mixture was sealed in glass tubes (after adding 1 mL H_2O to achieve a degree of filling 30%) with subsequent heating at 135°C for 90 days.

NaIn[BP$_2$O$_8$(OH)] also crystallizes in the monoclinic structure with the lattice parameters *a* = 517.7 ± 0.1, *b* = 1,681.5 ± 0.3, *c* = 768.4 ± 0.2 pm, and *β* = 94.10 ± 0.03° (Mao et al. 2004a) [*a* = 518.2 ± 0.3, *b* = 769.6 ± 0.5, *c* = 1,685 ± 1 pm, and *β* = 94.07 ± 0.05° and a calculated density of 3.513 ± 0.007 g cm^{-3} (Gurbanova et al. 2002)]. It was prepared from a mixture of In_2O_3, B_2O_3, and Na_3PO_4 (mass ratio 1:1:1) and 0.1 M HCl, which was heated in an autoclave at 280°C for 18–20 days (Gurbanova et al. 2002).

2.21 BORON–HYDROGEN–SODIUM–SCANDIUM–OXYGEN–PHOSPHORUS

The **NaSc[BP$_2$O$_6$(OH)$_3$][(HO)PO$_3$]** multinary compound, which crystallizes in the monoclinic structure with the lattice parameters *a* = 500.10 ± 0.01, *b* = 1,242.71 ± 0.09, *c* = 1,583.40 ± 0.14 pm, and *β* = 94.201 ± 0.004°, is formed in the B–H–Na–Sc–O–P system (Menezes et al. 2010a). Transparent, colorless single crystals of the title compound were prepared under hydrothermal conditions from a mixture of $Na_2B_4O_7 \cdot H_2O$ (2.7653 g), Sc_2O_3 (0.5002 g), and 85% H_3PO_4 (2.5079 g) in the molar ratio Na/Sc/B/P = 2:1:2:6. The mixture was heated whilst stirring together with water (10 mL). The pH was adjusted to 1 by the addition of 37% HCl (1.0 mL) and the system was heated at 90°C for 3 h. The reaction mixture was then filled into a Teflon-lined steel autoclave (degree of filling 45%) and heated under autogeneous pressure at 220°C for 7 weeks. The reaction product was separated by filtration, washed with hot water/acetone, and finally dried at 60°C in air. Single crystals were isolated from the mixture.

2.22 BORON–HYDROGEN–SODIUM–LEAD–OXYGEN–PHOSPHORUS

The $\mathbf{Na_3Pb[B(O_3POH)_4]}$ multinary compound, which crystallizes in the tetragonal structure with the lattice parameters $a = 691.82 \pm 0.08$ and $c = 2{,}730.9 \pm 0.3$ pm, is formed in the B–H–Na–Pb–O–P system (Hoffmann et al. 2008). To prepare the title compound, a total of 0.5 g of $PbB_2O_4{\cdot}H_2O$ and 1.145 g of $Na_2HPO_4{\cdot}2H_2O$ were dissolved in 8 mL of deionized water together with 1 mL of 85% H_3PO_4 and stirred at 80°C for 1 h. The pH of the solution was found to be 2. The white gel was transferred into a 10-mL Teflon autoclave (degree of filling 35%) and heated at 170°C for 5 days. The colorless crystals were separated from the mother liquor by vacuum filtration, washed with water/acetone, and dried in air at 60°C.

2.23 BORON–HYDROGEN–SODIUM–TITANIUM–OXYGEN–PHOSPHORUS

The $\mathbf{[Na_{0.5}(H_3O)_{0.5}]\{Ti_2[B(PO_4)_4]\}{\cdot}H_2O}$ multinary compound, which crystallizes in the tetragonal structure with the lattice parameters $a = 630.99 \pm 0.02$ and $c = 1{,}643.72 \pm 0.07$ pm, is formed in the B–H–Na–Ti–O–P system (Zhou et al. 2011b). The synthesis of the title compound was performed under hydrothermal conditions. First, Ti (1.966 g) was dissolved in concentrated HCl (100 mL) and kept in a sealed bottle for several days to get a clear blue solution, which is named afterwards Ti*. Then, the mixture of Ti* (5 mL), H_3BO_3 (1.236 g), NaCl (0.29 g), and 85% H_3PO_4 (1 mL) (Na/Ti/B/P molar ratio 2.5:1:10:10) was transferred to a Teflon autoclave. The resulting pH values were lower than 1. The synthesis was conducted at 180°C for 3 days. After that, the autoclave was directly taken out of the oven. The product was filtered and washed with deionized water for several times and dried at 50°C for 8 h.

2.24 BORON–HYDROGEN–SODIUM–VANADIUM–OXYGEN–PHOSPHORUS

The $\mathbf{NaV[BP_2O_7(OH)_3]}$ and $\mathbf{Na_2V[B_3P_2O_{12}(OH)]{\cdot}2.92H_2O}$ multinary compounds are formed in the B–H–Na–V–O–P system. The first of them crystallizes in the monoclinic structure with the lattice parameters $a = 1{,}041.52 \pm 0.03$, $b = 823.56 \pm 0.09$, $c = 919.63 \pm 0.06$ pm, and $\beta = 116.550 \pm 0.001°$ (Zhang et al. 2002). Green blocks of this compound were prepared under mild hydrothermal conditions from a mixture of VCl_3 (1.573 g), Na_2HPO_4 (1.419 g), H_3BO_3 (1.22 g), and 85% H_3PO_4 (2 mL) (molar ratio 1:1:2:3; concentrated solution; pH = 0–1). The reaction mixture was heated in a Teflon autoclave (20 mL; degree of filling about 50%) at 170°C under autogenous pressure for 3 days.

$Na_2V[B_3P_2O_{12}(OH)]{\cdot}2.92H_2O$ crystallizes in the cubic structure with the lattice parameter $a = 2{,}010.07 \pm 0.07$ and a calculated density of 2.222 g cm^{-3} (Yang et al. 2008c). The title compound was prepared by a boric acid flux method in the reaction system with molar compositions of $16.2H_3BO_3 + 1.0V_2O_5 + (1.0–1.4)Na_2HPO_4 + 12H_2O$. Typically, a mixture of H_3BO_3 (1 g), V_2O_5 (0.182 g), and $Na_2HPO_4{\cdot}12H_2O$ (0.5 g) was directly added into a 15-mL Teflon-lined stainless steel autoclave and heated at 200°C for 5 days. The final product containing large bright blue single crystals in the form of dodecahedron was washed with hot water (50°C) until the residual H_3BO_3 was completely removed and then dried in air. This compound could also be synthesized using $NaH_2PO_4{\cdot}2H_2O$ and NH_4VO_3 as phosphorus and vanadium sources, respectively, under similar conditions.

2.25 BORON–HYDROGEN–SODIUM–OXYGEN–PHOSPHORUS

The **$Na_2[BP_2O_7(OH)]$** quinary compound, which crystallizes in the orthorhombic structure with the lattice parameters $a = 682.36 \pm 0.01$, $b = 2{,}079.11 \pm 0.04$, and $c = 1{,}314.46 \pm 0.03$ pm and a calculated density of 2.64 g cm^{-3}, is formed in the B–H–Na–O–P system (Kniep and Engelhardt 1998). It was synthesized under hydrothermal conditions from a mixture of $NaNO_3$, $Na_2HPO_4{\cdot}2H_2O$, and $Na_2B_4O_7{\cdot}10H_2O$. The nitrate solution (pH = 1.5–2.0) was placed in a Teflon autoclave (20 mL; degree of filling 50%–60%) and heated at 165°C–170°C under autogenous pressure from a few days to 3 weeks. Colorless crystals of the title compound were obtained by filtration; then, they were washed with hot H_2O and dried at 50°C–70°C.

2.26 BORON–HYDROGEN–SODIUM–OXYGEN–CHROMIUM–PHOSPHORUS

The **$Na_8[Cr_4B_{12}P_8O_{44}(OH)_4][P_2O_7]{\cdot}nH_2O$** multinary compound, which crystallizes in the cubic structure with the lattice parameter $a = 2{,}002.42 \pm 0.03$ and a calculated density of 2.31 g cm^{-3}, is formed in the B–H–Na–O–Cr–P system (Yang et al. 2008b). It can be synthesized by a hydrothermal reaction method. Typically, $Cr(NO_3)_3{\cdot}9H_2O$ (2.5 mM) was dissolved in 14.6 M H_3PO_4 (3 mL) in a 50-mL Teflon reactor at 80°C. Then, $Na_2B_4O_7{\cdot}10H_2O$ (10 mM) was added, and the Teflon reactor was sealed in a stainless steel container. After heating at 220°C for 4 days and cooling to room temperature for 12 h, the solid sample was washed extensively with hot water (80°C) until the soluble components were completely removed. Dark green single crystals with a truncated cubic appearance were obtained. The reaction temperature and time are not constant and can vary from 180°C to 240°C and from 3 days to 2 weeks.

2.27 BORON–HYDROGEN–SODIUM–OXYGEN–TUNGSTEN–PHOSPHORUS

The **$K_5[W_4O_8(H_2BO_4)(HPO_4)_2(PO_4)_2]{\cdot}0.5H_2O$** multinary compound, which is stable in air up to about 380°C, has a continuous mass loss of total 4% to 1,000°C and crystallizes in the monoclinic structure with the lattice parameters $a = 2{,}698.7 \pm 0.5$, $b = 979.4 \pm 0.2$, $c = 954.07 \pm 0.19$ pm, and $\beta = 99.98 \pm 0.03°$ and a calculated density of 4.108 g cm^{-3}, is formed in the B–H–Na–O–W–P system (Mao et al. 2006). The synthesis of this compound can be successfully carried out under mild hydrothermal conditions. $Na_2WO_4{\cdot}2H_2O$ (1.652 g), KH_2PO_4 (4.080 g), and H_3BO_3 (1.850 g) were mixed in molar ratios of 1:6:6, and then 3 mL of H_2O and 4.631 g of KCl were added. The resulting solution was transferred into a Teflon-lined autoclave and heated at 124°C for 6 days. Colorless crystals of the title compound remained after filtration. The crystals were washed with water and air dried.

2.28 BORON–HYDROGEN–SODIUM–OXYGEN–CHLORINE–PHOSPHORUS

The **$Na_{13}(H_3O)_2[B_6P_{11}O_{42}(OH)_2]Cl_2{\cdot}H_2O$** multinary compound, which crystallizes in the cubic structure with the lattice parameter $a = 2{,}047.9 \pm 0.7$ and a calculated density of 2.378 g cm^{-3}, is formed in the B–H–Na–O–Cl–P system (Feng et al. 2013). It was synthesized from a mixture of H_3BO_3, $Na_2HPO_4{\cdot}12H_2O$, $SmCl_3{\cdot}6H_2O$, and H_2O (molar ratio 10:9:6:135) under hydrothermal conditions. The mixture was stirred for 30 min, and then transferred into a Teflon-lined stainless steel autoclave (50 mL) and heated at 170°C for 10 days. After the mixture was slowly cooled to room temperature, colorless polyhedral crystals were obtained. The products were filtered off, washed with deionized water and acetone, and dried at ambient temperature.

2.29 BORON–HYDROGEN–SODIUM–OXYGEN–CHLORINE–COBALT–PHOSPHORUS

The $\mathbf{Na_6Co_3B_2P_5O_{21}Cl{\cdot}H_2O}$ multinary compound, which crystallizes in the orthorhombic structure with the lattice parameters $a = 1{,}406.38 \pm 0.08$, $b = 988.13 \pm 0.07$, $c = 1{,}400.08 \pm 0.10$ pm and a calculated density of 2.541 g cm^{-3}, is formed in the B–H–Na–O–Cl–Co–P system (Su et al. 2011). The title compound was prepared under ionothermal conditions using the ionic liquid 1-ethyl-3-methylimidazolium bromide as a solvent (as ionic liquid 1-methyl-3-butylimidazolium bromide can be also used). The mixture of $CoCl_2{\cdot}6H_2O$ (0.238 g), NaH_2PO_4 (0.31 g), H_3BO_3 (0.062 g), and ionic liquid (1.0 g) (molar ratio 1:2:1:5.24) was heated in a 15-mL Teflon-lined autoclave under an autogeneous pressure of 200°C for 10–20 days followed by cooling to room temperature. The product was filtered off, washed with deionized water, rinsed with ethanol, purified ultrasonically, and dried in vacuum desiccators at ambient temperature. Single-phased dark-purple crystals were obtained.

2.30 BORON–HYDROGEN–SODIUM–OXYGEN–MANGANESE–PHOSPHORUS

Some multinary compounds are formed the B–H–Na–O–Mn–P system. $\mathbf{NaMn(H_2O)_2[(BP_2O_8)]{\cdot}H_2O}$ crystallizes in the hexagonal structure with the lattice parameters $a = 960.23 \pm 0.14$ and $c = 1{,}603.7 \pm 0.3$ pm and a calculated density of 2.57 g cm^{-3} at 100 K (Yakubovich et al. 2009a) [$a = 958.9 \pm 0.5$ and $c = 1{,}593.9 \pm 0.9$ and a calculated density of 2.57 g cm^{-3} (Kniep et al. 1997a, 1997b)]. Crystals of the title compound were synthesized under hydrothermal conditions at 280°C and 7 MPa in a Teflon-lined stainless steel autoclave (4 mL) (Yakubovich et al. 2009a). The mass ratio of the starting components of the system was $MnCl_2/Na_3PO_4/B_2O_3/H_2O$ = 2:1:4:30. The experiments were performed for 18 days. An excess amount of boron at pH = 6–7 promotes the crystallization of anionic borophosphate ribbons. Colorless transparent crystals were then obtained.

$NaMn(H_2O)_2[(BP_2O_8)]{\cdot}H_2O$ was also synthesized under mild hydrothermal conditions by Kniep et al. (1997a, 1997b). The experiment was carried out at 150°C in sealed glass ampoules (10–12 mL, degree of filling 25%–30%) by mixing $MnHPO_4$ (1 mM), $Na_2B_4O_7$ (2 mM), 18% HCl (0.85 mL), and H_2O (1 mL) within approximately 2 weeks. Solid reaction product (slightly pink crystals) was obtained after washing with hot water (~50% with respect to the amount of solid starting material).

$\mathbf{Na_2[MnB_3P_2O_{11}(OH)]{\cdot}0.67H_2O}$ also crystallizes in the hexagonal structure with the lattice parameters $a = 1{,}194.0 \pm 0.2$ and $c = 1{,}209.8 \pm 0.2$ pm and a calculated density of 2.670 g cm^{-3} (Yang et al. 2006c). It was synthesized under hydrothermal conditions. $Mn(CH_3COO)_2{\cdot}H_2O$ (5 mM), 3 mL of 14.6 M H_3PO_4, and $Na_2B_4O_7{\cdot}10H_2O$ (20 mM) were charged into a 50-mL Teflon reactor and heated at 200°C for 4 days. After cooling to room temperature, the solid products were washed extensively with hot water (80°C) until the soluble components were completely removed. Colorless crystals of the title compound were obtained.

$Na_5(H_3O)\{Mn_3[B_3O_3(OH)]_3(PO_4)_6\}·2H_2O$ also crystallizes in the hexagonal structure with the lattice parameters a = 1,196.83 ± 0.05 and c = 1,213.03 ± 0.06 pm and a calculated density of 2.642 g cm^{-3} (Yang et al. 2006a). The title compound can be prepared by a hydrothermal method. Typically, a mixture of H_3BO_3, $MnCl_2·4H_2O$, $Na_2HPO_4·12H_2O$, and 18% HCl (molar ratio 16.2:1.0:6.0:2.0–3.0) was added into a 15-mL Teflon-lined stainless steel autoclave and heated at 200°C for 5 days. The final product, containing large single crystals in the form of a hexagonal prism, was washed with hot water (50°C) until the residual H_3BO_3 was completely removed, and then it was dried in air. This compound has a high thermal stability and can keep its structure intact upon calcinations at 450°C.

Given the proximity of the chemical composition and parameters of the hexagonal cell, it is possible that $Na_2[MnB_3P_2O_{11}(OH)]·0.67H_2O$ and $Na_5(H_3O)\{Mn_3[B_3O_3(OH)]_3(PO_4)_6\}·2H_2O$ are in fact the same compound.

2.31 BORON–HYDROGEN–SODIUM–OXYGEN–IRON–PHOSPHORUS

Some multinary compounds are formed in the B–H–Na–O–Fe–P system. **$NaFe[BP_2O_7(OH)_3]$** crystallizes in the monoclinic structure with the lattice parameters a = 1,042.0 ± 0.2, b = 821.5 ± 0.1, c = 921.7 ± 0.1 pm, and β = 116.60 ± 0.01° and a calculated density of 2.962 g cm^{-3} (Boy et al. 1998a). It was synthesized under mild hydrothermal conditions. For this, $FeCl_3·6H_2O$ (1.54 mM), $Na_2B_4O_7·10H_2O$ (0.77 mM), $Na_2HPO_4·2H_2O$ (5.39 mM), and 18% HCl (1.5 mL) were placed into glass vials (degree of filling 30%). The synthesis was carried out at 170°C for 2 weeks. The resulting crystalline product was separated from the mother liquor, washed with hot H_2O, and dried at 60°C. Pale yellow crystals of the title compound were obtained.

$NaFe(H_2O)_2[(BP_2O_8)]·H_2O$ crystallizes in the hexagonal structure with the lattice parameters a = 946.7 ± 0.2 and c = 1,586.1 ± 0.2 pm (Boy et al. 2001d) [a = 949.9 ± 0.1 and c = 1,593.1 ± 0.2 (Kniep et al. 1997a, 1997b)]. It was prepared by hydrothermal treatment of a mixture of $Na_2B_4O_7·10H_2O$ (0.29 g), $NaH_2PO_4·2H_2O$ (0.96 g), and $FeCl_2·4H_2O$ (0.31 g) (molar ratio 2:1:7) (Boy et al. 2001d). A concentrated aqueous solution of pH = 1 (1.5 mL of 18% HCl) was held at 120°C for 2 weeks in a sealed glass ampoule (degree of filling 50%). Pale grey-purple, hexagonal bipyramids of the title compound were obtained.

$NaFe(H_2O)_2[(BP_2O_8)]·H_2O$ was also synthesized under mild hydrothermal conditions by Kniep et al. (1997a, 1997b). The experiment was carried out at 120°C in sealed glass ampoules (10–12 mL, degree of filling 25%–30%) by mixing $FeCl_2·4H_2O$ (2 mM), Na_2HPO_4 (7 mM), $Na_2B_4O_7·10H_2O$ (2 mM) 18% HCl (0.85 mL), and H_2O (0.7 mL) for approximately 2 weeks. A solid reaction product (light gray-violet crystals) was obtained after washing with hot water (~50% with respect to the amount of solid starting material).

$Na_2Fe[B_3P_2O_{11}(OH)]·0.67H_2O$ also crystallizes in the hexagonal structure with the lattice parameters a = 1,181.2 ± 0.2 and c = 1,206.7 ± 0.2 pm and a calculated density of 2.742 g cm^{-3} (Yang et al. 2006c). It was synthesized under hydrothermal conditions. $FeSO_4·7H_2O$ (5 mM),

3 mL of 14.6 M H_3PO_4, and $Na_2B_4O_7{\cdot}10H_2O$ (20 mM) were charged into a 50-mL Teflon reactor and heated at 200°C for 4 days. After cooling to room temperature, the solid products were washed extensively with hot water (80°C) until the soluble components were completely removed. Light yellow crystals of the title compound were obtained.

2.32 BORON–HYDROGEN–SODIUM–OXYGEN–COBALT–PHOSPHORUS

Some multinary compounds are formed in the B–H–Na–O–Co–P system. **$NaCo[BP_2O_8]{\cdot}H_2O$** crystallizes in the monoclinic structure with the lattice parameters $a = 654.7 \pm 0.1$, $b = 1{,}140.4 \pm 0.3$, $c = 965.0 \pm 0.1$ pm, and $\beta = 107.37 \pm 0.01°$ and a calculated density of 2.905 g cm^{-3} (Guesmi and Driss 2004). It was synthesized under mild hydrothermal conditions. For this, an aqueous solution of $NaBO_3{\cdot}10H_2O$, $Co(NO_3)_2{\cdot}6H_2O$, and 85% H_3PO_4 with an Na/Co/P molar ratio of 1:1:2 was prepared. A glass tube quarter filled with the mixture was kept at 90°C for about 2 days; it was then sealed and heated to 300°C for 3 days under autogeneous pressure. Normal cooling to room temperature produced pink crystals of the title compound with a colorless phase.

$NaCo(H_2O)_2[(BP_2O_8)]{\cdot}H_2O$ crystallizes in the hexagonal structure with the lattice parameters $a = 944.7 \pm 0.5$ and $c = 1{,}583 \pm 1$ pm and a calculated density of 2.742 g cm^{-3} (Shi et al. 2003a) [$a = 945.5 \pm 0.1$ and $c = 1{,}584.7 \pm 0.1$ (Kniep et al. 1997a, 1997b)]. A brilliant purple octahedral single crystals was hydrothermally synthesized by the reaction of $CoCl_2{\cdot}6H_2O$, H_3BO_3 and 85% H_3PO_4 in NaOH aqueous solution of $CH_3(CH_2)_{15}N(CH_3)_3Br$ (Shi et al. 2003a).

$NaCo(H_2O)_2[(BP_2O_8)]{\cdot}H_2O$ was also synthesized under mild hydrothermal conditions by Kniep et al. (1997a, 1997b). The experiment was carried out at 150°C in sealed glass ampoules (10–12 mL, degree of filling 25%–30%) by mixing $CoCl_2$ (1 mM), $Na_2B_4O_7$ (1.85 mM), Na_2HPO_4 (6 mM), 18% HCl (0.8 mL), and H_2O (1 mL) for approximately 2 weeks. A solid reaction product (violet crystals) was obtained after washing with hot water (~50% with respect to the amount of solid starting material).

$Na_2Co[B_3P_2O_{11}(OH)]{\cdot}0.67H_2O$ also crystallizes in the hexagonal structure with the lattice parameters $a = 1{,}175.9 \pm 0.2$ and $c = 1{,}209.9 \pm 0.2$ pm and a calculated density of 2.781 g cm^{-3} (Yang et al. 2006c). It was synthesized under hydrothermal conditions. $Co(CH_3COO)_2{\cdot}H_2O$ (5 mM), 3 mL of 14.6 M H_3PO_4, and $Na_2B_4O_7{\cdot}10H_2O$ (20 mM) were charged into a 50-mL Teflon reactor and heated at 200°C for 4 days. After cooling to room temperature, the solid products were washed extensively with hot water (80°C) until the soluble components were completely removed. Purple crystals of the title compound were obtained.

$Na_5(H_3O)\{Co_3[B_3O_3(OH)]_3(PO_4)_6\}{\cdot}2H_2O$ also crystallizes in the hexagonal structure with the lattice parameters $a = 1{,}176.91 \pm 0.15$ and $c = 1{,}211.2 \pm 0.2$ pm and a calculated density of 2.764 g cm^{-3} (Yang et al. 2006a). The title compound can be prepared by hydrothermal method. Typically, a mixture of H_3BO_3, $CoCl_2{\cdot}6H_2O$, $Na_2HPO_4{\cdot}12H_2O$, and 18% HCl (molar ratio 16.2:1.0:6.0:2.0–3.0) was added into a 15-mL Teflon-lined stainless steel autoclave and heated at 200°C for 5 days. The final product, containing large single crystals in the form of a hexagonal prism, was washed with hot water (50°C) until the residual

H_3BO_3 was completely removed, and then it was dried in air. This compound has a high thermal stability and can keep their structure intact upon calcinations at 500°C.

Given the proximity of the chemical composition and parameters of the hexagonal cell, it is possible that $Na_2Co[B_3P_2O_{11}(OH)]{\cdot}0.67H_2O$ and $Na_5(H_3O)\{Co_3[B_3O_3(OH)]_3(PO_4)_6\}{\cdot}2H_2O$ are in fact the same compound.

2.33 BORON–HYDROGEN–SODIUM–OXYGEN–NICKEL–PHOSPHORUS

Some multinary compounds are formed in the B–H–Na–O–Ni–P system. **$NaNi(H_2O)_2[(BP_2O_8)]{\cdot}H_2O$** crystallizes in the hexagonal structure with the lattice parameters $a = 937.7 \pm 0.1$ and $c = 1{,}584.8 \pm 0.1$ pm at 20°C and 935.8 ± 0.1 and $c = 1{,}583.3 \pm 0.1$ pm at –75°C (Boy et al. 2001e) [$a = 937.1 \pm 0.1$ and $c = 1{,}583.1 \pm 0.1$ (Kniep et al. 1997a, 1997b)]. Yellow-green hexagonal bipyramids of the title compound were prepared by hydrothermal treatment of a mixture of $Na_2B_4O_7{\cdot}10H_2O$ (0.29 g), $NaH_2PO_4{\cdot}2H_2O$ (0.96 g), and $NiCl_2{\cdot}6H_2O$ (0.37 g) (molar ratio 2:1:7). A concentrated aqueous solution of pH = 1 (1.5 mL of 18% HCl was added) was held at 120°C for 2 weeks in a sealed glass ampoule (degree of filling 50%) (Boy et al. 2001e).

$NaNi(H_2O)_2[(BP_2O_8)]{\cdot}H_2O$ was also synthesized under mild hydrothermal conditions by Kniep et al. (1997a, 1997b). The experiment was carried out at 150°C in sealed glass ampoules (10–12 mL, degree of filling 25%–30%) by mixing $NiHPO_4$ (1 mM), $Na_2B_4O_7$ (2 mM), 18% HCl (0.85 mL), and H_2O (1 mL) for approximately 2 weeks. A solid reaction product (yellow-green crystals) was obtained after washing with hot water (~50% with respect to the amount of solid starting material).

$Na_2Ni[B_3P_2O_{11}(OH)]{\cdot}0.67H_2O$ also crystallizes in the hexagonal structure with the lattice parameters $a = 1{,}172.8 \pm 0.3$ and $c = 1{,}207.4 \pm 0.1$ pm and a calculated density of 2.799 g cm^{-3} (Yang et al. 2006c). It was synthesized under hydrothermal conditions. $Ni(CH_3COO)_2{\cdot}H_2O$ (5 mM), 3 mL of 14.6 M H_3PO_4, and $Na_2B_4O_7{\cdot}10H_2O$ (20 mM) were charged into a 50-mL Teflon reactor and heated at 200°C for 4 days. After cooling to room temperature, the solid products were washed extensively with hot water (80°C) until the soluble components were completely removed. Yellow crystals of the title compound were obtained.

$Na_5(H_3O)\{Ni_3[B_3O_3(OH)]_3(PO_4)_6\}{\cdot}2H_2O$ also crystallizes in the hexagonal structure with the lattice parameters $a = 1{,}171.71 \pm 0.05$ and $c = 1{,}207.59 \pm 0.07$ pm and a calculated density of 2.795 g cm^{-3} (Yang et al. 2006a). The title compound can be prepared by a hydrothermal method. Typically, a mixture of H_3BO_3, $NiCl_2{\cdot}6H_2O$, $Na_2HPO_4{\cdot}12H_2O$, and 18% HCl (molar ratio 16.2:1.0:6.0:2.0–3.0) was added into a 15-mL Teflon-lined stainless steel autoclave and heated at 200°C for 5 days. The final product, containing large single crystals in the form of a hexagonal prism, was washed with hot water (50°C) until the residual H_3BO_3 was completely removed, and then it was dried in air. This compound has a high thermal stability and can keep their structure intact upon calcinations at 550°C.

Given the proximity of the chemical composition and parameters of the hexagonal cell, it is possible that $Na_2Ni[B_3P_2O_{11}(OH)]{\cdot}0.67H_2O$ and $Na_5(H_3O)\{Ni_3[B_3O_3(OH)]_3(PO_4)_6\}{\cdot}2H_2O$ are in fact the same compound.

2.34 BORON–HYDROGEN–POTASSIUM–COPPER–OXYGEN–PHOSPHORUS

The $\mathbf{K_6Cu_2[B_4P_8O_{28}(OH)_6]}$ multinary compound, which crystallizes in the monoclinic structure with the lattice parameters $a = 961.8 \pm 0.1$, $b = 1{,}755.0 \pm 0.1$, $c = 942.0 \pm 0.1$ pm, and $\beta = 112.29 \pm 0.01°$, is formed in the B–H–K–Cu–O–P system (Boy et al. 1998c). It was synthesized under mild hydrothermal conditions. For this, $CuCl_2{\cdot}2H_2O$ (1.31 g), $K_2B_4O_7{\cdot}4H_2O$ (2.35 g), and KH_2PO_4 (8.38 g) (molar ratio 1:1:8) were dissolved in H_2O (20 mL), heated to 80°C with dropwise addition of 18% HCl (4.5 mL), and concentrated to 10 mL. The resulting highly viscous gel (pH = 1.5) was loaded in a Teflon autoclave (degree of filling 50%) and heated at 160°C for a week. The resulting crystalline products were separated from the mother liquor by filtration, washed with hot H_2O, and dried at 60°C. Turquoise crystals of the title compound were obtained.

2.35 BORON–HYDROGEN–POTASSIUM–BERYLLIUM–OXYGEN–PHOSPHORUS

The $\mathbf{KBe[BP_2O_8]{\cdot}0.33H_2O}$ multinary compound, which is thermally stable up to 800°C and crystallizes in the cubic structure with the lattice parameters $a = 1{,}242.7 \pm 0.6$ pm and a calculated density of 2.640 g cm^{-3}, is formed in the B–H–K–Be–O–P system (Zhang et al. 2003). In a typical reaction, a mixture of $Be(OH)_2$ (0.172 g), H_3BO_3 (0.26 g), 85% H_3PO_4 (6 g), KOH (0.236 g), and distilled water (5 mL) (molar ratio 1:2:7.2:1:70) were mixed with constant stirring for 1 h, until the components were dissolved completely, and the pH of the mixture was 2.0. After stirring at room temperature for 2 h, the clear solution was transferred to a Teflon-coated steel autoclave and heated at 160°C for 5 days. The crystalline products were recovered by filtration and washed with distilled water. Transparent, nearly globular crystals of the title compound were obtained.

2.36 BORON–HYDROGEN–POTASSIUM–MAGNESIUM–OXYGEN–PHOSPHORUS

The $\mathbf{KMg(H_2O)_2[BP_2O_8]{\cdot}H_2O}$ multinary compound, which crystallizes in the hexagonal structure with the lattice parameters $a = 946.3 \pm 0.1$ and $c = 1{,}581.5 \pm 0.1$, is formed in the B–H–K–Mg–O–P system (Kniep et al. 1997a, 1997b). It was synthesized under hydrothermal conditions. The experiment was carried out at 150°C in sealed glass ampoules (10–12 mL, degree of filling 25%–30%) by mixing $MgHPO_4$ (1 mM), $K_2B_4O_7$ (2 mM), 18% HCl (0.85 mL), and H_2O (1 mL) for approximately 2 weeks. Solid reaction product (colorless crystals) was obtained after washing with hot water (~50% with respect to the amount of solid starting material).

2.37 BORON–HYDROGEN–POTASSIUM–CALCIUM–OXYGEN–COBALT–PHOSPHORUS

The $\mathbf{K_{0.17}Ca_{0.42}Co(H_2O)_2BP_2O_8{\cdot}H_2O}$ multinary phase, which crystallizes in the hexagonal structure with the lattice parameters $a = 947.20 \pm 0.15$ and $c = 1{,}587.2 \pm 0.6$ pm and a calculated density of 2.722 g cm^{-3}, is formed in the B–H–K–Ca–O–Co–P system (Guesmi and Driss 2012). This compound was obtained during the preparation of $KCo(H_2O)_2BP_2O_8{\cdot}0.48H_2O$ (the reaction container was the source of calcium).

2.38 BORON–HYDROGEN–POTASSIUM–ALUMINUM–OXYGEN–PHOSPHORUS

The **$KAl[BP_2O_8(OH)]$** multinary compound, which crystallizes in the monoclinic structure with the lattice parameters a = 925.5 ± 0.4, b = 819.0 ± 0.1, c = 932.3 ± 0.1 pm, and β = 102.89 ± 0.03°, is formed in the B–H–K–Al–O–P system (Kniep et al. 2002b). It was prepared under mild hydrothermal conditions from a mixture of $Al(H_2PO_4)_3$ (2.54 g), $K_2B_4O_7{\cdot}4H_2O$ (2.44 g), K_2HPO_4 (2.35 g), and 85% H_3PO_4 (5.01 g, pH = 1.5). The molar ratio of K/Al/B/P amounted to 5:1:4:10. The syntheses were carried out in Teflon autoclaves (degree of filling 90%) at 165°C for 6 days.

2.39 BORON–HYDROGEN–POTASSIUM–ALUMINUM–OXYGEN–IRON–PHOSPHORUS

The **$K(Fe_{0.69}Al_{0.31})[BP_2O_8(OH)]$** multinary compound, which crystallizes in the triclinic structure with the lattice parameters a = 513.9 ± 0.2, b = 806.5 ± 0.4, and c = 829.0 ± 0.4 pm and α = 86.841 ± 0.008°, β = 80.346 ± 0.008°, and γ = 86.622 ± 0.008° and a calculated density of 2.984 g cm^{-3} at 100 K, is formed in the B–H–K–Al–O–Fe–P system (Yakubovich et al. 2010). Crystals of the title compound were synthesized under hydrothermal conditions at 280°C and 7 MPa in a standard 4-mL Teflon-lined autoclave. The mass ratio of the initial components of $AlCl_3/FeCl_3/K_3PO_4/B_2O_3/H_2O$ was 1:1:1:4:30. The experiment lasted for 18 days. Excess boron at pH = 6–7 facilitated crystallization. The crystals that were synthesized were colorless transparent triangular plates.

2.40 BORON–HYDROGEN–POTASSIUM–GALLIUM–OXYGEN–PHOSPHORUS

The **$KGa[BP_2O_7(OH)_3]$** multinary compound, which crystallizes in the monoclinic structure with the lattice parameters a = 1,086.3 ± 0.1, b = 816.23 ± 0.07, c = 930.5 ± 0.1 pm, and β = 116.587 ± 0.004°, is formed in the B–H–K–Ga–O–P system (Mi et al. 2002d). Colorless prisms of the title compound were prepared under mild hydrothermal conditions. The reaction was carried out with a mixture of Ga (0.139 g) dissolved in 1 mL of 18% HCl with KCl (0.447 g), H_3BO_3 (0.371 g), and $K_2HPO_4{\cdot}3H_2O$ (0.817 g) (molar ratio of Ga/P/B = 2:6:10) in aqueous solution. The mixture was sealed in glass tubes (after adding 1 mL H_2O to achieve a degree of filling 30%) with subsequent heating at 135°C for 14 days.

2.41 BORON–HYDROGEN–POTASSIUM–INDIUM–OXYGEN–PHOSPHORUS

The **$KIn[BP_2O_8(OH)]$** multinary compound, which crystallizes in the triclinic structure with the lattice parameters a = 526.38 ± 0.04, b = 847.91 ± 0.05, and c = 814.69 ± 0.09 pm and α = 91.741 ± 0.007°, β = 93.061 ± 0.007°, and γ = 79.823 ± 0.005°, is formed in the B–H–K–In–O–P system (Mao et al 2002b, 2004a). It was synthesized under mild hydrothermal conditions. The reaction was carried out with a mixture of In (0.057 g) dissolved in 1 mL of 18% HCl, H_3BO_3 (0.062 g), and K_2HPO_4 (0.522 g) (molar ratio of In/P/B = 1:2:3) in aqueous solution. The mixture was sealed in glass tubes (after adding 1 mL of H_2O to achieve a degree of filling 30%) with subsequent heating at 145°C for 14 days.

2.42 BORON–HYDROGEN–POTASSIUM–SCANDIUM–OXYGEN–PHOSPHORUS

The **$KSc[BP_2O_8(OH)]$** multinary compound, which crystallizes in the triclinic structure with the lattice parameters $a = 526.96 \pm 0.02$, $b = 827.39 \pm 0.08$, and $c = 838.90 \pm 0.05$ pm and $\alpha = 88.22 \pm 0.01°$, $\beta = 79.101 \pm 0.006°$, and $\gamma = 86.67 \pm 0.01°$, is formed in the B–H–K–Sc–O–P system (Menezes et al. 2006b). To prepare the title compound, the reaction was carried out with a mixture of Sc_2O_3 (0.25 g), $K_2B_4O_7{\cdot}4H_2O$ (0.2769 g), and K_2HPO_4 (1.1051 g) (molar ratio of K/Sc/B/P = 4:1:1:1.75). HCl (37%, 2.0 mL) was added to adjust the pH value to 1. The mixture was filled in a 10-mL Teflon autoclave (degree of filling 30%) and heated at 170°C under autogenous pressure for 2 weeks. The reaction product was washed with water and dried in air at 60°C.

2.43 BORON–HYDROGEN–POTASSIUM–YTTERBIUM–OXYGEN–PHOSPHORUS

The **$K_3Yb[OB(OH)_2]_2[HOPO_3]_2$** multinary compound, which crystallizes in the trigonal structure with the lattice parameters $a = 568.09 \pm 0.02$ and $c = 3{,}659.4 \pm 0.5$ pm and a calculated density of 2.942 g cm^{-3}, is formed in the B–H–K–Yb–O–P system (Zhou et al. 2011a). To synthesize the title compound, Yb_2O_3 (0.591g) was dissolved in 1 mL of concentrated HCl. Then, $K_2B_4O_7{\cdot}4H_2O$ (1.8334 g), K_2HPO_4 (3.658 g), and H_2O (4 mL) (molar ratio of K/Yb/B/P = 18:1:8:7) were mixed with the earlier solution, and the pH value was adjusted to 7 using 0.75 mL of concentrated HCl. The mixture was transferred to a Teflon vessel with a cover, which was placed in a steel autoclave. The synthesis was conducted at 180°C for 5 days. After that time, the autoclave was directly taken out of the hot oven. The product was filtered and washed with distilled water several times and dried at 50°C for 8 h.

2.44 BORON–HYDROGEN–POTASSIUM–LUTETIUM–OXYGEN–PHOSPHORUS

The **$K_3Lu[OB(OH)_2]_2[HOPO_3]_2$** multinary compound, which crystallizes in the trigonal structure with the lattice parameters $a = 566.68 \pm 0.02$ and $c = 3{,}669.2 \pm 0.2$ pm and a calculated density of 2.958 g cm^{-3}, is formed in the B–H–K–Lu–O–P system (Zhou et al. 2011a). To prepare this compound, $LuCl_3{\cdot}6H_2O$ (1.7340 g), $K_2B_4O_7{\cdot}4H_2O$ (1.8336 g), and K_2HPO_4 (3.6372 g) were dissolved in water (4 mL) and the pH value was adjusted to 6–7 using concentrated HCl. The suspension was transferred into a Teflon autoclave and maintained at 180°C for 3 days. Finally, the autoclave was directly taken out of the oven. Crystals settled at the bottom of the autoclave were washed with distilled water several times and dried at 50°C for 8 h.

2.45 BORON–HYDROGEN–POTASSIUM–TITANIUM–OXYGEN–PHOSPHORUS

The **$K\{Ti_2[B(PO_4)_4]\}{\cdot}H_2O$** multinary compound, which crystallizes in the tetragonal structure with the lattice parameters $a = 631.40 \pm 0.02$ and $c = 1{,}638.08 \pm 0.06$ pm and a calculated density of 2.765 g cm^{-3}, is formed in the B–H–K–Ti–O–P system (Zhou et al. 2011b). The synthesis of the title compound was performed under hydrothermal conditions. First, Ti (1.966 g) was dissolved in concentrated HCl (100 mL) and kept in a sealed bottle for several days to get a clear blue solution, which is named afterwards Ti*. Then, the mixture of Ti* (5 mL), H_3BO_3 (1.236 g), and KH_2PO_4 (1.02 g) (K/Ti/B/P molar ratio 3.75:1:10:3.75) was transferred to a Teflon autoclave. The resulting pH values were lower than 1. The synthesis was conducted at 180°C for 3 days. After that, the autoclave was directly taken out of the oven. The product was filtered and washed with deionized water for several times and dried at 50°C for 8 h. Light bluish white, bipyramid-shaped crystals of $K\{Ti_2[B(PO_4)_4]\}\cdot H_2O$ were obtained.

2.46 BORON–HYDROGEN–POTASSIUM–VANADIUM–OXYGEN–PHOSPHORUS

The **$KV[BP_2O_8(OH)]$** multinary compound, which crystallizes in the triclinic structure with the lattice parameters $a = 516.47 \pm 0.01$, $b = 805.85 \pm 0.02$, and $c = 835.67 \pm 0.02$ pm and $\alpha = 86.847 \pm 0.004°$, $\beta = 80.196 \pm 0.003°$, and $\gamma = 86.330 \pm 0.003°$, is formed in the B–H–K–V–O–P system (Lin et al. 2008a). Greenish platelets of the title compound were synthesized under mild hydrothermal conditions. The reaction was performed with a mixture of $K_2B_4O_7\cdot 4H_2O$ (3.06 g), VCl_3 (0.20 g), 85% H_3PO_4 (0.25 mL), and H_2O (2 mL) heated at 280°C for 5 days in a Teflon-lined stainless steel autoclave (25 mL). After the reaction, the autoclave was moved out of the oven and cooled down to ambient temperature. The reaction product was washed with hot water (60°C) until soluble substances were completely removed. Finally, the reaction product was dried in air at 60°C.

2.47 BORON–HYDROGEN–POTASSIUM–OXYGEN–PHOSPHORUS

Some quinary compounds are formed in the B–H–K–O–P system. **$K_3B(HPO_4)_3$** was obtained at the synthesis of $KMn[BP_2O_7(OH)_2]$ (Wang et al. 2012a).

$K_3[BP_3O_9(OH)_3]$ crystallizes in the monoclinic structure with the lattice parameters $a = 2{,}454.6 \pm 0.8$, $b = 736.3 \pm 0.2$, $c = 1{,}406.2 \pm 0.4$ pm, and $\beta = 118.35 \pm 0.02°$ and a calculated density of 2.471 g cm^{-3} (Ewald et al. 2005a). The title compound was prepared solvothermally in a Teflon-lined steel autoclave (28 mL; degree of filling 30%–40%) from a mixture of $K_2B_4O_7\cdot 4H_2O$ (4.910 mM) and KH_2PO_4 (58.920 mM) that were homogenized by grinding before H_3PO_4 (9.820 mM) was added. Ethanol was used as a solvent. After a reaction time of typically 4 days at 220°C, the autoclave was removed from the furnace and allowed to cool to room temperature. The raw product was separated from the mother liquor by vacuum filtration. The transparent and colorless crystalline samples were washed with hot water, filtered, and washed again with acetone before they were dried in air at 60°C.

$K_3[B_5PO_{10}(OH)_3]$ also crystallizes in the monoclinic structure with the lattice parameters $a = 710.8 \pm 0.4$, $b = 668.2 \pm 0.4$, $c = 1{,}227.6 \pm 0.5$ pm, and $\beta = 95.71 \pm 0.01°$ (Hauf and Kniep 1996b). It was synthesized by hydrothermal treatment of a concentrated solution of KOH, P_2O_5, and B_2O_3 at 150°C.

$K[B_6PO_{10}(OH)_4]$ crystallizes in the tetragonal structure with the lattice parameters $a = 1{,}209.66 \pm 0.13$ and $c = 759.05 \pm 0.07$ pm and a calculated density of 2.171 g cm^{-3}

(Boy and Kniep 1999). It was prepared under mild hydrothermal conditions (170°C) from a concentrated solution of $K_2B_4O_7{\cdot}4H_2O$, KH_2PO_4, and 18% HCl. The title compound is formed under strong acidic conditions (pH = 0.5). For this purpose, $K_2B_4O_7{\cdot}4H_2O$ (18.80 g) and KH_2PO_4 (1.05 g) (molar ratio 8:1) in 20 mL of 18% HCl were heated at 80°C and were then treated in a Teflon autoclave (degree of filling 80%) at 170°C for 2 days. The crystalline product was separated by filtration, washed with demineralized water, and dried at 60°C.

2.48 BORON–HYDROGEN–POTASSIUM–OXYGEN–MANGANESE–PHOSPHORUS

Three multinary compounds are formed in the B–H–K–O–Mn–P system. **$KMn[BP_2O_7(OH)_2]$** crystallizes in the monoclinic structure with the lattice parameters a = 665.94 ± 0.13, b = 1,204.9 ± 0.2, c = 979.0 ± 0.2 pm, and β = 109.12 ± 0.03° (Wang et al. 2012a). This compound can be made by ionothermally treating a mixture of $MnCl_2{\cdot}4H_2O$ (0.2985 g), KH_2PO_4 (0.2036 g), H_3BO_3 (0.03 g), and 1-butyl-1-methylpyrrolidinium bromide (0.5 g). The mixture was placed into a Teflon-lined stainless steel autoclave and allowed to react at 200°C for several days by cooling to room temperature. The products were filtered off, washed with deionization water and acetone, filtered by suction, and dried at 60°C for a day. Pure $KMn[BP_2O_7(OH)_2$ was obtained only for $MnCl_2{\cdot}4H_2O/KH_2PO_4$ molar ratios <3 and >1/3. If this ratio was >3, crystallization of another dense phase occurred; if it was less than 1/3, crystallization of $K_3B(HPO_4)_3$ occurred. $MnBr_2{\cdot}4H_2O$ can also be used as the manganese source.

$K[Mn(H_2O)_2(BP_2O_8)]{\cdot}H_2O$ crystallizes in the hexagonal structure with the lattice parameters a = 968.3 ± 0.4 and c = 1,613.9 ± 0.6 pm (Wang 2012) [a = 963.9 ± 0.1 and c = 1,593.1 ± 0.2 pm (Kniep et al. 1997a, 1997b)]. This compound was prepared as follows (Wang 2012). A mixture of $K_2B_4O_7{\cdot}4H_2O$ (1.5285 g), Ga_2O_3 (0.0371 g), MnO_2 (0.4366 g), $MnCO_3$ (0.5743 g), and 85% H_3PO_4 (3 mL) was sealed in a 25-mL Teflon-lined stainless steel vessel and heated at 175°C for about 20 days under autogeneous pressure and then cooled to room temperature. The resulting columnar colorless crystals of the title compound were collected and dried in air at ambient temperature.

$KMn(H_2O)_2[(BP_2O_8)]{\cdot}H_2O$ was also synthesized under mild hydrothermal conditions by Kniep et al. (1997a, 1997b). The experiment was carried out at 150°C in sealed glass ampoules (10–12 mL, degree of filling 25%–30%) by mixing $MnHPO_4$ (1 mM), $K_2B_4O_7$ (2 mM), 18% HCl (0.85 mL), and H_2O (1 mL) for approximately 2 weeks. A solid reaction product (slightly pink crystals) was obtained after washing with hot water (~50% with respect to the amount of solid starting material).

$K_5Mn_2B_2P_5O_{19}(OH)_2$ crystallizes in the monoclinic structure with the lattice parameters a = 1,449.39 ± 0.03, b = 925.39 ± 0.03, c = 1,480.31 ± 0.04 pm, and β = 101.460 ± 0.001° and a calculated density of 2.792 g cm^{-3} (Su et al. 2011). The title compound was prepared under ionothermal conditions using the ionic liquid 1-ethyl-3-methylimidazolium bromide as a solvent (1-methyl-3-butylimidazolium bromide can be also used as an ionic liquid). The mixture of $MnCl_2{\cdot}4H_2O$ (0.20 g), $K_2HPO_4{\cdot}3H_2O$ (0.20 g), H_3BO_3 (0.06 g), and ionic liquid (1.0 g) (molar ratio 1:0.9:1:5.24) was heated in a 15-mL Teflon-lined autoclave under autogeneous pressure at 200°C for 5 days followed by cooling to room temperature. The product was filtered off, washed with deionized water, rinsed with ethanol, purified ultrasonically, and dried in vacuum desiccators at ambient temperature. Light pink plate-like crystals were obtained.

2.49 BORON–HYDROGEN–POTASSIUM–OXYGEN–IRON–PHOSPHORUS

Some multinary compounds are formed in the B–H–K–O–Fe–P system. **$KFe[BP_2O_8(OH)]$** crystallizes in the monoclinic structure with the lattice parameters a = 937.2 ± 0.2, b = 814.6 ± 0.2, c = 958.7 ± 0.2 pm, and β = 101.18 ± 0.03° (Wang and Mudring 2011). It was prepared under ionothermal synthesis conditions using the ionic liquid 1-butyl-1-methylpyrrolidinium bromide as the solvent. A mixture of $FeCl_3{\cdot}4H_2O$, KH_2PO_4, H_3BO_3, and 1-butyl-1-methylpyrrolidinium bromide (molar ratio 3:3:1:1.5) was reacted in a 3-mL Teflon-lined stainless steel container at 200°C for 5 days followed by cooling to room temperature. The products were filtered off, washed with deionized water and acetone, filtered by suction, and dried at 60°C for a day.

$KFe(H_2O)_2[BP_2O_8]{\cdot}0.5H_2O$ crystallizes in the hexagonal structure with the lattice parameters a = 952.3 ± 0.1 and c = 1,599.8 ± 0.1 pm (Boy et al. 2001d) [951.0 ± 0.1 and c = 1,595.2 ± 0.4 pm for **$KFe(H_2O)_2[BP_2O_8]{\cdot}H_2O$** (Kniep et al. 1997a, 1997b)]. $KFe(H_2O)_2[BP_2O_8]{\cdot}0.5H_2O$ was prepared by hydrothermal treatment of mixtures of $K_2B_4O_7{\cdot}4H_2O$ (0.12 g), KH_2PO_4 (0.47 g), and $FeCl_2{\cdot}4H_2O$ (0.15 g) (molar ratio 2:1:7) (Boy et al. 2001d). A concentrated aqueous solution of pH = 0.5–1 (1.5 mL of 9% HCl was added) was held at 120°C for 2 weeks in a sealed glass ampoule (degree of filling 50%).

$KFe(H_2O)_2[(BP_2O_8)]{\cdot}H_2O$ was synthesized under mild hydrothermal conditions by Kniep et al. (1997a, 1997b). The experiment was carried out at 120°C in sealed glass ampoules (10–12 mL, degree of filling 25%–30%) by mixing $FeCl_2{\cdot}4H_2O$ (2 mM), K_2HPO_4 (7 mM), $K_2B_4O_7{\cdot}4H_2O$ (2 mM), 18% HCl (0.85 mL), and H_2O (0.7 mL) for approximately 2 weeks. A solid reaction product (light gray-violet crystals) was obtained after washing with hot water (~50% with respect to the amount of solid starting material).

$K_2Fe_2[B_2P_4O_{16}(OH)_2]$ crystallizes in the triclinic structure with the lattice parameters a = 516.7 ± 0.1, b = 808.9 ± 0.1, and c = 834.0 ± 0.1 pm and α = 87.06 ± 0.01°, β = 80.21 ± 0.01°, and γ = 86.59 ± 0.01° and a calculated density of 3.031 g cm^{-3} (Boy et al. 1998a). It was synthesized under mild hydrothermal conditions. For this, $FeCl_3{\cdot}6H_2O$ (1.54 mM), $K_2B_4O_7{\cdot}4H_2O$ (0.77 mM), K_2HPO_4 (5.39 mM), and 18% HCl (1.5 mL) were placed into glass vials (degree of filling 30%). The synthesis was carried out at 170°C for 2 weeks. The resulting crystalline product was separated from the mother liquor, washed with hot H_2O, and dried at 60°C. Pink crystals of the title compound were obtained.

2.50 BORON–HYDROGEN–POTASSIUM–OXYGEN–COBALT–PHOSPHORUS

The **$KCo(H_2O)_2BP_2O_8{\cdot}0.48H_2O$** multinary compound, which crystallizes in the hexagonal structure with the lattice parameters a = 952.31 ± 0.14 and c = 1,563.3 ± 0.2 pm and a calculated density of 2.787 g cm^{-3} (Guesmi and Driss 2012) [948.3 ± 0.1 and c = 1,582.7 ± 0.2 pm for **$KCo(H_2O)_2[BP_2O_8]{\cdot}H_2O$** (Kniep et al. 1997a, 1997b)], is formed in the B–H–K–O–Co–P system. $KCo(H_2O)_2BP_2O_8{\cdot}0.48H_2O$ was synthesized hydrothermally, and the reaction was carried out in thick-walled Pyrex tubes (Guesmi and Driss 2012). A mixture of KNO_3 (1.49 g), $Co(CH_3COO)_2$ (1.84 g), H_3BO_3 (0.91 g), and 85% H_3PO_4 (2 mL) was dissolved in deionized water (25 mL). The mixture was then heated at 180°C for 4 weeks, and the degree of filling (30%–50%) was modified to vary the autogeneous pressure. Hexagonal pyramidal and bipyramidal purple crystals of $KCo(H_2O)_2BP_2O_8{\cdot}0.48H_2O$ as the major phase, accompanied by some pink crystals of $K_{0.17}Ca_{0.42}Co(H_2O)_2BP_2O_8{\cdot}H_2O$, were obtained in the most filled tube.

$KCo(H_2O)_2[(BP_2O_8)]{\cdot}H_2O$ was synthesized under mild hydrothermal conditions by Kniep et al. (1997a, 1997b). The experiment was carried out at 150°C in sealed glass ampoules (10–12 mL, degree of filling 25%–30%) by mixing $CoCl_2$ (1 mM), $K_2B_4O_7$ (1.85 mM), K_2HPO_4 (6 mM), 18% HCl (0.8 mL), and H_2O (1 mL) for approximately 2 weeks. A solid reaction product (violet crystals) was obtained after washing with hot water (~50% with respect to the amount of solid starting material).

2.51 BORON–HYDROGEN–POTASSIUM–OXYGEN–NICKEL–PHOSPHORUS

Two multinary compounds are formed in the B–H–K–O–Ni–P system. $\mathbf{KNi(H_2O)_2[(BP_2O_8)]{\cdot}H_2O}$ crystallizes in the hexagonal structure with the lattice parameters $a = 939.2 \pm 0.1$ and $c = 1{,}584.2 \pm 0.2$ (Kniep et al. 1997a, 1997b). It was synthesized under mild hydrothermal conditions. The experiment was carried out at 150°C in sealed glass ampoules (10–12 mL, degree of filling 25%–30%) by mixing $NiHPO_4$ (1 mM), $K_2B_4O_7$ (2 mM), 18% HCl (0.85 mL), and H_2O (1 mL) for approximately 2 weeks. A solid reaction product (yellow-green crystals) was obtained after washing with hot water (~50% with respect to the amount of solid starting material).

$\mathbf{KNi_5[P_6B_6O_{23}(OH)_{13}]}$ crystallizes in the cubic structure with the lattice parameters $a = 1{,}346.7 \pm 0.2$ pm at 100 K (Yakubovich et al. 2009b). The title compound was synthesized under mild hydrothermal conditions. The starting materials $NiCl_2$, K_3PO_4, B_2O_3, and K_2CO_3 were mixed in distilled water (mass ratio 4:2:8:1:45) and placed in a Teflon-lined stainless steel autoclave. The pH of the initial solution was 5–6. The experiment was performed at 280°C and a pressure of 7 MPa over a period of 20 days. Light-green crystals of an octahedral shape were filtered off, washed with water, and dried in air.

2.52 BORON–HYDROGEN–RUBIDIUM–ALUMINUM–OXYGEN–PHOSPHORUS

The $\mathbf{RbAl[BP_2O_8(OH)]}$ multinary compound, which crystallizes in the monoclinic structure with the lattice parameters $a = 924.4 \pm 0.1$, $b = 840.3 \pm 0.1$, $c = 939.4 \pm 0.2$ pm, and $\beta = 103.388 \pm 0.007°$, is formed in the B–H–Rb–Al–O–P system (Mi et al. 2002f). Colorless prisms of the title compound were synthesized under mild hydrothermal conditions. The reaction was carried out with a mixture of RbCl (1.813 g), $Al(H_2PO_4)_3$ (3.179 g), H_3BO_3 (0.618 g), and 85% H_3PO_4 (4 mL) (molar ratio of Rb/Al/B/P = 1.5:1:1:3.6) in an aqueous solution. The mixture was filled in a Teflon autoclave (20 mL; degree of filling about 50%). The autoclave was placed in an oven with subsequent heating at 140°C for 7 days.

2.53 BORON–HYDROGEN–RUBIDIUM–ALUMINUM–OXYGEN–IRON–PHOSPHORUS

The **$Rb(Al_{0.68}Fe_{0.32})[BP_2O_8(OH)]$** multinary compound, which crystallizes in the monoclinic structure with the lattice parameters $a = 938.1 \pm 0.6$, $b = 839.8 \pm 0.5$, and $c = 957.9 \pm 0.6$ pm and $\beta = 102.605 \pm 0.010$ and a calculated density of 3.062 ± 0.003 g cm^{-3} at 100 K, is formed in the B–H–Rb–Al–O–Fe–P system (Yakubovich et al. 2010). Crystals of the title compound were synthesized under hydrothermal conditions at 280°C and 7 MPa in a standard 4-mL Teflon-lined autoclave. The mass ratio of the initial components of $AlCl_3$/$FeCl_3$/Rb_3PO_4/B_2O_3/H_2O was 2:1:1:4:30. The experiment lasted for 18 days. Excess boron at pH = 6–7 facilitated the crystallization. The crystals that were synthesized were colorless transparent parallelepipeds.

2.54 BORON–HYDROGEN–RUBIDIUM–GALLIUM–OXYGEN–PHOSPHORUS

The **$RbGa[BP_2O_8(OH)]$** multinary compound, which crystallizes in the monoclinic structure with the lattice parameters $a = 932.07 \pm 0.08$, $b = 836.86 \pm 0.07$, and $c = 953.71 \pm 0.09$ pm and $\beta = 102.527 \pm 0.003$, is formed in the B–H–Rb–Ga–O–P system (Mi et al. 2003d). It was synthesized under mild hydrothermal conditions. Reaction was carried out with mixtures of RbOH (1.538 g), $GaCl_3$ (0.35 g Ga dissolved in 2 mL of 37% HCl), H_3BO_3 (0.309 g), and 3 mL of 85% H_3PO_4 (molar ratio of Rb/Ga/B/P = 3:1:1:8). The mixture was filled in a 20-mL Teflon autoclave (degree of filling 50%). The autoclave was placed in an oven with subsequent heating at 140°C for 7 days.

2.55 BORON–HYDROGEN–RUBIDIUM–INDIUM–OXYGEN–PHOSPHORUS

The **$RbIn[BP_2O_8(OH)]$** multinary compound, which crystallizes in the triclinic structure with the lattice parameters $a = 531.57 \pm 0.05$, $b = 832.09 \pm 0.08$, and $c = 848.40 \pm 0.07$ pm and $\alpha = 87.35 \pm 0.01°$, $\beta = 80.23 \pm 0.01°$, and $\gamma = 86.78 \pm 0.01°$, is formed in the B–H–Rb–In–O–P system (Huang et al. 2002b). The title compound was synthesized under mild hydrothermal conditions. The reaction was carried out with a mixture of RbOH (6.15 g), $InCl_3$ (1.1058 g), H_3BO_3 (0.9277 g), and 1.5 mL of 85% H_3PO_4. HCl (1.0 mL of 37%) was added to adjust the pH value to 1.5. The mixture was filled in a 10-mL Teflon autoclave (degree of filling 60%) and heated at 170°C for 10 days under autogeneous pressure.

2.56 BORON–HYDROGEN–RUBIDIUM–SCANDIUM–OXYGEN–PHOSPHORUS

The **$RbSc[BP_2O_8(OH)]$** multinary compound, which crystallizes in the triclinic structure with the lattice parameters $a = 532.96 \pm 0.02$, $b = 839.19 \pm 0.08$, and $c = 843.19 \pm 0.05$ pm and $\alpha = 87.27 \pm 0.01°$, $\beta = 80.124 \pm 0.006°$, and $\gamma = 86.60 \pm 0.01°$, is formed in the B–H–Rb–Sc–O–P system (Menezes et al. 2006c). Single crystals of the title compound for X-ray diffraction (XRD) were isolated from a sample synthesized hydrothermally by using a mixture of Sc_2O_3 (0.25 g), RbOH (1.009 g), H_3BO_3 (0.6043 g), and 85% H_3PO_4 (2.8126 g) (molar ratio of Rb/Sc/B/P = 3:1:4:10). HCl (1.0 mL 37%) was added to adjust the pH value to 1. The mixture was placed in a 10-mL Teflon-lined autoclave (degree of filling 30%), treated under autogeneous pressure at 170°C for 10 days, cooled down to 60°C, and kept there for 12 h before cooling down to room temperature. The resulting product was filtered off, washed with water, and dried in air.

2.57 BORON–HYDROGEN–RUBIDIUM–TITANIUM–OXYGEN–PHOSPHORUS

The **$Rb\{Ti_2[B(PO_4)_4]\}\cdot H_2O$** multinary compound, which crystallizes in the tetragonal structure with the lattice parameters $a = 631.16 \pm 0.02$ and $c = 1{,}641.44 \pm 0.07$ pm, is formed in the B–H–Rb–Ti–O–P system (Zhou et al. 2011b). The synthesis of the title compound was performed under hydrothermal conditions. First, Ti (1.966 g) was dissolved in concentrated HCl (100 mL) and kept in a sealed bottle for several days to get a clear blue solution, which is named afterwards Ti*. Then, the mixture of Ti* (5 mL), H_3BO_3 (0.618 g), 85% H_3PO_4 (1 mL), and $RbOH{\cdot}xH_2O$ (0.375 g) (Rb/Ti/B/P molar ratio 1.85:1:5:10) was transferred to a Teflon autoclave. The resulting pH values were lower than 1. The synthesis was conducted at 180°C for 3 days. After that, the autoclave was directly taken out of the oven. The product was filtered and washed with deionized water for several times and dried at 50°C for 8 h. Crystals of $Rb\{Ti_2[B(PO_4)_4]\}\cdot H_2O$ were obtained.

2.58 BORON–HYDROGEN–RUBIDIUM–VANADIUM–OXYGEN–PHOSPHORUS

The **$RbV[BP_2O_8(OH)]$** multinary compound, which crystallizes in the monoclinic structure with the lattice parameters $a = 937.89 \pm 0.05$, $b = 832.96 \pm 0.04$, and $c = 967.01 \pm 0.03$ pm and $\beta = 102.0781 \pm 0.0008$, is formed in the B–H–Rb–V–O–P system (Engelhardt et al. 2000a). This compound was prepared under mild hydrothermal conditions from a mixture of 50% RbOH, VCl_3, H_3BO_3, and 85% H_3PO_4 (molar ratio 1:1:1:2; concentrated solution; pH = 0–1; 7.5–16 g total mass). The synthesis was carried in a Teflon autoclave (165°C, 3 days) under autogeneous pressure.

2.59 BORON–HYDROGEN–RUBIDIUM–OXYGEN–PHOSPHORUS

Two quinary compounds are formed the B–H–Rb–O–P system. **$Rb[B_2P_2O_8(OH)]$** is stable up to 430°C and crystallizes in the monoclinic structure with the lattice parameters $a = 676.2 \pm 0.3$, $b = 1{,}217.3 \pm 0.5$, $c = 885.0 \pm 0.4$ pm, and $\beta = 97.28 \pm 0.01°$ and a calculated density of 2.887 g cm^{-3} (Hauf and Kniep 1997b). The crystals of the title compound were grown under hydrothermal conditions from a mixture of P_2O_5, B_2O_3, and RbCl (molar ratio 3:3:8), which was dissolved in H_2O (15 mL), heated, and concentrated to a volume of about 10 mL. The highly viscous gel was kept in a Teflon-lined autoclave (degree of filling about 50%) for 4 days at 150°C. The crystalline product was isolated from the mother liquor, washed with H_2O, and dried at 60°C.

$Rb_3[B_2P_3O_{11}(OH)_2]$ also crystallizes in the monoclinic structure with the lattice parameters $a = 781.6 \pm 0.2$, $b = 667.3 \pm 0.2$, $c = 2{,}424.8 \pm 0.5$ pm, and $\beta = 92.88 \pm 0.01°$ and a calculated density of 3.055 g cm^{-3} (Ewald et al. 2004a, 2005c). The title compound was prepared solvothermally in a Teflon-lined steel autoclave (28 mL; degree of filling 30%–40%) from a mixture of RbOH (32.35 mM), H_3BO_3 (16.17 mM), and 85% H_3PO_4 (32.35 mM) that was homogenized before the reaction. Ethanol was used as a solvent. After a reaction time of typically 4 days at 170°C, the autoclave was removed from the furnace and allowed to cool to room temperature. The raw product was separated from the mother liquor by vacuum filtration. The transparent and colorless crystalline samples were washed with hot water, filtered, and washed again with acetone before they were dried in air at 60°C.

2.60 BORON–HYDROGEN–RUBIDIUM–OXYGEN–IRON–PHOSPHORUS

The **$RbFe[BP_2O_8(OH)]$** multinary compound, which crystallizes in the monoclinic structure with the lattice parameters $a = 935.8 \pm 0.5$, $b = 833.9 \pm 0.6$, $c = 956.6 \pm 0.5$ pm, and $\beta = 101.69 \pm 0.04°$ and a calculated density of 3.232 g cm^{-3}, is formed the B–H–Rb–O–Fe–P system (Kniep et al. 1999a). It was prepared under mild hydrothermal conditions (165°C–170°C) from a mixture of RbOH (aqueous solution), $FeCl_2·4H_2O$, H_3BO_3, and 85% H_3PO_4 (molar ratio 1:1:1:2).

2.61 BORON–HYDROGEN–RUBIDIUM–OXYGEN–COBALT–PHOSPHORUS

The **$Rb_2Co_3(H_2O)_2[B_4P_6O_{24}(OH)_2]$** multinary compound, which crystallizes in the orthorhombic structure with the lattice parameters $a = 950.1 \pm 0.1$, $b = 1{,}227.2 \pm 0.2$, and $c = 2{,}007.4 \pm 0.2$ pm and a calculated density of 2.925 g cm^{-3}, is formed the B–H–Rb–O–Co–P system (Engelhardt et al. 2000c). This compound was prepared under mild hydrothermal conditions from a mixture of RbOH (2.62 g), $CoCl_2$ (1.66 g), H_3BO_3 (0.79 g), and 85% H_3PO_4 (2.95 g) (molar ratio 1:1:1:2). The starting reactants were mixed with stirring in approximately 10 mL of H_2O. The deep dark violet solution (pH = 0–1) was concentrated to about 8 mL and heated in a Teflon autoclave (10 mL; degree of filling 75%–80%) at 165°C for 3 days. The resulting rose- to blood-red colored crystalline reaction products were filtered with suction, washed several times with water, and dried at 80°C.

2.62 BORON–HYDROGEN–CESIUM–ALUMINUM–OXYGEN–PHOSPHORUS

The **$CsAl[BP_2O_8(OH)]$** multinary compound, which crystallizes in the monoclinic structure with the lattice parameters $a = 920.8 \pm 0.2$, $b = 869.57 \pm 0.09$, $c = 946.9 \pm 0.1$ pm, and $\beta = 104.189 \pm 0.006°$, is formed in the B–H–Cs–Al–O–P system (Mi et al. 2002c). Colorless octahedra of the title compound were synthesized under mild hydrothermal conditions. The reaction was carried out with a mixture of CsOH (1.679 g), $Al(H_2PO_4)_3$ (3.179 g), H_3BO_3 (0.618 g), and 85% H_3PO_4 (2 mL) (molar ratio of Cs/Al/B/P = 1:1:1:3.3). The mixture was filled in a Teflon autoclave (20 mL; degree of filling ~50%). The autoclave was placed in an oven with subsequent heating at 140°C for 7 days.

2.63 BORON–HYDROGEN–CESIUM–GALLIUM–OXYGEN–PHOSPHORUS

The **$CsGa[BP_2O_8(OH)]$** multinary compound, which crystallizes in the monoclinic structure with the lattice parameters $a = 925.9 \pm 0.1$, $b = 864.62 \pm 0.09$, and $c = 961.5 \pm 0.1$ pm and $\beta = 103.059 \pm 0.006$, is formed in the B–H–Cs–Ga–O–P system (Mi et al. 2003a). Colorless transparent prisms of this compound were synthesized under mild hydrothermal conditions. The reaction was carried out with mixtures of $CsOH·H_2O$ (1.679 g), $GaCl_3$ (0.35 g Ga dissolved in 2 mL of 37% HCl), H_3BO_3 (0.618 g), LiH_2PO_4 (3.118 g), and 2 mL of 85% H_3PO_4 (molar ratio of Cs/Ga/B/Li/P = 2:1:2:6:12). The mixture was filled in a 20-mL Teflon autoclave. The autoclave was placed in an oven with subsequent heating at 170°C for 7 days.

2.64 BORON–HYDROGEN–CESIUM–SCANDIUM–OXYGEN–PHOSPHORUS

The **$CsSc[B_2P_3O_{11}(OH)_3]$** multinary compound, which crystallizes in the orthorhombic structure with the lattice parameters a = 1,305.29 ± 0.15, b = 1,834.03 ± 0.17, and c = 1,038.38 ± 0.12 pm and a calculated density of 2.773 g cm^{-3} (Menezes et al. 2007b) [a = 1,306.41 ± 0.07, b = 1,833.33 ± 0.08, and c = 1,038.86 ± 0.05 pm (Menezes et al. 2006d)], is formed in the B–H–Cs–Sc–O–P system. Single crystals of the title compound for XRD were isolated from a sample synthesized hydrothermally by using a mixture of Sc_2O_3 (0.25 g), CsOH (1.009 g), H_3BO_3 (0.6043 g), and 85% H_3PO_4 (2.8126 g) (molar ratio of Cs/Sc/B/P = 3:1:4:10). The pH value was adjusted to 1 by the addition of 1.0 mL of 37% HCl. The mixture was then transferred to a 10-mL Teflon autoclave (degree of filling 30%), treated under autogenous pressure at 170°C for 10 days, cooled down to 60°C and kept there for 12 h before cooling down to room temperature. The reaction product was separated by filtration, washed with water/acetone, and dried in air at 60°C. The product consisted of transparent platy crystals of the title compound.

2.65 BORON–HYDROGEN–CESIUM–TITANIUM–OXYGEN–PHOSPHORUS

The **$[Cs_{0.5}(H_3O)_{0.5}]\{Ti_2[B(PO_4)_4]\}$** multinary compound, which crystallizes in the tetragonal structure with the lattice parameters a = 631.48 ± 0.01 and c = 1,647.92 ± 0.04 pm, is formed in the B–H–Cs–Ti–O–P system (Zhou et al. 2011b). The synthesis of the title compound was performed under hydrothermal conditions. First, Ti (1.966 g) was dissolved in concentrated HCl (100 mL) and kept in a sealed bottle for several days to get a clear blue solution, which is named afterwards Ti*. Then, the mixture of Ti* (5 mL), H_3BO_3 (1.236 g), 85% H_3PO_4 (1 mL), and CsCl (0.84 g) (Cs/Ti/B/P molar ratio 5:1:10:10) was transferred to a Teflon autoclave. The resulting pH values were lower than 1. The synthesis was conducted at 180°C for 3 days. After that, the autoclave was directly taken out of the oven. The product was filtered and washed with deionized water for several times and dried at 50°C for 8 h.

2.66 BORON–HYDROGEN–CESIUM–VANADIUM–OXYGEN–PHOSPHORUS

The **$CsV_3(H_2O)_2[B_2P_4O_{16}(OH)_4]$** multinary compound, which crystallizes in the monoclinic structure with the lattice parameters a = 958.82 ± 0.15, b = 1,840.8 ± 0.4, c = 503.49 ± 0.03 pm, and β = 110.675 ± 0.004° and a calculated density of 3.161 g cm^{-3}, is formed in the B–H–Cs–V–O–P system (Engelhardt et al. 2000b). It was synthesized under hydrothermal conditions (165°C) from a mixture of CsOH (3.2 g), VCl_3 (1.67 g), H_3BO_3 (0.66 g), and 85% H_3PO_4 (2.46 g) (molar ratio 1:1:1:2). The starting materials were dissolved with stirring and heating in approximately 10 mL of H_2O. The obtained dark-green solution (pH = 0–1) was concentrated to about 8 mL and added to a 10-mL Teflon-lined stainless steel autoclave (degree of filling 85%–90%). After 3–4 days at 165°C, a green crystalline product was separated by suction, washed several times with H_2O, and dried at 80°C.

2.67 BORON–HYDROGEN–CESIUM–OXYGEN–PHOSPHORUS

The **$Cs[B_2P_2O_8(OH)]$** quinary compound, which is stable up to 430°C and crystallizes in the monoclinic structure with the lattice parameters a = 682.7 ± 0.3, b = 1,236.8 ± 0.5, c = 900.6 ± 0.5 pm, and β = 98.1 ± 0.1° and a calculated density of 3.189 g cm^{-3}, is formed in the B–H–Cs–O–P system (Hauf and Kniep 1997b). The crystals of the title compound were grown under hydrothermal conditions from a mixture of P_2O_5, B_2O_3, and CsCl (molar ratio 3:2:10), which was dissolved in H_2O (15 mL), heated, and concentrated to a volume of about 10 mL. The highly viscous gel was kept in a Teflon-lined autoclave (degree of filling of about 50%) for 4 days at 150°C. The crystalline product was isolated from the mother liquor, washed with H_2O, and dried at 60°C.

2.68 BORON–HYDROGEN–CESIUM–OXYGEN–IRON–PHOSPHORUS

The **$CsFe[BP_2O_8(OH)]$** multinary compound, which is stable up to 370°C (Kritikos et al. 2001) and crystallizes in the monoclinic structure with the lattice parameters a = 935.1 ± 0.2, b = 863.1 ± 0.4, c = 976.3 ± 0.2 pm, and β = 102.58 ± 0.02° (Engelhardt and Kniep 1999), is formed in the B–H–Cs–O–Fe–P system. Pale violet prisms of the title compound were obtained by hydrothermal reaction of 50% CsOH (2.78 g), $FeCl_3{\cdot}6H_2O$ (2.51 g), H_3BO_3 (0.57 g), and 85% H_3PO_4 (2.14 g) (molar ratio 1:1:1:2; concentrated solution; pH = 1) in a Teflon autoclave at 160°C–170°C for 6 days (Engelhardt and Kniep 1999).

This compound could also be prepared from a mixture of $FeCl_3$ (1.00 mmol), 85% H_3PO_4 (6.88 mM), H_3BO_3 (2.01 mM), CsCl (3.02 mM), and H_2O (2.0 mL) (Kritikos et al. 2001). The mixture was placed in a 5-mL Teflon-lined stainless steel autoclave. The pH was adjusted to 0.2 with concentrated HCl. The autoclave was heated under autogeneous pressure at 180°C for 6 days. Transparent, pale pink, rod-shaped crystals of $CsFe[BP_2O_8(OH)]$ were isolated.

2.69 BORON–HYDROGEN–CESIUM–OXYGEN–COBALT–PHOSPHORUS

The **$Cs_2Co_3(H_2O)_2[B_4P_6O_{24}(OH)_2]$** multinary compound, which crystallizes in the orthorhombic structure with the lattice parameters a = 955.26 ± 0.04, b = 1,231.90 ± 0.04, and c = 2,011.23 ± 0.08 pm, is formed in the B–H–Cs–O–Co–P system (Menezes et al. 2009a). The preparation of this compound was carried out hydrothermally by treating a mixture of CsOH (1.2037 g), $Co(CH_3COO)_2{\cdot}4H_2O$ (2.0003 g), B_2O_3 (1.1180 g), and 85% H_3PO_4 (5.5542 g) (molar ratio of Cs/Co/B/P = 1:1:2:6). Ten milliliters of deionized water was added, and the system was stirred at 100°C. The mixture was then filled into a 20-mL Teflon-lined autoclave (degree of filling 60%) and heated at 170°C under autogeneous pressure for 12 days. The reaction product was filtered, washed with deionized water/acetone, and finally dried in air at 60°C. Pink platelet crystals of the title compound were obtained.

2.70 BORON–HYDROGEN–COPPER–OXYGEN–PHOSPHORUS

Two quinary compounds are formed in the B–H–Cu–O–P system. **$Cu(H_2O)_2[B_2P_2O_8(OH)_2]$** crystallizes in the orthorhombic structure with the lattice parameters a = 1,348.7 ± 0.7,

$b = 833.1 \pm 0.4$, and $c = 782.7 \pm 0.4$ pm and a calculated density of 2.607 g cm^{-3} at 193 K (Shi et al. 2005). Investigation of phase changes between 50°C and 850°C showed that the sample was still crystalline below 300°C, then became amorphous between 300°C and 500°C, and above 600°C formed a new phase. This compound was synthesized by heating a mixture of $Cu(CH_3COO)_2{\cdot}H_2O$, H_3BO_3, 85% H_3PO_4, and H_2O (molar ratio 1:2:3:11) at 110°C for 3 days in a sealed thick-walled Pyrex tube. Blue polyhedral crystals of the title compound were obtained.

$Cu_2[B_2P_3O_{12}(OH)_3]$ crystallizes in the monoclinic structure with the lattice parameters $a = 618.95 \pm 0.06$, $b = 1{,}362.09 \pm 0.01$, $c = 1{,}193.73 \pm 0.11$ pm, and $\beta = 97.62°$ and a calculated density of 3.6501 g cm^{-3} (Duan et al. 2014). The title compound was synthesized using boric acid as flux in a closed system. A typical example was to transfer the homogenized mixture of $CuCl_2{\cdot}2H_2O$, $NH_4H_2PO_4$, and H_3BO_3 (molar ratio 1:1:10) in an agate mortar. Then, put the mixtures into a 23-mL Teflon-lined stainless steel autoclave and keep the autoclave at 170°C for 7 days. After cooling to room temperature, the opaque sticky liquid and pale-blue rod crystals were obtained. The pure single crystals were recovered by vacuum filtration, washed with water, and dried at 80°C.

2.71 BORON–HYDROGEN–SILVER–MAGNESIUM–OXYGEN–PHOSPHORUS

The **$AgMg(H_2O)_2[BP_2O_8]{\cdot}H_2O$** multinary compound, which crystallizes in the hexagonal structure with the lattice parameters $a = 945.77 \pm 0.04$ and $c = 1{,}583.01 \pm 0.13$ pm and a calculated density of 3.144 g cm^{-3}, is formed in the B–H–Ag–Mg–O–P system (Zouihri et al. 2011a). It was hydrothermally synthesized at 180°C for 7 days in a 25-mL Teflon-lined steel autoclave from a mixture of MgO, H_3BO_3, 85% H_3PO_4, $AgNO_3$, and 5 mL of distilled water (molar ratio 1:4:6:1:165). The brilliant colorless octahedral crystals were recovered and washed with hot water and then dried in air.

2.72 BORON–HYDROGEN–SILVER–OXYGEN–COBALT–PHOSPHORUS

The **$(Ag_{0.79}Co_{0.11})Co(H_2O)_2[BP_2O_8]{\cdot}0.67H_2O$** multinary phase, which crystallizes in the hexagonal structure with the lattice parameters $a = 943.21 \pm 0.11$ and $c = 1{,}575.0 \pm 0.4$ pm and a calculated density of 3.274 g cm^{-3}, is formed in the B–H–Ag–O–Co–P system (Zouihri et al. 2012). It was hydrothermally synthesized at 180°C for 7 days in a 25-mL Teflon-lined steel autoclave from the mixture of $CoCO_3$, H_3BO_3, 85% H_3PO_4, $AgNO_3$, and 5 mL of distilled water (molar ratio 1:4:6:1:165). The reaction product was separated by filtration, washed with hot water, and dried in air. Pink hexagonal bipyramidal crystals were obtained.

2.73 BORON–HYDROGEN–SILVER–OXYGEN–NICKEL–PHOSPHORUS

The **$(Ag_{0.57}Ni_{0.22})Ni(H_2O)_2[BP_2O_8]{\cdot}0.67H_2O$** multinary phase, which crystallizes in the hexagonal structure with the lattice parameters $a = 938.48 \pm 0.06$ and $c = 1{,}584.11 \pm 0.18$ pm and a calculated density of 3.145 g cm^{-3}, is formed in the B–H–Ag–O–Ni–P system (Zouihri et al. 2011b). The title compound was hydrothermally synthesized at 180°C for 7 days in a 25-mL Teflon-lined steel autoclave from the mixture of $NiCO_3$, H_3BO_3, 85% H_3PO_4, $AgNO_3$, and 5 mL of distilled water (molar ratio 1:4:6:1:165). The brilliant colorless octahedral crystals were recovered and washed with hot water and then dried in air.

2.74 BORON–HYDROGEN–MAGNESIUM–OXYGEN–PHOSPHORUS

Some quinary compounds are formed in the B–H–Mg–O–P system.

$Mg[BPO_4(OH)_2]$ crystallizes in the trigonal structure with the lattice parameters $a = 746.072 \pm 0.003$, $c = 1{,}263.283 \pm 0.005$ pm and a calculated density of 2.652 g cm^{-3} (Yang et al. 2008a). It was synthesized under hydrothermal conditions. Typically, $Mg(NO_3)_2 \cdot 6H_2O$ (4 mM), H_3BO_3 (20 mM), and 85% H_3PO_4 (3 mL) were charged into a 50-mL Teflon autoclave and heated at 220°C for 7 days. After cooling to the room temperature, the solid products were washed extensively with hot water (80°C). Small, colorless, needle-like crystals that often aggregate as balls were obtained.

$Mg[BPO_4(OH)_2](H_2O)_2$ crystallizes in the trigonal structure with the lattice parameters $a = 1{,}495.18 \pm 0.09$ and $c = 1{,}380.9 \pm 0.2$ pm (Shi et al. 2003b). To prepare it, a mixture of $MgCl_2 \cdot 6H_2O$, 85% H_3PO_4, H_3BO_3, C_5H_5N, $NH_2NHCSNH_2$, and H_2O (molar ratio 1:1.5:8:2.5:1:22) was loaded in a sealed thick wall Pyrex tube and then placed in an oven at 110°C for 6 days. Colorless cubic crystals were obtained after washing with deionized water.

$Mg_2[BP_2O_7(OH)_3]$ crystallizes in the triclinic structure with the lattice parameters $a = 645.2 \pm 0.1$, $b = 645.5 \pm 0.2$, and $c = 836.0 \pm 0.2$ pm and $\alpha = 82.50 \pm 0.01°$, $\beta = 82.56 \pm 0.02°$, and $\gamma = 80.98 \pm 0.02°$ (Hauf et al. 1999). Colorless platelets of this compound were prepared by hydrothermal reactions of H_3BO_3, $MgHPO_4 \cdot 3H_2O$, and $MgCl_2 \cdot 6H_2O$ (molar ratio 1:1:2; concentrated solution; pH = 0.5 by adding of HCl) at 160°C.

$(H_3O)Mg(H_2O)_2[BP_2O_8] \cdot H_2O$ crystallizes in the hexagonal structure with the lattice parameters $a = 944.62 \pm 0.07$ and $c = 1{,}575.9 \pm 0.2$ pm and a calculated density of $2.439 \pm 0{,}003$ g cm^{-3} (Yang et al. 2011b). The title compound was synthesized under mild hydrothermal conditions. $MgCl_2 \cdot 6H_2O$ (20 mM) and B_2O_3 (40 mM) were mixed in deionized water (40 mL) whilst stirring. H_3PO_4 (85%, 5.2 mL) and pyridine (4 mL) were added dropwise to the mixture in sequence. After being stirred homogeneously, the mixture was evaporated and condensed to about 20 mL in a water bath. Pyridine (2 mL) and 37% HCl (2 mL) were added to the mixture, which resulted in a clear gel with a pH value of 1. The gel was sealed in a Teflon-lined stainless steel autoclave and heated to 170°C for 3 days. The resulting colorless transparent crystals with a hexagonal bipyramids shape were recovered by filtration and dried for 1 day at 60°C.

$Mg_2(H_2O)[BP_3O_9(OH)_4]$ crystallizes in the orthorhombic structure with the lattice parameters $a = 709.44 \pm 0.05$, $b = 859.70 \pm 0.04$, and $c = 1{,}635.1 \pm 0.1$ pm and a calculated density of 2.547 g cm^{-3} (Ewald et al. 2005c). Single crystals of this compound were obtained under hydrothermal conditions from aqueous solutions of $Mg(OH)_2$ (25.720 mM), H_3BO_3 (12.859 mM), and 85% H_3PO_4 (38.580 mM), which was additionally acidified with 1 mL of HCl. This solution was concentrated by evaporation of water to form a transparent gel. Filled in a Teflon-lined autoclave (23 mL; degree of filling 10%–20%), the mixture was kept at 190°C in a furnace for 7 days. The raw product was separated from the mother liquor by vacuum filtration, washed with water and acetone, and finally dried at 60°C in air.

$Mg(H_2O)_2[B_2P_2O_8(OH)_2]·H_2O$ crystallizes in the monoclinic structure with the lattice parameters $a = 776.04 \pm 0.05$, $b = 1{,}464.26 \pm 0.09$, $c = 824.10 \pm 0.04$ pm, and $\beta = 90.25 \pm 0.01°$ and a calculated density of 2.298 g cm^{-3} (Ewald et al. 2005c). Single crystals of the title compound were obtained in the same way as $Mg_2(H_2O)[BP_3O_9(OH)_4]$ was prepared starting from a composition of $Mg(OH)_2$ (25.720 mM), H_3BO_3 (25.721 mM), and 85% H_3PO_4 (51.441 mM), which was additionally acidified with 1 mL of HCl.

$2MgO·P_2O_5·B_2O_3·7H_2O$ or **$Mg_2[B_2P_2O_{10}]·7H_2O$** is formed from aqueous solutions of boric acid and magnesium phosphates after 2 days at 60°C–80°C (Portnova et al. 1969). It is thermally unstable and decomposes at 250°C.

$Mg_3(H_2O)_6[B_2(OH)_6(PO_4)_2]$ (mineral lüneburgite) (Biltz and Marcus 1912) crystallizes in the triclinic structure with the lattice parameters $a = 634.75 \pm 0.06$, $b = 980.27 \pm 0.11$, and $c = 629.76 \pm 0.05$ pm and $\alpha = 84.46 \pm 0.01°$, $\beta = 106.40 \pm 0.01°$, and $\gamma = 96.40 \pm 0.01°$ and the calculated and experimental densities of 2.204 and 2.05 g cm^{-3}, respectively (Gupta et al. 1991). Synthetic $Mg_3(H_2O)_6[B_2(OH)_6(PO_4)_2]$ was prepared from the solutions of MgO and $MgHPO_4·3H_2O$ in boric acid, which were boiled for a few days (Berdesinski 1951).

2.75 BORON–HYDROGEN–MAGNESIUM–OXYGEN–COBALT–PHOSPHORUS

The **$Mg_{0.77}Co_{0.23}(H_2O)_2[B_2P_2O_8(OH)_2]·H_2O$** multinary compound, which crystallizes in the monoclinic structure with the lattice parameters $a = 775.91 \pm 0.05$, $b = 1{,}465.4 \pm 0.1$, $c = 823.82 \pm 0.05$ pm, and $\beta = 90.26 \pm 0.01°$, is formed in the B–H–Mg–O–Co–P system (Ewald et al. 2005b). A single phase product of the title compound was obtained hydrothermally from an aqueous solution of $Mg(OH)_2$, $CoCl_2$, H_3BO_3, and H_3PO_4. A mixture of $Mg(OH)_2$ (8.573 mM), $CoCl_2$ (4.287 mM), H_3BO_3 (45.687 mM), and 85% H_3PO_4 (68.589 mM) was placed in a 20-mL Teflon autoclave that was then filled with water up to a filling degree of 50%. The reaction was performed at 170°C. After reaction duration of 13 days, the autoclave was removed from the furnace and allowed to cool down to room temperature. The raw product was separated from the mother liquor by vacuum filtration and washed with water and acetone before drying in air at 60°C.

2.76 BORON–HYDROGEN–MAGNESIUM–OXYGEN–NICKEL–PHOSPHORUS

The **$(Ni_{3-x}Mg_x)[B_3P_3O_{12}(OH)_6]·6H_2O$** ($x \approx 1.5$) multinary phase, which is stable up to about 250°C and crystallizes in the trigonal structure with the lattice parameters $a = 1{,}495.7 \pm 1.0$ and $c = 1{,}381.2 \pm 0.6$ pm and a calculated density of 2.428 g cm^{-3}, is formed in the B–H–Mg–O–Ni–P system (Boy et al. 2001c). The title compound was prepared hydrothermally from a mixture of $NiCl_2·6H_2O$ (7.7 mM), $Mg(OH)_2$ (30.8 mM), and B_2O_3 (38.5 mM). The mixture was heated under constant stirring together with 20 mL of deionized water, and the volume was reduced by heating to approximately 10 mL. After adding 5 mL of 85% H_3PO_4, the resulting clear green gel (pH = 1) was filled into a Teflon autoclave (20 mL; degree of filling ~80%) and was allowed to react at 170°C for 3 days. The pale green reaction product was separated by filtration, washed with deionized water, and dried for 1 day at 60°C.

2.77 BORON–HYDROGEN–CALCIUM–OXYGEN–IRON–PHOSPHORUS

Two multinary compounds are formed in the B–H–Ca–O–Fe–P system.

$Ca_{0.5}Fe(H_2O)_2[BP_2O_8]\cdot H_2O$ crystallizes in the hexagonal structure with the lattice parameters $a = 950.91 \pm 0.04$ and $c = 1{,}573.4 \pm 0.1$ pm (Menezes et al. 2008d). To prepare the title compound, a mixture of $Ca(OH)_2$ (0.7453 g), $FeCl_2\cdot 4H_2O$ (1.000 g), B_2O_3 (0.7003 g), 85% H_3PO_4 (3.4793 g), and H_2O (10 mL) (molar ratio of Ca/Fe/B/P = 1:2:2:6) was slowly heated to form a gel. The pH of the solution was noted to be 1. The acidic mixture was then transferred into a 10-mL Teflon-lined autoclave (degree of filling about 30%). The autoclave was treated hydrothermally under autogenous pressure at 170°C for 10 days. The resulting product was filtered off, washed with hot water/acetone and dried at 70°C in air. Colorless hexagonal bipyramidal crystals were obtained.

$CaFe[BP_2O_7(OH)_3]$ crystallizes in the monoclinic structure with the lattice parameters $a = 1{,}023.3 \pm 0.2$, $b = 823.91 \pm 0.08$, $c = 915.9 \pm 0.1$ pm, and $\beta = 117.069 \pm 0.006°$ (Menezes et al. 2008c). To obtain it, the reaction was carried out hydrothermally by treating a mixture of $Ca(OH)_2$ (0.7453 g), $FeCl_2\cdot 4H_2O$ (1.0000 g), B_2O_3 (0.7003 g), and 85% H_3PO_4 (3.4793 g) (molar ratio of Ca/Fe/B/P = 1:2:2:6). The mixture was heated with 10 mL of deionized water for 3 h and filled into a 20-mL Teflon-lined autoclave (degree of filling about 30%). The autoclave was placed in an oven with subsequent heating at 170°C for 8 days. The reaction product was washed with deionized water and dried in air at 60°C. The prismatic colorless transparent crystals were obtained.

2.78 BORON–HYDROGEN–CALCIUM–OXYGEN–COBALT–PHOSPHORUS

The **$CaCo(H_2O)[BP_2O_8(OH)]\cdot H_2O$** multinary compound, which crystallizes in the triclinic structure with the lattice parameters $a = 657.93 \pm 0.03$, $b = 783.20 \pm 0.01$, and $c = 881.72 \pm 0.01$ pm and $\alpha = 68.785 \pm 0.007°$, $\beta = 82.719 \pm 0.10°$, and $\gamma = 73.985 \pm 0.09°$, is formed in the B–H–Ca–O–Co–P system (Menezes et al. 2009b). It was synthesized under mild hydrothermal conditions. A mixture of $Ca(OH)_2$ (4.01 mM), $Co(CH_3COO)_2\cdot 4H_2O$ (4.01 mM), B_2O_3 (8.02 mM), 85% H_3PO_4 (24.08 mM) (molar ratio of Ca/Co/B/P = 1:1:2:6), and deionized water (10 mL) was heated up to 100°C whilst stirring. Meanwhile, 37% HCl (1 mL) was added to adjust the pH value to 1. Afterwards, the reaction mixture was filled into a Teflon-lined autoclave (degree of filling 40%) and heated under autogenous pressure at 170°C for 7 days. Pink platelets of the title compound were isolated as a by-product.

2.79 BORON–HYDROGEN–CALCIUM–OXYGEN–NICKEL–PHOSPHORUS

Two multinary compounds are formed in the B–H–Ca–O–Ni–P system. **$Ca_{0.5}Ni(H_2O)_2[BP_2O_8]\cdot H_2O$** crystallizes in the hexagonal structure with the lattice parameters $a = 937.15 \pm 0.03$ and $c = 1{,}572.61 \pm 0.06$ pm (Menezes et al. 2007a). This compound was prepared by mild hydrothermal synthesis. A mixture of $Ca(OH)_2$ (0.2480 g), NiO (0.2500 g), B_2O_3 (0.4606 g), and 85% H_3PO_4 (2.3153 g) (molar ratio of Ca/Ni/B/P = 1:1:4:6) was treated hydrothermally. The mixture was placed in a 10-mL Teflon-lined autoclave (degree of filling 30%) and treated hydrothermally under autogenous pressure at 170°C for 8 days.

The resulting product was filtered off, washed with hot water, and dried in air. The yellow prisms of the title compound were obtained.

$CaNi[BP_2O_7(OH)_3]$ crystallizes in the monoclinic structure with the lattice parameters $a = 1{,}025.15 \pm 0.09$, $b = 833.64 \pm 0.05$, $c = 917.5 \pm 0.1$ pm, and $\beta = 116.34 \pm 0.04°$ (Menezes et al. 2006a). It was synthesized under mild hydrothermal conditions. The reaction was carried out with a mixture of $Ca(OH)_2$ (0.2480 g), NiO (0.2500 g), B_2O_3 (0.4606 g), and 85% H_3PO_4 (2.3153 g) (molar ratio of Ca/Ni/B/P = 1:1:4:6). The mixture was filled in a 10-mL Teflon-lined autoclave (degree of filling of about 30%). The autoclave was placed in an oven with subsequent heating at 170°C for 8 days. The product was washed with H_2O and dried in air at 60°C. The greenish crystals were obtained.

2.80 BORON–HYDROGEN–STRONTIUM–OXYGEN–IRON–PHOSPHORUS

The **$SrFe[BP_2O_8(OH)_2]$** multinary compound, which crystallizes in the triclinic structure with the lattice parameters $a = 667.04 \pm 0.12$, $b = 669.27 \pm 0.13$, and $c = 938.91 \pm 0.19$ pm and $\alpha = 109.829 \pm 0.005°$, $\beta = 102.068 \pm 0.006°$, and $\gamma = 103.151 \pm 0.003°$, is formed in the B–H–Sr–O–Fe–P system (Menezes et al. 2009c). Colorless and transparent single crystals of the title compound were prepared hydrothermally from a mixture of $Sr(OH)_2{\cdot}8H_2O$ (2.88 mM), $FeC_2O_4{\cdot}2H_2O$ (2.74 mM), B_2O_3 (5.5 mM), and 85% H_3PO_4 (16.57 mM) (molar ratio Sr/Fe/B/P = 1:1:2:6). The mixture was heated while stirring together with deionized water (10 mL) at 90°C for 3 h. The pH was adjusted to 1 by the addition of 37% HCl (0.5 mL). Afterwards, the reaction mixture was filled into a Teflon-lined autoclave (degree of filling 70%) and heated under autogenous pressure at 170°C for 8 days. The reaction product was separated by filtration, washed with hot water, filtered, washed again with acetone, and finally dried at 60°C in air.

2.81 BORON–HYDROGEN–BARIUM–URANIUM–OXYGEN–PHOSPHORUS

The **$Ba_5[(UO_2)(PO_4)_3(B_5O_9)]{\cdot}nH_2O$** multinary compound, which crystallizes in the tetragonal structure with the lattice parameters $a = 2{,}493.4 \pm 0.4$ and $c = 677.32 \pm 0.06$ pm, is formed in the B–H–Ba–U–O–P system (Wu et al. 2012). The crystals of this compound were obtained from a high-temperature solid-state reaction. The following compounds were used as initial components: H_3BO_3 (61.38 mg), BPO_4 (52.89 mg), $UO_2(NO_3)_2{\cdot}6H_2O$ (251.05 mg), and $Ba(CO_3)_2$ (197.34 mg). The components were ground in an agate mortar, placed into a platinum crucible, heated to 1,000°C, and then slowly (5°C h^{-1}) cooled to room temperature. The resulting mixture consisted of crystals of the title compound and a glassy mass.

2.82 BORON–HYDROGEN–BARIUM–OXYGEN–PHOSPHORUS

The **$Ba_{11}B_{26}O_{44}(PO_4)_2(OH)_6$** quinary compound, which crystallizes in the monoclinic structure with the lattice parameters $a = 689.09 \pm 0.14$, $b = 1{,}362.9 \pm 0.3$, $c = 2{,}585.1 \pm 0.2$ pm, and $\beta = 90.04 \pm 0.03°$ and a calculated density of 3.854 g cm^{-3}, is formed in the B–H–Ba–O–P system (Heyward et al. 2013). Crystals of this compound were grown through a direct hydrothermal synthesis using $Ba(OH)_2{\cdot}H_2O$ (0.21 mM), $(NH_4)_2B_{10}O_{16}{\cdot}8H_2O$ (0.66 mM), $BaHPO_4$ (0.17 mM), and 0.1 M NaOH (0.8 mL). All powders were sealed in welded silver ampoules and loaded into a cold seal autoclave (575°C, 0.1 GPa) for 6 days. The autoclave was cooled to room temperature and the crystalline products were washed with deionized water and air dried. This reaction produced the target phase as colorless needles in low yield.

2.83 BORON–HYDROGEN–BARIUM–OXYGEN–IRON–PHOSPHORUS

The **$BaFe[BP_2O_8(OH)]$** multinary compound, which crystallizes in the triclinic structure with the lattice parameters $a = 533.93 \pm 0.04$, $b = 801.79 \pm 0.09$, and $c = 834.4 \pm 0.1$ pm and $\alpha = 89.05 \pm 0.02°$, $\beta = 79.21 \pm 0.01°$, and $\gamma = 87.45 \pm 0.01°$, is formed in the B–H–Ba–O–Fe–P system (Menezes et al. 2008b). It was synthesized under mild hydrothermal conditions. The reaction was carried out by a mixture of $Ba(OH)_2{\cdot}8H_2O$ (0.8768 g), $FeC_2O_4{\cdot}2H_2O$ (1.0000 g), B_2O_3 (0.7740 g), and 85% H_3PO_4 (3.8451 g) (molar ratio Ba/Fe/B/P = 0.5:1:2:6). Ten milliliters of deionized water was added and the solution was stirred at 100°C for 3 h. The pH value was adjusted to 1 by the addition of 1.0 mL of 37% HCl. The mixture was filled into a 20-mL Teflon-lined autoclave (degree of filling about 30%). The autoclave was heated at 170°C under autogenous pressure for 8 days. The reaction product was washed with water/acetone and dried in air at 60°C. The colorless platelets of the title compound were obtained.

2.84 BORON–HYDROGEN–BARIUM–OXYGEN–COBALT–PHOSPHORUS

The **$BaCo[BP_2O_8(OH)]$** multinary compound, which crystallizes in the triclinic structure with the lattice parameters $a = 531.47 \pm 0.03$, $b = 799.26 \pm 0.08$, and $c = 829.36 \pm 0.08$ pm and $\alpha = 88.63 \pm 0.01°$, $\beta = 79.46 \pm 0.01°$, and $\gamma = 87.14 \pm 0.01°$, is formed in the B–H–Ba–O–Co–P system (Menezes et al. 2008a). The preparation of this compound was carried out hydrothermally by treating a mixture of $Ba(OH)_2{\cdot}8H_2O$ (0.6332 g), $Co(CH_3COO)_2{\cdot}4H_2O$ (1.0000 g), B_2O_3 (0.5590 g), and 85% H_3PO_4 (2.7771) (molar ratio of Ba/Co/B/P = 0.5:1:2:6). Ten milliliters of deionized water was added and the solution was stirred at 100°C. Meanwhile, 1.0 mL of 37% HCl was added to adjust the pH value to 1. The mixture was filled into a 20-mL Teflon-lined autoclave (degree of filling 40%) and heated at 170°C under autogenous pressure for 2 weeks. The reaction product was filtered, washed with deionized water/acetone, and finally dried in air at 60°C. The reaction product contained pink platelets of the title compound.

2.85 BORON–HYDROGEN–ZINC–LEAD–OXYGEN–PHOSPHORUS

The **$PbZn[BP_2O_8(OH)]$** multinary compound, which crystallizes in the triclinic structure with the lattice parameters $a = 521.72 \pm 0.02$, $b = 784.32 \pm 0.07$, and $c = 832.05 \pm 0.04$ pm and $\alpha = 89.429 \pm 0.008°$, $\beta = 79.239 \pm 0.005°$, and $\gamma = 87.536 \pm 0.008°$, is formed in the B–H–Zn–Pb–O–P system (Tanh Jeazet et al. 2006). This compound was synthesized under mild hydrothermal conditions from a mixture of ZnO (0.2618 g), $PbB_2O_4{\cdot}H_2O$ (0.5 g), H_3BO_3 (0.1987 g), and 85% H_3PO_4 (0.6304 g) (molar ratio 1:1:2:4). The pH of the solution was adjusted to 1.5 by adding 37% HCl. The mixture was filled into a 10-mL autoclave (degree of filling 30%) and heated at 170°C for 3 weeks under autogenous pressure. Colorless platelets of the title compound were obtained.

2.86 BORON–HYDROGEN–ZINC–OXYGEN–PHOSPHORUS

Two quinary compounds are formed in the B–H–Zn–O–P system. **Zn[BPO$_4$(OH)$_2$]** crystallizes in the orthorhombic structure with the lattice parameters a = 915.07 ± 0.03, b = 897.22 ± 0.03, and c = 1,059.19 ± 0.03 pm and a calculated density of 3.13 g cm^{-3} (Huang et al. 2008). The title compound was prepared from a mixture of ZnO (1.628 g), B_2O_3 (0.624 g), P_2O_5 (2.840 g), and H_2O (7.5 mL) by hydrothermal treatment. The mixture was stirred at 100°C till a homogeneous colorless gel was obtained, which was then transferred into a Teflon autoclave (20 mL; degree of filling 30%) and treated at 170°C for 3 days. After that, the autoclave was directly removed from the oven and cooled down to room temperature. Colorless platelet crystals were separated from the mother liquid by vacuum filtration followed by washing and drying at 60°C.

(H$_3$O)Zn(H$_2$O)$_2$BP$_2$O$_8$·H$_2$O crystallizes in the hexagonal structure with the lattice parameters a = 960.4 ± 0.4 and c = 1,529.7 ± 0.6 pm and a calculated density of 2.912 g cm^{-3} (Shi et al. 2003c). Single crystals of this compound were synthesized by a reaction of $Zn(NO_3)_2 \cdot 6H_2O$ and H_3BO_3 with 85% H_3PO_4 (molar ratio 1:3:2) in a mixed solvent of ethylene glycol (0.15 mL) and 18% HCl (0.1 mL). The starting materials were loaded in a thick-walled Pyrex tube, which was sealed in vacuum and held at 110°C for a week. The single-phase colorless octahedral crystals were obtained after washing with deionized water.

2.87 BORON–HYDROGEN–ALUMINUM–OXYGEN–PHOSPHORUS

The **Al[B$_2$P$_2$O$_7$(OH)$_5$]·H$_2$O** quinary compound, which crystallizes in the monoclinic structure with the lattice parameters a = 1,899.4 ± 0.4, b = 670.4 ± 0.1, c = 691.0 ± 0.1 pm, and β = 99.03 ± 0.03°, is formed in the B–H–Al–O–P system (Kniep et al. 2002a). Transparent prisms of the title compound was synthesized under mild hydrothermal conditions from a mixture of $Al(H_2PO_4)_3$ (2.12 g), H_3BO_3 (1.66 g), and 85% H_3PO_4 (2.36 g). The mixture was treated in a Teflon autoclave (degree of filling 70%) at 165°C for 6 days.

2.88 BORON–HYDROGEN–GALLIUM–OXYGEN–PHOSPHORUS

The **Ga[B$_2$P$_2$O$_7$(OH)$_5$]** quinary compound, which crystallizes in the monoclinic structure with the lattice parameters a = 1,764.04 ± 0.03, b = 670.735 ± 0.008, c = 699.525 ± 0.008 pm, and β = 109.157 ± 0.001° and a calculated density of 2.976 g cm^{-3}, is formed in the B–H–Ga–O–P system (Mao et al. 2004b). It was prepared under mild hydrothermal conditions. The reaction was carried out with a mixture of $GaCl_3$ (0.523 g Ga dissolved in 2 mL of 37% HCl), H_3BO_3 (1.237 g), and 85% H_3PO_4 (2.5 mL) (molar ratio of Ga/B/P = 1.5:4:7). The mixture was treated in a Teflon autoclave (degree of filling about 50%) at 170°C for 7 days.

2.89 BORON–HYDROGEN–INDIUM–OXYGEN–PHOSPHORUS

The **In(H$_2$O)$_2$[BP$_2$O$_8$]·H$_2$O** quinary compound, which crystallizes in the hexagonal structure with the lattice parameters a = 957.00 ± 0.06 and c = 1,587.0 ± 0.1 pm (Ewald et al. 2004d) [$a = 957.2 \pm 0.1$ and $c = 1{,}589.4 \pm 0.3$ pm (Ewald et al. 2004b); $a = 958.1 \pm 0.06$ and $c = 1590 \pm 2$ pm and a calculated density of 2.60 g cm^{-3} for **$In[BP_2O_8]{\cdot}0.8H_2O$** (Belokoneva et al. 2001b)], is formed in the B–H–In–O–P system. The hydrothermal preparation of $In(H_2O)_2[BP_2O_8]{\cdot}H_2O$ was performed with acidic aqueous suspensions of In_2O_3 (10.730 mM), H_3BO_3 (32.345 mM), and 85 % H_3PO_4 (53.904 mM) in a Teflon autoclave (20 mL) (Ewald et al. 2004b, 2004d). The mixture was homogenized in H_2O (5 mL) and then treated hydrothermally at 170°C for 20 days (pH = 1; degree of filling 70%). The product was washed with water and dried in air at 60°C. $In[BP_2O_8]{\cdot}0.8H_2O$ was synthesized hydrothermally at 260°C–280°C and ~7 MPa in an autoclave (Belokoneva et al. 2001b). A mixture of In_2O_3, B_2O_3, and P_2O_5 (molar ratio 1:3–4:2–3) was used. As a mineralizer, chlorine and lithium ions were present in the solution; pH of the solution was equal to 6.

2.90 BORON–HYDROGEN–SCANDIUM–OXYGEN–PHOSPHORUS

Two quinary compounds are formed in the B–H–Sc–O–P system. **$Sc(H_2O)_2[BP_2O_8]{\cdot}H_2O$** crystallizes in the hexagonal structure with the lattice parameters $a = 957.52 \pm 0.03$ and $c = 1{,}581.45 \pm 0.06$ pm and a calculated density of 2.378 g cm^{-3} (Ewald et al. 2006) [$a = 957.6 \pm 0.1$ and $c = 1{,}582.5 \pm 0.3$ pm (Ewald et al. 2004b)]. Single-phase samples of the title compound were yielded hydrothermally in a Teflon autoclave (20 mL) from acidic suspensions of In_2O_3, H_3BO_3, and 85% H_3PO_4 (Ewald et al. 2006). A mixture of Sc_2O_3 (5.44 mM), H_3BO_3 (14.51 mM), and 85% H_3PO_4 (21.76 mM) (molar ratio of Sc/B/P = 3:4:6) was homogenized in 10 mL of water and acidified with 1.5 mL of concentrated HCl. By evaporating the water, the mixture was concentrated to form a colorless nontransparent gel that was then treated hydrothermally at 170°C (degree of filling 30%). After a reaction time of 14 days, the autoclave was allowed to cool to room temperature, and the raw product was separated from the mother liquor by vacuum filtration. The transparent and colorless crystalline sample was washed with hot water, filtered, and washed again with acetone before they were dried in air at 60°C.

To characterize the dehydrated phase, $Sc(H_2O)_2[BP_2O_8]{\cdot}H_2O$ was heated to 300°C (Ewald et al. 2006). It was held at that temperature for 60 min, and was subsequently kept under inert conditions after being cooled to room temperature to avoid any rehydration. **$Sc(H_2O)_2[BP_2O_8]$** was obtained as a result of dehydration. This compound also crystallizes in the hexagonal structure with the lattice parameters $a = 953.5 \pm 0.1$ and $c = 1{,}576.8 \pm 0.2$. Rehydration appears as soon as the samples were exposed to air moisture for 24 h (Ewald et al. 2004b).

2.91 BORON–HYDROGEN–LEAD–OXYGEN–CHLORINE–COBALT–PHOSPHORUS

The **$Pb^{II}_4\{Co_2[B(OH)_2P_2O_8](PO_4)_2\}Cl$** multinary compound, which crystallizes in the trigonal structure with the lattice parameters $a = 975.13 \pm 0.07$ and $c = 9{,}106.0 \pm 1.3$ pm and a calculated density of 5.607 g cm^{-3}, is formed in the B–H–Pb–O–Cl–Co–P system (Yang et al. 2010a). The title compound was prepared under mild hydrothermal conditions from a mixture of $Co(CH_3COO)_2{\cdot}4H_2O$, $PbB_2O_4{\cdot}H_2O$, H_3BO_3, H_3PO_4, HCl, and H_2O (molar ratio 1:1:2:5:1.23:174). The mixture was filled into a 20-mL Teflon-lined stainless steel autoclave and heated for 5 days at 180°C under autogenous pressure. Blue crystals of this compound were obtained and separated by filtration, washed with deionized water, and dried at 60°C in air. It was found that the type of acid and molar ratio of Co/Pb have considerable effects on the formation of final products.

2.92 BORON–HYDROGEN–LEAD–OXYGEN–COBALT–PHOSPHORUS

The **$PbCo[BP_2O_8(OH)]$** multinary compound, which crystallizes in the triclinic structure with the lattice parameters $a = 522.08 \pm 0.02$, $b = 784.67 \pm 0.07$, and $c = 834.22 \pm 0.03$ pm and $\alpha = 89.473 \pm 0.008°$, $\beta = 79.219 \pm 0.004°$, and $\gamma = 87.521 \pm 0.007°$, is formed in the B–H–Pb–O–Co–P system (Tanh Jeazet et al. 2006). This compound was synthesized under mild hydrothermal conditions from a mixture of $Co(CH_3COO)_2{\cdot}4H_2O$ (0.4007 g), $PbB_2O_4{\cdot}H_2O$ (0.5 g), H_3BO_3 (0.1987 g), and 85% H_3PO_4 (0.7880 g) (molar ratio 1:1:2:5). One milliliter of 37% HCl was added to adjust the pH value to 1. The mixture was filled into a 10-mL autoclave (degree of filling 30%) and heated at 170°C for 3 weeks under autogenous pressure. Red platelets of the title compound were obtained.

2.93 BORON–HYDROGEN–TITANIUM–OXYGEN–PHOSPHORUS

Two quinary compounds are formed in the B–H–Ti–O–P system. **$(H_3O)\{Ti_2[B(PO_4)_4]\}{\cdot}H_2O$** crystallizes in the tetragonal structure with the lattice parameters $a = 632.10 \pm 0.02$ and $c = 1{,}643.6 \pm 0.1$ pm and a calculated density of 2.643 g cm^{-3} (Zhou et al. 2011b). The synthesis of the title compound was performed under hydrothermal conditions. First, Ti (1.966 g) was dissolved in concentrated HCl (100 mL) and kept in a sealed bottle for several days to get a clear blue solution, which is named afterwards Ti*. Then, the mixture of Ti* (5 mL), H_3BO_3 (1.236 g), and 85% H_3PO_4 (1 mL) (Ti/B/P molar ratio 1:10:1) was transferred to a Teflon autoclave. The resulting pH values were lower than 1. The synthesis was conducted at 180°C for 3 days. After that, the autoclave was directly taken out of the oven. The product was filtered and washed with deionized water for several times and dried at 50°C for 8 h. Colorless crystals of $(H_3O)\{Ti_2[B(PO_4)_4]\}{\cdot}H_2O$ were obtained.

$Ti[BP_2O_7(OH)_3]$ also crystallizes in the tetragonal structure with the lattice parameters $a = 631.3 \pm 0.2$ and $c = 1{,}644.6 \pm 0.6$ pm and a calculated density of 2.872 g cm^{-3} (Shi et al. 2011). It was synthesized from a mixture of $Ti[O(CH_2)_3Me]_4$, B_2O_3, H_3PO_4, and H_2O (molar ratio 2:3:9:75) by hydrothermal reaction (pH = 1.0). The mixture was stirred for 0.5 h, and then transferred into a Teflon-lined stainless steel autoclave (50 mL) and treated at 170°C for 18 days. After the mixture was slowly cooled to room temperature, colorless bipyramid-shaped crystals were obtained. The products were filtered off, washed with deionized water, purified ultrasonically, and dried in vacuum desiccators at ambient temperature.

Given the proximity of the chemical composition and parameters of the tetragonal cell, it is possible that $(H_3O)\{Ti_2[B(PO_4)_4]\}{\cdot}H_2O$ and $Ti[BP_2O_7(OH)_3]$ are in fact the same compound.

2.94 BORON–HYDROGEN–OXYGEN–MANGANESE–PHOSPHORUS

Three quinary compounds are formed in the B–H–O–Mn–P system. **$Mn[BPO_4(OH)_2]$** crystallizes in the tetragonal structure with the lattice parameters $a = 757.50 \pm 0.04$

and c = 1,292.68 ± 0.11 pm (Huang et al. 2006). This compound was synthesized under mild hydrothermal conditions. The starting materials $MnCl_2$ (0.76 g), H_3BO_3 (2.97 g), 1,4-diazobicyclo[2,2,2]octane ($C_6H_{12}N_2$, 2.69 g), and 85% H_3PO_4 (2.52 g) were heated at 60°C in deionized water (5 mL) under constant stirring for 2 h. Meanwhile, 37% HCl was added to adjust the pH value to 1.5. The clear solution was transferred into a Teflon autoclave (20 mL; degree of filling ~40%) and treated at 170°C for 3 days. Crystals were separated from the mother liquor by vacuum, washed with deionized water, and dried at 60°C in air. The title compound could also be synthesized from MnO, B_2O_3, P_2O_5, and H_2O.

$Mn^{II}_3B(PO_4)(OH)_6$ (mineral seamanite) crystallizes in the orthorhombic structure with the lattice parameters a = 782.31 ± 0.09, b = 1,514.05 ± 0.14, and c = 669.99 ± 0.07 pm (Huminicki and Hawthorne 2002) [a:b:c = 0.5195:1:0.4508 and a calculated density of 3.128 g cm^{-3} (Kraus et al. 1930); a = 783 ± 2, b = 1514 ± 2, and c = 671 ± 2 pm and the calculated and experimental densities of 3.09 and 3.08 g cm^{-3}, respectively (McConnell and Pondrom 1941); a = 786, b = 1483, and c = 670 pm (Kurkutova et al. 1971); a = 781.1 ± 0.5, b = 1,511.4 ± 1.0, and c = 669.1 ± 0.5 pm and the calculated and experimental densities of 3.132 and 3.128 g cm^{-3}, respectively (Moore and Ghose 1971)].

$(H)_{0.5}Mn_{1.25}(H_2O)_{1.5}[BP_2O_8]\cdot H_2O$ crystallizes in the hexagonal structure with the lattice parameters a = 965.47 ± 0.12 and c = 1,579.1 ± 0.3 pm and a calculated density of 2.462 g cm^{-3} (Yilmaz et al. 2005). The title compound was synthesized under mild hydrothermal conditions: $MnCl_2$ (3.580 g), H_3BO_3 1.758 g, $(NH_4)_2HPO_4$ 7.550 g, and 85% H_3PO_4 (7.712 g) were mixed in 5 mL of distilled water and treated under constant stirring with 85% H_3PO_4, until the components dissolved completely. The pH of the initial solution was 2. The clear solutions were heated without boiling to concentrate to about 15 mL, then transferred to Teflon-coated steel autoclaves (degree of filling 65%), and heated at 180°C for 2 days. The single-phase, pale-pink product was filtered off in a vacuum, washed with distilled water, and dried at 60°C.

2.95 BORON–HYDROGEN–OXYGEN–MANGANESE–COBALT–PHOSPHORUS

The **$(Co_{0.6}Mn_{0.4})_2(H_2O)[BP_3O_9(OH)_4]$** multinary phase, which crystallizes in the orthorhombic structure with the lattice parameters a = 713.55 ± 0.06, b = 873.21 ± 0.08, and c = 1,640.5 ± 0.2 pm, is formed in the B–H–O–Mn–Co–P system (Huang et al. 2009). To prepare this compound, the reaction was carried out by mixing MnO (0.710 g), CoO (0.750.g), P_2O_5 (4.260 g), and B_2O_3 (0.624 g) with 5 mL of deionized water. The mixture was stirred at 100°C for evaporation of water. The resulting violet gel was transferred into a Teflon autoclave (20 mL; degree of filling 30%; pH < 0.5) and was held at 170°C for 3 days. After vacuum filtration and washing, the purple crystalline compound was obtained.

2.96 BORON–HYDROGEN–OXYGEN–IRON–PHOSPHORUS

Some quinary compounds are formed in the B–H–O–Fe–P system. **$Fe[BPO_4(OH)_2]$** crystallizes in the tetragonal structure with the lattice parameters a = 748.44 ± 0.04 and c = 1,284.4 ± 1.1 pm (Huang et al. 2006). This compound was synthesized by the same way as $Mn[BPO_4(OH)_2]$ was obtained using $FeCl_2\cdot 4H_2O$ (1.18 g) instead of $MnCl_2$.

$Fe(H_2O)_2BP_2O_8·H_2O$ crystallizes in the hexagonal structure with the lattice parameters $a = 945.83 \pm 0.08$ and $c = 1{,}570.7 \pm 0.2$ pm and a calculated density of 2.543 g cm^{-3} (Yilmaz et al. 2000). The title compound was synthesized under mild hydrothermal conditions: $FeCl_2·4H_2O$ (5.675 g), H_3BO_3 (1.758 g), $(NH_4)_2HPO_4$ (7.550 g), and 85% H_3PO_4 (7.712 g) were mixed in 5 mL of distilled water and treated under constant stirring with 85% H_3PO_4, until the components dissolved completely; the pH of the initial solution was 2. The clear solution was heated without boiling to concentrate to about 15 mL, then transferred to Teflon-coated steel autoclaves (degree of filling 65%) and heated at 180°C for 2 days. The crystalline product was filtered off in a vacuum, washed with distilled water, and dried at 60°C.

$Fe[B_2P_2O_7(OH)_5]$ crystallizes in the monoclinic structure with the lattice parameters $a = 1{,}774.7 \pm 0.5$, $b = 672.0 \pm 0.2$, $c = 705.9 \pm 0.2$ pm, and $\beta = 109.01 \pm 0.02°$ and a calculated density of 2.808 g cm^{-3} (Boy et al. 1998d). It was prepared under mild hydrothermal conditions. A mixture of $FeCl_3·6H_2O$ (4.16 g), $Li_2B_4O_7$ (1.3 g), and LiH_2PO_4 (5.6 g) (molar ratio 2:1:7) was heated in 20 mL of H_2O and dissolved by dropwise addition of 18% HCl (5 mL). The reaction mixture was then concentrated to ~10 mL. The obtained highly viscous gel (pH = 1) was placed in a Teflon autoclave (degree of filling 50%) and stored for 2 weeks at 170°C. The resulting crystalline product was separated by filtration, washed with hot H_2O, and dried at 60°C. $Fe[B_2P_2O_7(OH)_5]$ was obtained as pale pink crystals.

$Fe(H_2O)_2[B_2P_2O_8(OH)_2]·H_2O$ crystallizes in the monoclinic structure with the lattice parameters $a = 774.49 \pm 0.04$, $b = 1{,}479.00 \pm 0.10$, $c = 824.29 \pm 0.06$ pm, and $\beta = 90.304 \pm 0.004°$ and a calculated density of 2.500 g cm^{-3} (Menezes et al. 2010b). Transparent colorless single crystals of this compound were prepared from a mixture of $FeC_2O_4·2H_2O$ (1.0003 g), B_2O_3 (0.7740 g), and 85% H_3PO_4 (3.8451 g) (molar ratio of Fe/B/P = 1:2:6). Five milliliters of water was added, and the mixture was transferred to a 20-mL Teflon-lined autoclave (degree of filling 40%), which was then treated at 170°C for 7 days. The reaction product was separated by vacuum filtration, washed with hot H_2O, filtered, and washed again with acetone, and finally dried at 60°C in air.

2.97 BORON–HYDROGEN–OXYGEN–COBALT–PHOSPHORUS

Three quinary compounds are formed in the B–H–O–Co–P system. **$Co[BPO_4(OH)_2]$** crystallizes in the tetragonal structure with the lattice parameters $a = 745.54 \pm 0.02$ and $c = 1{,}273.97 \pm 0.10$ pm (Huang et al. 2006). This compound was synthesized by the same way as $Mn[BPO_4(OH)_2]$ using $CoCl_2$ (0.78 g) instead of $MnCl_2$. It could also be synthesized from CoO, B_2O_3, P_2O_5, and H_2O.

$(H)_{0.5}Co_{1.25}(H_2O)_{1.5}[BP_2O_8]·H_2O$ crystallizes in the hexagonal structure with the lattice parameters $a = 949.60 \pm 0.06$ and $c = 1{,}562.30 \pm 0.13$ pm and a calculated density of 2.613 g cm^{-3} (Yilmaz et al. 2005). This compound was synthesized by the same way as $(H)_{0.5}Mn_{1.25}(H_2O)_{1.5}[BP_2O_8]·H_2O$ and was prepared using $MnCl_2$ (3.580 g) instead of $CoCl_2·6H_2O$. Purple crystals of the title compound were obtained.

$Co(H_2O)_2[B_2P_2O_8(OH)_2]·H_2O$ crystallizes in the monoclinic structure with the lattice parameters $a = 774.08 \pm 0.04$, $b = 1{,}469.75 \pm 0.05$, $c = 822.19 \pm 0.04$ pm, and $\beta = 90.26 \pm 0.02°$ and a calculated density of 2.546 g cm^{-3} (Menezes et al. 2010b). Violet single crystals of this compound were prepared from a mixture of $CoCO_3$ (0.5002 g), B_2O_3 (0.5853 g), and 85% H_3PO_4 (2.9078 g) (molar ratio of Co/B/P = 1:2:6). Ten milliliters of water was added, and the mixture was transferred into a Teflon-lined autoclave (degree of filling 30%) followed by treatment at 170°C for 10 days. The reaction product was separated by vacuum filtration, washed with hot H_2O, filtered and washed again with acetone, and finally dried at 60°C in air.

2.98 BORON–HYDROGEN–OXYGEN–NICKEL–PHOSPHORUS

Two quinary compounds are formed in the B–H–O–Ni–P system. **$Ni[BPO_4(OH)_2]$** crystallizes in the trigonal structure with the lattice parameters $a = 741.862 \pm 0.003$, $c = 1{,}258.982 \pm 0.006$ pm, and a calculated density of 3.295 g cm^{-3} (Yang et al. 2008a). It was prepared under hydrothermal conditions. Typically, $Ni(CH_3COO)_2{\cdot}4H_2O$ (4 mM), H_3BO_3 (20 mM), and 85% H_3PO_4 (3 mL) were charged into a 50-mL Teflon autoclave and heated at 220°C for 7 days. After cooling to room temperature, the solid product was washed extensively with hot water (80°C). Yellow, small needle-like crystals that often aggregate as balls were obtained.

$Ni(H_2O)_2[B_2P_2O_8(OH)_2]{\cdot}H_2O$ crystallizes in the monoclinic structure with the lattice parameters $a = 773.89 \pm 0.07$, $b = 1{,}457.35 \pm 0.11$, $c = 820.60 \pm 0.05$ pm, and $\beta = 90.30 \pm 0.04°$ and a calculated density of 2.572 g cm^{-3} (Menezes et al. 2010b). Greenish single crystals of this compound were prepared from a mixture of NiO (0.2501 g), B_2O_3 (0.4660 g), and H_3PO_4 (2.3153 g) (molar ratio of Ni/B/P = 1:2:6). Ten milliliters of water was added and the mixture was filled to a 20-mL Teflon-lined autoclave followed by treatment under autogeneous pressure at 170°C for 7 days. The reaction product was separated by vacuum filtration, washed with hot H_2O, filtered and washed again with acetone, and finally dried at 60°C in air.

2.99 BORON–LITHIUM–SODIUM–OXYGEN–PHOSPHORUS

The **$Li_2NaBP_2O_8$** and **$LiNa_2B_5P_2O_{14}$** quinary compounds are formed in the B–Li–Na–O–P system (Hasegawa and Yamane 2015). The first of them crystallizes in the triclinic structure with the lattice parameters $a = 543.44 \pm 0.03$, $b = 737.93 \pm 0.04$, and $c = 798.40 \pm 0.04$ pm and $\alpha = 103.243 \pm 0.003°$, $\beta = 109.270 \pm 0.004°$, and $\gamma = 87.391 \pm 0.002°$ and a calculated density of 2.684 g cm^{-3}. The second compound crystallizes in the monoclinic structure with the lattice parameters $a = 820.8 \pm 0.3$, $b = 915.1 \pm 0.3$, $c = 834.9 \pm 0.3$ pm, and $\beta = 115.709 \pm 0.007°$ and a calculated density of 2.309 g cm^{-3}. To prepare these compounds, the powders of Li_2CO_3, Na_2CO_3, H_3BO_3, and $NH_4H_2PO_4$ (molar ratio 1:1:2:2) were mixed and pressed into a pellet. The pellet was placed on a Pt plate and heated at 200°C for 9 h in air. After cooling, the product was powdered, pelletized, and heated at 550°C for 12 h. The obtained polycrystalline sample was heated again at 600°C for 1 h and then cooled to 500°C at a rate of 2°C h^{-1}. After cooling from 500°C to room temperature in the furnace, the product was crushed and a $Li_2NaBP_2O_8$ single crystal for crystal structure analysis was chosen. Furthermore, the polycrystalline sample (0.0310 g) and H_3BO_3 (0.0308 g) were mixed and pressed into a pellet and heated at 610°C for 1 h, followed by cooling to 510°C at a rate of 2°C h^{-1}. The product was crushed, and a single crystal of $LiNa_2B_5P_2O_{14}$ was chosen for crystal structure analysis.

2.100 BORON–LITHIUM–POTASSIUM–OXYGEN–PHOSPHORUS

The $\mathbf{LiK_2BP_2O_8}$ and $\mathbf{Li_3K_2BP_4O_{14}}$ quinary compounds are formed in the B–Li–K–O–P system (Wang et al 2012b). $LiK_2BP_2O_8$ melts at 627°C and crystallizes in the monoclinic structure with the lattice parameters $a = 835.44 \pm 0.03$, $b = 879.12 \pm 0.03$, $c = 986.26 \pm 0.04$ pm and $\beta = 95.043 \pm 0.002°$, a calculated density of 2.632 g cm^{-3}, and an energy gap of 3.74 eV. It is hard to conclusively prove whether this compound melts congruently as its melt had difficulty crystallizing due to the high viscosity. Single crystals of $LiK_2BP_2O_8$ were grown from a high-temperature solution. This solution was prepared in a Pt crucible by melting a mixture of LiF, KF, Cs_2CO_3, H_3BO_3, and $NH_4H_2PO_4$ (molar ratio 1:1:0.75:1:2). The addition of Cs_2CO_3 acted as the flux and helped the crystallization of the title compound. The mixture (10 g) was kept at 750°C for 24 h in a furnace. The homogenized solution was then cooled (20°C h^{-1}) to the initial crystallization temperature (600°C). It was further slowly cooled to 500°C at a rate of 1°C h^{-1}, and finally cooled to room temperature at a rate of 50°C h^{-1}. Colorless block crystals were separated from the crucible.

$Li_3K_2BP_4O_{14}$ melts at 695°C and crystallizes in the orthorhombic structure with the lattice parameters $a = 2{,}404.4 \pm 0.9$, $b = 1{,}390.2 \pm 0.5$, and $c = 687.4 \pm 0.3$ pm, a calculated density of 2.646 g cm^{-3} and an energy gap of 3.88 eV (Wang et al 2012b). It is hard to conclusively prove whether this compound melts congruently as its melt had difficulty crystallizing due to the high viscosity. Single crystals of $Li_3K_2BP_4O_{14}$ were grown under the same conditions as single crystals of $LiK_2BP_2O_8$ were obtained at a molar ratio of LiF/K_2CO_3/H_3BO_3/$NH_4H_2PO_4$ = 1:0.5:1:2 using LiF/K_2CO_3/H_3BO_3 as the flux.

2.101 BORON–LITHIUM–RUBIDIUM–OXYGEN–PHOSPHORUS

The $\mathbf{Li_3Rb_2BP_4O_{14}}$ quinary compound, which melts at 694°C and crystallizes in the orthorhombic structure with the lattice parameters $a = 2{,}390.4 \pm 1.7$, $b = 1{,}413.2 \pm 1.0$, and $c = 707.6 \pm 0.5$ pm, a calculated density of 3.059 g cm^{-3} and an energy gap of 3.82 eV, is formed in the B–Li–Rb–O–P system (Wang et al 2012b). It is hard to conclusively prove whether this compound melts congruently as its melt had difficulty crystallizing due to the high viscosity. Single crystals of $Li_3Rb_2BP_4O_{14}$ were grown under the same conditions as single crystals of $LiK_2BP_2O_8$ were obtained at a molar ratio of LiF/Rb_2CO_3/H_3BO_3/$NH_4H_2PO_4$ = 1:0.5:1:2 by using LiF/Rb_2CO_3/H_3BO_3 as the flux.

2.102 BORON–LITHIUM–CESIUM–OXYGEN–PHOSPHORUS

The $\mathbf{Li_2Cs_2B_2P_4O_{15}}$ quinary compound, which melts congruently at 736°C and crystallizes in the triclinic structure with the lattice parameters $a = 709.2 \pm 0.4$, $b = 778.8 \pm 0.4$, and $c = 1{,}397.4 \pm 0.7$ pm and $\alpha = 86.287 \pm 0.006°$, $\beta = 75.948 \pm 0.006°$, and $\gamma = 66.363 \pm 0.006°$, a calculated density of 3.223 g cm^{-3} and an energy gap of 3.58 eV, is formed in the B–Li–Cs–O–P system (Wang et al 2012b). Single crystals of this compound were grown by spontaneous crystallization from a stoichiometric composition melt of Li_2CO_3, Cs_2CO_3, H_3BO_3, and $NH_4H_2PO_4$. After grinding, the mixture (10 g) was kept at 750°C for 24 h in a furnace. The homogenized melt was then cooled slowly to 500°C at a rate of 5°C h^{-1}, and finally decreased to room temperature at a rate of 50°C h^{-1}. Several colorless block crystals were mechanically separated from the crucible. Subsequently, a pure powder sample of the title compound was prepared by solid-state reactions of the stoichiometric starting components of Li_2CO_3, Cs_2CO_3, H_3BO_3, and $NH_4H_2PO_4$. The mixture was ground and then calcined at 680°C for 24 h with several intermediate grindings.

2.103 BORON–SODIUM–STRONTIUM–OXYGEN–PHOSPHORUS

The **$Sr_{9.402}Na_{0.209}(PO_4)_6B)_{0.996}O_2$** quinary phase, which crystallizes in the trigonal structure with the lattice parameters $a = 973.4 \pm 0.4$, $c = 727.9 \pm 0.2$ pm, and a calculated density of 4.006 g cm^{-3}, is formed in the B–Na–Sr–O–P system (Calvo et al. 1975). Crystals of this phase were grown in a Pt crucible from a mixture of $(NH_4)_2HPO_4$, $SrCO_3$, and $Na_2B_4O_7 \cdot 10H_2O$ (molar ratio 2:3:6) heated to 1,450°C and then cooled to room temperature. The product, washed with hot water, contained clear colorless crystals.

2.104 BORON–SODIUM–CADMIUM–OXYGEN–PHOSPHORUS

The **$Na_3Cd_3B(PO_4)_4$** quinary compound, which crystallizes in the orthorhombic structure with the lattice parameters $a = 1{,}368.54 \pm 0.03$, $b = 533.46 \pm 0.11$, and $c = 1{,}821.69 \pm 0.04$ pm and a calculated density of 3.980 g cm^{-3}, is formed in the B–Na–Cd–O–P system (Shi et al. 2012). Polycrystalline $Na_3Cd_3B(PO_4)_4$ was prepared by solid-state reaction techniques. A mixture of Na_2CO_3, CdO, H_3BO_3, and $NH_4H_2PO_4$ (molar ratio 1.5:3:1:4) was ground and loaded into a fused silica crucible. The mixture was preheated at 200°C for 2 h to decompose the $NH_4H_2PO_4$ and eliminate the water. Then, the temperature was raised to 400°C to decompose the carbonate, and the products were cooled to room temperature and ground. Finally, the sample was calcined at 680°C for 3 days with several intermediate grindings until a single-phase powder was obtained.

Single crystals of the title compound were grown by a spontaneous crystallization method (Shi et al. 2012). The solution was prepared in a Pt crucible by melting a mixture of $Na_2CO_3/H_3BO_3/CdO/NH_4H_2PO_4$ (molar ratio 1:1:1:2). The Pt crucible was placed in the center of a vertical furnace. It was held at 850°C for 10 h until the solution became transparent and clear and then quickly cooled to 700°C. The temperature was further decreased to 650°C at a rate of 1°C h^{-1}; then the temperature was allowed to cool to room temperature at a rate of 10°C h^{-1}. Some colorless, transparent block crystals were obtained.

2.105 BORON–SODIUM–CADMIUM–CHLORINE–PHOSPHORUS

BP-NaCl-$CdCl_2$. According to the data of metallography and XRD, the solubility of BP in the NaCl–$CdCl_2$ eutectic is less than 0.01 mass% (Goryunova et al. 1964).

2.106 BORON–POTASSIUM–STRONTIUM–OXYGEN–PHOSPHORUS

The **$KSrBP_2O_8$** quinary compound, which is stable up to 1,100°C and crystallizes in the tetragonal structure with the lattice parameters $a = 710.95 \pm 0.18$, $c = 1{,}388.2 \pm 0.5$ pm, a calculated density of 3.100 g cm^{-3} and an energy gap of 3.97 eV, is formed in the B–K–Sr–O–P system (Zhao et al. 2009). Single crystals of this compound were initially obtained by the high-temperature solution growth method. At room temperature, K_2CO_3 (6.468 mM), $SrCO_3$ (5.390 mM), H_3BO_3 (32.34 mM), and $NH_4H_2PO_4$ (5.390 mM) (molar ratio of K/Sr/B/P = 12:5:30:5) were mixed. Then the reaction mixture was thoroughly ground and pressed into a pellet, which was put into a Pt crucible. Subsequently, the crucible was put into an oven and heated at 1,050°C in the air for 2 days. Afterward, it was allowed to cool at a rate of 0.05°C min^{-1} to 500°C before switching off the furnace. After boiling for 24 h in water, prism-shaped colorless crystals were obtained in very low yield. The polycrystalline samples of the title compound were synthesized by the solid-state reactions of analytical reagents with stoichiometric amounts (K/Sr/B/P = 1:1:1:2). The powdered mixtures were ground and then calcined at 900°C for 7 days with several intermediate grindings.

2.107 BORON–POTASSIUM–BARIUM–OXYGEN–PHOSPHORUS

The **$KBaBP_2O_8$** quinary compound, which is stable up to 1,100°C and crystallizes in the tetragonal structure with the lattice parameters $a = 720.2 \pm 0.2$, $c = 1,430.0 \pm 0.6$ pm, and a calculated density of 3.378 g cm^{-3}, is formed in the B–K–Ba–O–P system (Zhao et al. 2009). A few prism-shaped colorless crystals of this compound was obtained by the same way as $KSrBP_2O_8$ was synthesized using $BaCO_3$ instead of $SrCO_3$.

2.108 BORON–POTASSIUM–ZINC–OXYGEN–PHOSPHORUS

The **$KZnBP_2O_8$** quinary compound, which crystallizes in the monoclinic structure with the lattice parameters $a = 1,261.7 \pm 0.5$, $b = 1,277.3 \pm 0.6$, $c = 841.5 \pm 0.3$ pm and $\beta = 91.25 \pm 0.03°$ and a calculated density of 2.991 g cm^{-3}, is formed in the B–K–Zn–O–P system (Kniep et al. 1999b, 1999c). It was obtained under mild hydrothermal conditions in a Teflon autoclave (20 mL). For this, ZnO (0.65 g), $K_2B_4O_7{\cdot}4H_2O$ (2.44 g), and K_2HPO_4 (2.2 g) (molar ratio 1:1:2), were heated to about 80°C in demineralized water (10 mL) and treated under stirring with 85% H_3PO_4 (5.5 mL), until the components had dissolved completely. The clear solutions were concentrated to about 15–20 mL to give a highly viscous gel. The pH value was between 1 and 1.5. The highly viscous gel was transferred to the Teflon autoclave (degree of filling about 60%) and stored for 3–4 days at about 170°C. The crystalline reaction product was filtered off in vacuum, washed with demineralized water, and dried at 60°C. Single crystals grow to a length of about 0.8 mm under these conditions.

2.109 BORON–POTASSIUM–LEAD–OXYGEN–PHOSPHORUS

The **$KPbBP_2O_8$** quinary compound, which melts congruently at 865°C and crystallizes in the tetragonal structure with the lattice parameters $a = 714.64 \pm 0.07$, $c = 1,389.17 \pm 0.16$ pm, and a calculated density of 4.185 g cm^{-3}, is formed in the B–K–Pb–O–P system (Li et al. 2013). Polycrystalline samples of this compound were synthesized by solid-state reaction techniques. A mixture of KNO_3, PbO, H_3BO_3, and $NH_4H_2PO_4$ (molar ratio 1:1:1:2) was ground thoroughly and packed into a Pt crucible. The temperature was raised to 400°C at a rate of 3°C min^{-1}, and after 4 h at 400°C, the products were cooled to room temperature and ground up again. The temperature was raised to 750°C at the same rate as earlier and held at this temperature for 48 h with several intermediate grinds and mixes to ensure completion of the solid-state reaction. Large crystals can be grown from a stoichiometric melt.

2.110 BORON–RUBIDIUM–ZINC–OXYGEN–PHOSPHORUS

The **$RbZnBP_2O_8$** quinary compound, which crystallizes in the triclinic structure with the lattice parameters $a = 743.9 \pm 0.2$, $b = 763.9 \pm 0.2$, $c = 786.1 \pm 0.2$ pm and $\alpha = 118.82 \pm 0.02°$, $\beta = 101.73 \pm 0.02°$, and $\gamma = 103.51 \pm 0.02°$ and a calculated density of 3.304 g cm^{-3}, is formed in the B–Rb–Zn–O–P system (Kniep et al. 1999b, 1999c). The title compound was obtained under mild hydrothermal conditions in a Teflon autoclave (20 mL). For this, RbOH (7.62 g), ZnO (3.02 g), and B_2O_3 (1.29 g) (molar ratio 2:2:1) were heated to about 80°C in demineralized water (10 mL) and treated under constant stirring with 85% H_3PO_4 (9 mL), until the components had dissolved completely. The clear solutions were concentrated to about 15–20 mL to give a highly viscous gel. The pH value was between 1 and 1.5. The highly viscous gel was transferred to a Teflon autoclave (degree of filling about 65%) and stored for 3–4 days at about 170°C. The crystalline reaction product was filtered off in vacuum, washed with demineralized water, and dried at 60°C. Single crystals grow to a length of about 0.8 mm under these conditions.

2.111 BORON–RUBIDIUM–LEAD–OXYGEN–PHOSPHORUS

The **$RbPbBP_2O_8$** quinary compound, which melts congruently at 750°C and crystallizes in the tetragonal structure with the lattice parameters $a = 721.3 \pm 0.3$, $c = 1{,}398.7 \pm 1.0$ pm, and a calculated density of 4.503 g cm^{-3}, is formed in the B–Rb–Pb–O–P system (Wang et al. 2013). Polycrystalline samples of the title compound were synthesized via conventional solid-state reactions of stoichiometric amounts of Rb_2CO_3, PbO, H_3BO_3, and $NH_4H_2PO_4$. The mixture was initially ground thoroughly and packed into an alumina crucible. The temperature was raised to 600°C at a rate of 2°C min^{-1}. After preheating, the mixture was ground up again, heated to 700°C, and then dwelled at this temperature for 48 h with several intermediate grindings and mixings.

Large size crystals were grown by the Kyropoulos method from its stoichiometric composition melt without any flux (Wang et al. 2013). The pure crystalline powders of $RbPbBP_2O_8$ were melted at 800°C in a Pt crucible that was placed into a vertical, programmable temperature furnace. It was held at that temperature for 24 h. In the first run of growth, a Pt wire was dipped into the melt, and the temperature was decreased to 720°C at a rate of 2°C h^{-1}. Thus, a few colorless, transparent crystals crystallized on the Pt wire, and were singled out to be used as seeds for growth of the single crystals of this compound.

2.112 BORON–CESIUM–ZINC–OXYGEN–PHOSPHORUS

The **$CsZnBP_2O_8$** quinary compound, which crystallizes in the triclinic structure with the lattice parameters $a = 750.6 \pm 0.2$, $b = 791.4 \pm 0.2$, $c = 803.8 \pm 0.2$ pm and $\alpha = 118.05 \pm 0.02°$, $\beta = 102.96 \pm 0.02°$, and $\gamma = 104.50 \pm 0.02°$ and a calculated density of 3.545 g cm^{-3}, is formed in the B–Cs–Zn–O–P system (Kniep et al. 1999b, 1999c). The title compound was obtained under mild hydrothermal conditions in a Teflon autoclave (20 mL). For this, CsOH (8.6 g), ZnO (2.33 g), and B_2O_3 (0.99 g) (molar ratio 2:2:1) were heated to about 80°C in demineralized water (10 mL) and treated under stirring with 85% H_3PO_4 (7.1 mL), until the components had dissolved completely. The clear solutions were concentrated to about 15–20 mL to give a highly viscous gel. The pH value was between 1 and 1.5. The highly viscous gel was transferred to the Teflon autoclave (degree of filling about 90%) and stored for 3–4 days at about 170°C. The crystalline reaction product was filtered off in vacuum, washed with demineralized water, and dried at 60°C. Single crystals grow to a length of about 0.8 mm under these conditions.

2.113 BORON–CESIUM–OXYGEN–CHROMIUM–PHOSPHORUS

The $\mathbf{Cs_2Cr_3(BP_4O_{14})(P_4O_{13})}$ quinary compound, which is thermally stable up to 1,200°C and crystallizes in the monoclinic structure with the lattice parameters $a = 1{,}479.18 \pm 0.14$, $b = 1{,}581.90 \pm 0.15$, $c = 970.37 \pm 0.08$ pm and $\beta = 92.450 \pm 0.006°$ and a calculated density of 3.257 g cm^{-3}, is formed in the B–Cs–O–Cr–P system (Zhang et al. 2010). This compound was synthesized by a solid-state reaction. A powder mixture of Cs_2CO_3 (0.81 g), Cr_2O_3 (0.073 g), H_3BO_3 (0.77 g), and $NH_4H_2PO_4$ (1.01 g) was first ground in an agate mortar and then transferred to a Pt crucible. The sample was gradually heated in air at 300°C for 6 h to decompose H_3BO_3 and $NH_4H_2PO_4$ and finally heated at 1,050°C for 24 h. The intermediate product was slowly cooled to 800°C at a rate of 2°C h^{-1}, where it was kept for 10 h and then quenched to room temperature. Some bottle-green crystals were selected carefully from the solidified flux with hot water. After structural analysis, a green crystalline powder sample of the title compound was then obtained quantitatively by the reaction of a mixture of Cs_2CO_3, Cr_2O_3, H_3BO_3, and $NH_4H_2PO_4$ (molar ratio 2:3:2:16) at 850°C for 30 h.

2.114 BORON–CESIUM–OXYGEN–IRON–PHOSPHORUS

The $\mathbf{CsFe(BP_3O_{11})}$ quinary compound, which is thermally stable up to 1,200°C and crystallizes in the orthorhombic structure with the lattice parameters $a = 853.75 \pm 0.07$, $b = 1{,}278.29 \pm 0.12$, and $c = 833.46 \pm 0.05$ pm and a calculated density of 3.434 g cm^{-3}, is formed in the B–Cs–O–Fe–P system (Zhang et al. 2010). The synthesis of single crystals of this compound was carried out using a powder mixture of Cs_2CO_3 (0.17 g), Fe_2O_3 (0.05 g), H_3BO_3 (0.50 g), and $NH_4H_2PO_4$ (0.63 g) (molar ratio of Cs/Fe/B/P = 2:1:20:12); the excess amounts of $NH_4H_2PO_4$ and H_3BO_3 are present to promote crystal growth. The mixture was first ground in an agate mortar and then introduced into a Pt crucible. The crucible was heated at 300°C for 1 day, then heated for 2 days at 850°C, cooled to 650°C at a rate of 3°C h^{-1}, and finally quenched to room temperature. Some transparent block crystals were selected carefully from the sintered product with hot water. Subsequently, a pure powder sample of $CsFe(BP_3O_{11})$ was prepared quantitatively by the reaction of a mixture of Cs_2CO_3, Fe_2O_3, H_3BO_3, and $NH_4H_2PO_4$ (molar ratio 1:1:2:6) at 880°C.

2.115 BORON–COPPER–GALLIUM–CARBON–CHROMIUM–MOLYBDENUM–IRON–PHOSPHORUS

The multinary $\mathbf{B_{5.5}Fe_{65.5}Cr_{4-x}Mo_{4-y}Cu_{x+y}Ga_4C_5P_{12}}$ ($x, y = 0, 0.5, 1$) master alloy ingots were obtained by induction melting starting from Fe-B, Fe-C, and Fe–P prealloys and pure Mo, Cr, and Fe metals and crystalline B (Borrego et al. 2006). The substitution of Cr and Mo by Cu decreases the glass-forming ability of the initial alloy, and the maximum crystalline volume fraction achieved at the end of the nanocrystallization process was found to be only around 20%.

2.116 BORON–COPPER–SILICON–IRON–PHOSPHORUS

Quinary alloys in the B–Cu–Si–Fe–P system were prepared by arc-melting (Cui et al. 2009; He et al. 2009; Guo et al. 2011; Kubota et al. 2011) or induction melting (Fu et al. 2017) of pure Fe, Si, B, Cu metals, and prealloyed Fe–P (Cui et al. 2009; He et al. 2009; Fu et al. 2017) [of pure Fe, Si, B, Cu metals, and premelted Fe_3P (Kubota et al. 2011)] in a high-purity Ar atmosphere. A single-roller melt-spinning method in air atmosphere was used to produce the rapidly solidified ribbons. The nanocrystalline $\mathbf{B_{12}Fe_{81}Si_4Cu_1P_2}$ alloy composed of a nanocrystalline phase embedded in an amorphous matrix was obtained by annealing the as-quenched amorphous ribbons in suitable heat-treatment conditions (Guo et al. 2011). The melt-spun alloy of $\mathbf{B_8Fe_{85}Si_2Cu_1P_4}$ forms a heteroamorphous structure (Kubota et al. 2011) and the annealed alloy of $\mathbf{B_8Fe_{83.3}Si_4Cu_{0.7}P_4}$ is composed of a uniform nanocrystalline structure with a grain size of about 20 nm (Cui et al. 2009). According to the data of He et al. (2009), the simultaneous additions of P and Cu decrease the grain size of the α-Fe particles.

Annealing effect of amorphous $B_8Fe_{83.3}Si_4Cu_{0.7}P_4$ precursor alloys on the formation of the nanoporous structure has been investigated by Fu et al. (2017). It was shown that the annealing of amorphous alloy causes the formation of nanocrystalline heteroamorphous microstructure coexisting α-Fe grains and a residual amorphous phase. Preferential dissolution of α-Fe grains in the solution of H_2SO_4 leads to the formation of nanoporous structure.

2.117 BORON–BARIUM–FLUORINE–CHLORINE–PHOSPHORUS

BP–$BaCl_2$–BaF_2. According to the data of metallography and XRD, the solubility of BP in the $BaCl_2$–BaF_2 eutectic was not detected (Goryunova et al. 1964).

2.118 BORON–ALUMINUM–GALLIUM–CARBON–SILICON–IRON–PHOSPHORUS

The effect of Si addition on the thermal stability of the supercooled liquid before crystallization and glass-forming ability was examined for amorphous alloy series $\mathbf{B_4Al_5Ga_2C_6Si_xFe_{72-x}P_{11}}$, $\mathbf{B_4Al_{5-x}Ga_2C_6Si_xFe_{72}P_{11}}$, $\mathbf{B_4Al_5Ga_2C_6Si_xFe_{72}P_{11-x}}$, and $\mathbf{B_4Al_5Ga_2C_{6-x}Si_xFe_{72}P_{11}}$ (Inoue et al. 1997). The increases in the thermal stability and glass-forming ability were recognized in the replacements of P by 1–2 at% Si and of C by 1 at% Si. The supercooled liquid region (ΔT_x) increases from 53°C for $\mathbf{B_4Al_5Ga_2C_6Fe_{72}P_{11}}$ to 58°C for $\mathbf{B_4Al_5Ga_2C_5Si_1Fe_{72}P_{11}}$. Multinary alloys were prepared by induction melting a mixture of pure Fe, Al, and Ga metals, prealloyed Fe–C and Fe–P, and pure B and Si crystals in an Ar atmosphere.

2.119 BORON–ALUMINUM–GALLIUM–CARBON–VANADIUM–COBALT–PHOSPHORUS

The $\mathbf{B_4Al_5Ga_2C_1V_3Co_{70}P_{15}}$ multinary alloy was prepared by induction melting a mixture of pure Co, Al, Ga, and V metals, prealloyed Co–P and Co–C, and B in an Ar atmosphere (Inoue and Katsuya 1996). The glass transition temperature and the temperature interval of the supercooled region for this alloy are equal to 449°C and 47°C, respectively.

2.120 BORON–ALUMINUM–GALLIUM–CARBON–NIOBIUM–IRON–PHOSPHORUS

The dissolution of 2 at% Nb for the $\mathbf{B_4Al_5Ga_2C_6Nb_xFe_{72-x}P_{11}}$ glassy alloys was found to be effective for the extension of the supercooled liquid region (ΔT_x) (Inoue and Cook 1996). The ΔT_x value is 60°C for the $\mathbf{B_4Al_5Ga_2C_6Fe_{72}P_{11}}$ alloy and increases to 66°C for 2 at% Nb. The ingots ($x = 0, 2, 4, 6$ at%) were prepared by induction melting a mixture of pure Fe, Al, Ga, and Nb metals, prealloyed Fe–P and Fe–C, and pure B crystal in an Ar atmosphere.

2.121 BORON–ALUMINUM–GALLIUM–CARBON–CHROMIUM–IRON–PHOSPHORUS

The dissolution of 4 at% Cr for the $\mathbf{B_4Al_5Ga_2C_6Cr_xFe_{72-x}P_{11}}$ glassy alloys was found to be effective for the extension of the supercooled liquid region (ΔT_x) (Inoue and Cook 1996). The ΔT_x value is 60°C for the $B_4Al_5Ga_2C_6Fe_{72}P_{11}$ alloy and increases to 61°C for 4 at% Cr. The ingots ($x = 0, 2, 4, 6$ at%) were prepared by induction melting a mixture of pure Fe, Al, Ga, and Cr metals, prealloyed Fe–P and Fe–C, and pure B crystal in an Ar atmosphere.

2.122 BORON–ALUMINUM–GALLIUM–CARBON–CHROMIUM–IRON–COBALT–PHOSPHORUS

The $\mathbf{B_4Al_5Ga_2C_1Cr_3Fe_3Co_{67}P_{15}}$ multinary alloy was prepared by induction melting a mixture of Co, Al, Fe, Ga, and Cr, prealloyed Co–P and Co–C, and B in an Ar atmosphere (Inoue and Katsuya 1996). The glass transition temperature and the temperature interval of the supercooled region for this alloy are equal to 448°C and 45°C, respectively.

2.123 BORON–ALUMINUM–GALLIUM–CARBON–CHROMIUM–COBALT–PHOSPHORUS

The $\mathbf{B_4Al_5Ga_2C_1Cr_3Co_{70}P_{15}}$ multinary alloy was prepared by induction melting a mixture of Co, Al, Ga, and Cr, prealloyed Co–P and Co–C, and B in an Ar atmosphere (Inoue and Katsuya 1996). The glass transition temperature and the temperature interval of the supercooled region for this alloy are equal to 444°C and 51°C, respectively.

2.124 BORON–ALUMINUM–GALLIUM–CARBON–MOLYBDENUM–IRON–PHOSPHORUS

The dissolution of 5 at% Mo for the $\mathbf{B_4Al_5Ga_2C_6Mo_xFe_{72-x}P_{11}}$ glassy alloys was found to be effective for the extension of the supercooled liquid region (ΔT_x) (Inoue and Cook 1996). The ΔT_x value is 60°C for the $B_4Al_5Ga_2C_6Fe_{72}P_{11}$ alloy and increases to 62°C for 4 at% Mo. The ingots ($x = 0, 2, 4, 6$ at%) were prepared by induction melting a mixture of pure Fe, Al, Ga, and Mo metals, prealloyed Fe–P and Fe–C, and pure B crystal in an Ar atmosphere.

2.125 BORON–ALUMINUM–GALLIUM–CARBON–IRON–PHOSPHORUS

The multinary $\mathbf{B_4Al_5Ga_2C_6Fe_{72}P_{11}}$ alloy was found to form a glassy phase with a wide supercooled liquid region before crystallization (Inoue and Cook 1995a). The glass transition temperature (T_g), crystallization temperature (T_x), and temperature interval of supercooled liquid (ΔT_x) for this glassy alloy are 459°C, 520°C, and 61°C, respectively [ΔT_x = 53°C for the $\mathbf{B_4Al_5Ga_2C_5Fe_{73}P_{11}}$ alloy ingot (Inoue et al. 1995)]. These ingots were prepared by induction melting the mixtures of pure Fe, Al, and Ga metals, prealloyed Fe–P and Fe–C, and pure B crystal powder in an Ar atmosphere (Inoue and Cook 1995a; Inoue et al. 1995).

2.126 BORON–ALUMINUM–GALLIUM–CARBON–IRON–COBALT–PHOSPHORUS

The dissolution of 6 at% Co for the $\mathbf{B_4Al_5Ga_2C_6Co_xFe_{72-x}P_{11}}$ glassy alloys was found to be effective for the extension of the supercooled liquid region (ΔT_x) (Inoue and Cook 1996). The ΔT_x value is 60°C for the $B_4Al_5Ga_2C_6Fe_{72}P_{11}$ alloy and increases to 61°C for 6 at% Co. The ingots ($x = 0, 2, 4, 6$ at%) were prepared by induction melting a mixture of pure Fe, Al, Ga, and Co metals, prealloyed Fe–P and Fe–C, and pure B crystal in an Ar atmosphere.

2.127 BORON–ALUMINUM–GALLIUM–SILICON–IRON–PHOSPHORUS

The multinary $\mathbf{B_4Al_4Ga_2Si_4Fe_{74}P_{12}}$ alloy was prepared by induction melting a mixture of pure Fe, Al, and Ga metals, prealloyed Fe–P ingots, and pure B and Si crystal in an Ar atmosphere (Inoue and Park 1996). The ΔT_x value is 49°C for this alloy.

2.128 BORON–ALUMINUM–CARBON–GERMANIUM–IRON–PHOSPHORUS

The multinary $\mathbf{B_4Al_5C_6Ge_xFe_{A-x}P_{11}}$ alloys ($A = 72, 73, 74$, and 76; $x = 0, 2, 4$, and 6) were prepared by induction melting a mixture of pure Fe and Al metals, prealloyed Fe–P and Fe–C, and pure B and Ge in an Ar atmosphere (Inoue and Cook 1995b). The $\mathbf{B_4Al_5C_6Ge_2Fe_{72}P_{11}}$ glassy was found to cause the extension of the supercooled liquid region to 60°C.

2.129 BORON–ALUMINUM–CARBON–IRON–PHOSPHORUS

Fe-based bulk metallic glasses have been found in the B–Al–C–Fe–P system over a wide compositional range, and the quinary $\mathbf{B_2Al_3C_9Fe_{77}P_9}$ alloy shows the best glass-forming ability yielding the formation of an amorphous cylinder rod with a diameter of 3.0 mm (Wang et al. 2015). The ΔT_x value is 36°C for the $\mathbf{B_4Al_5C_5Fe_{75}P_{11}}$ alloy (Inoue et al. 1995). Alloy ingots with nominal composition $\mathbf{Fe_{77}Al_3(P, C, B)_{20}}$ were obtained by induction melting a mixture of pure Fe, Al, C, and Si, prealloyed Fe–P (22.41 mass% P) and Fe–B (17.24 mass% P) in an Ar atmosphere (Wang et al. 2015). Quinary $\mathbf{B_4Al_5C_5Fe_{75}P_{11}}$ alloy was prepared by induction melting the mixtures of pure Fe and Al metals, prealloyed Fe–P and Fe–C ingots and pure B crystal powder in an Ar atmosphere (Inoue et al. 1995).

2.130 BORON–ALUMINUM–SILICON–IRON–PHOSPHORUS

The multinary $\mathbf{B_4Al_4Si_4Fe_{76}P_{12}}$ alloy was prepared by induction melting a mixture of pure Fe and Al metals, prealloyed Fe–P ingots, and pure B and Si crystal in an Ar atmosphere (Inoue and Park 1996). The ΔT_x value is 46°C for this alloy.

2.131 BORON–GALLIUM–CARBON–SILICON–CHROMIUM–MOLYBDENUM–IRON–PHOSPHORUS

The multinary $\mathbf{B_4Ga_2C_4Si_2Cr_xMo_2Fe_{76-x}P_{10}}$ ($x = 0$–6) alloys were prepared by induction melting under Ar atmosphere by using Fe, Cr, Mo, Ga, and Si; P and C were alloyed by adding prealloyed Fe–P and Fe–C (Shen et al. 2006). The glass transition (T_g) and crystallization temperature (T_x) increase gradually from 465°C to 467°C and 515°C to 532°C, respectively, with increasing Cr content combined with a slight increase of ΔT_x from 50°C to 55°C. The substitution of a small amount of Fe with Cr was found to be effective in enhancing the glass-forming ability.

2.132 BORON–GALLIUM–CARBON–SILICON–IRON–PHOSPHORUS

The alloys with compositions of $\mathbf{B_4Ga_3C_4Si_xFe_{77}P_{12-x}}$ and $\mathbf{B_4Ga_2C_4Si_xFe_{78}P_{12-x}}$ were prepared by induction melting a mixture of pure Fe, Ga, and Si, prealloyed Fe–C and Fe–P ingots, and pure B crystal in an Ar atmosphere (Shen and Inoue 2002). The glass transition temperature (T_g) and ΔT_x of the $\mathbf{B_4Ga_3C_4Si_{2.5}Fe_{77}P_{9.5}}$ and $\mathbf{B_4Ga_2C_4Si_{3.5}Fe_{78}P_{9.5}}$ alloys are equal to 477°C and 48°C, and 462°C and 40°C, respectively.

2.133 BORON–GALLIUM–CARBON–ANTIMONY–CHROMIUM–MOLYBDENUM–IRON–PHOSPHORUS

The multinary $\mathbf{B_5Ga_4C_5Sb_2Cr_4Mo_4Fe_{65}P_{11}}$ bulk amorphous alloy was synthesized using the following techniques: (a) mechanical alloying the various elements; (b) melting the alloyed powders in a fused silica tube; (c) purifying the melt with a flux; and (d) water quenching the molten alloy (Shen and Schwarz 1999). The glass transition temperature (T_g) and ΔT_x of this alloy are equal to 446°C and 56°C, respectively.

2.134 BORON–GALLIUM–CARBON–CHROMIUM–MOLYBDENUM–IRON–PHOSPHORUS

Bulk amorphous Fe-based quinary alloys with the nominal composition $\mathbf{B_{5.5}Ga_4C_5Cr_4Mo_4Fe_{65.5}P_{12}}$ have been obtained by copper mold casting in different shapes (Stoica et al. 2007). This bulk metallic glass shows a high glass transition temperature as well as a high crystallization temperature, with an extension of the supercooled liquid region of around 65°C. This amorphous alloy was obtained by induction melting a mixture of pure Fe, Cr, and Mo, prealloyed Fe-B, Fe–C, Fe–Ga, and Fe–P ingots, and pure crystalline boron in an Ar atmosphere.

Multinary $\mathbf{B_5Ga_4C_5Cr_4Mo_4Fe_{67}P_{11}}$ and $B_{5.5}Ga_4C_5Cr_4Mo_4Fe_{65.5}P_{12}$ bulk amorphous alloys were also synthesized using the following techniques: (a) mechanical alloying the various elements; (b) melting the alloyed powders in a fused silica tube; (c) purifying the melt with a flux; and (d) water quenching the molten alloy (Shen and Schwarz 1999). The glass transition temperature (T_g) and ΔT_x of these alloys are equal to 457°C and 56°C, and 450°C and 61°C, respectively.

2.135 BORON–GALLIUM–CARBON–CHROMIUM–MOLYBDENUM–IRON–COBALT–PHOSPHORUS

The multinary $\mathbf{B_5Ga_4C_5Cr_4Mo_4Fe_{62}Co_5P_{11}}$ bulk amorphous alloy was synthesized using the following techniques: (a) mechanical alloying the various elements; (b) melting the alloyed powders in a fused silica tube; (c) purifying the melt with a flux; and (d) water quenching the molten alloy (Shen and Schwarz 1999). The glass transition temperature (T_g) and ΔT_x of this alloy are equal to 458°C and 55°C, respectively.

2.136 BORON–GALLIUM–CARBON–MOLYBDENUM–IRON–PHOSPHORUS

The effect of the addition of Mo on the glass-forming ability of $\mathbf{B_5Ga_2C_4Mo_xFe_{78-x}P_{12}}$ glassy alloys was investigated by Shen et al. (2000b). The addition of 4 to 6 at% Mo was found to be effective for the extension of supercooled region: it is 30°C for $\mathbf{B_5Ga_2C_4Fe_{78}P_{12}}$ alloy and increases to 60°C for the alloy containing 4 at% Mo. The ingots were prepared by induction melting a mixture of Fe, Mo, and Ga metals, prealloyed Fe–C and Fe–P ingots, and pure B in an Ar atmosphere. The melt-spun $\mathbf{B_5Ga_2C_4Mo_xFe_{78-x}P_{12}}$ alloys were composed of a glassy phase without any crystallinity over the whole composition range up to 10 at% Mo.

2.137 BORON–GALLIUM–CARBON–IRON–PHOSPHORUS

The quinary $\mathbf{B_4Ga_2C_5Fe_{78}P_{11}}$ and $\mathbf{B_4Ga_xC_4Fe_{80-x}P_{12}}$ ($x \leq 8$) alloys were prepared by induction melting the mixtures of pure Fe and Ga metals, prealloyed Fe–C and Fe–P ingots, and pure B crystal powder in a flowing Ar atmosphere (Inoue et al. 1995; Shen et al. 2000c). The glass transition temperature (T_g) and ΔT_x of $\mathbf{B_4Ga_5C_4Fe_{75}P_{12}}$ alloy are equal to 458°C and 37°C, respectively (Shen et al. 2000c). The ΔT_x value for $\mathbf{B_4Ga_2C_5Fe_{78}P_{11}}$ alloy is 31°C (Inoue et al. 1995).

2.138 BORON–GALLIUM–CARBON–IRON–COBALT–PHOSPHORUS

The addition of 4–8 at% Co (10–15 at% Co) was found to be effective for the extension of ΔT_x of the multinary $\mathbf{B_4Ga_2C_4Fe_{78-x}Co_xP_{12}}$ ($\mathbf{B_4Ga_5C_4Fe_{75-x}Co_xP_{12}}$) alloy (Shen et al. 2000a; Shen et al. 2001). The ΔT_x value for the $\mathbf{B_4Ga_2C_4Fe_{78}P_{12}}$ ($\mathbf{B_4Ga_5C_4Fe_{75}P_{12}}$) is 26°C (37°C) and increases to 37°C (48°C) for the alloy containing 4 at% Co (10, 12.5, and 15 at% Co) alloys. The ingots were obtained by induction melting a mixture of Fe, Co, and Ga metals, prealloyed Fe–C and Fe–P ingots, and pure crystalline B in an Ar atmosphere.

2.139 BORON–CARBON–SILICON–MOLYBDENUM–IRON–PHOSPHORUS

The alloy ingots of $\mathbf{B_{2.5}C_{7.5}Si_xMo_{6-x}Fe_{74}P_{10}}$ ($x = 1$, 2, and 3 at%) and $\mathbf{B_{2.5}C_{7.5}Si_2Mo_2Fe_{76}P_{10}}$ were prepared by induction melting the mixtures of pure Fe, Mo, C, B, and Si, and prealloyed Fe–P ingots under an Ar atmosphere (Liu et al. 2009). From the master alloys, cylindrical rod and ring-shaped samples were prepared by copper mold casting under an Ar atmosphere. The addition of Si to the quinary B–C–Mo–Fe–P alloy is favorable to glass-forming ability.

2.140 BORON–CARBON–SILICON–IRON–PHOSPHORUS

The glass-forming region was observed at the limited $\mathbf{Fe_{76}P_5(Si_xB_yC_z)_{19}}$ ($x+y+z = 1$) compositions in the amorphous-forming range (Figure 2.1) and the large ΔT_x of over 50°C was observed in the range of $x = 0.3$–0.4, $y = 0.4$–0.66, and $z = 0$–0.2 (Makino et al. 2009). The maximum value of ΔT_x is 54°C for the $\mathbf{Fe_{76}P_5(Si_{0.3}B_{0.5}C_{0.2})_{19}}$ alloy. The ingots were prepared by induction melting the mixtures of pure Fe metal and prealloyed Fe–P, Fe–C, and pure crystalline powder of B and Si under an Ar atmosphere.

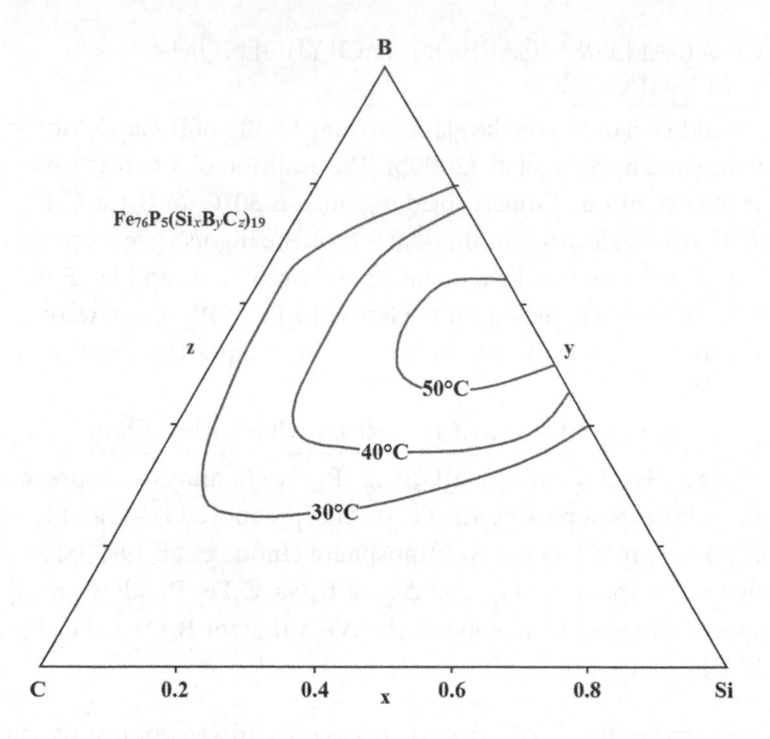

FIGURE 2.1 Glass-forming region and ΔT_x in the B–C–Si–Fe–P quinary system at 76 at% Fe and 5 at% P [$Fe_{76}P_5(Si_xB_yP_z)_{19}$]. (With permission from Makino, A., et al., *J. Alloys Compd.*, 483(1–2), 616, 2009.)

2.141 BORON–CARBON–CHROMIUM–MOLYBDENUM–IRON–PHOSPHORUS

The multinary $\mathbf{B_5C_5Cr_4Mo_4Fe_{71}P_{11}}$ bulk amorphous alloy was synthesized using the following techniques: (a) mechanical alloying the various elements; (b) melting the alloyed powders in a fused silica tube; (c) purifying the melt with a flux; and (d) water quenching the molten alloy (Shen and Schwarz 1999). The glass transition temperature (T_g) and ΔT_x of this alloy are equal to 445°C and 35°C, respectively.

2.142 BORON–CARBON–MOLYBDENUM–IRON–PHOSPHORUS

The quinary $\mathbf{B_{2.5}C_{7.5}Mo_5Fe_{75}P_{10}}$ alloy ingots were produced by induction melting the mixtures of pure Fe, Mo, C, and B and prealloyed Fe–P under an Ar atmosphere (Zhang et al. 2007). This alloy shows a glass transition at 435°C, followed by a supercooled liquid region of 32°C before crystallization at 467°C.

2.143 BORON–CARBON–IRON–COBALT–NICKEL–PHOSPHORUS

Substitution of Co and Ni for Fe in a $Fe_{75}P_{10}C_{10}B_5$ metallic glass effectively increases the stability of supercooled liquid and glass-forming ability (Miao et al. 2015, Zhang et al. 2017). The bulk metallic glasses with large supercooled liquid region (ΔT_x) exceeding 50°C and the critical sample diameter (d_c) of over 1 mm were obtained in a wide composition range from 30 to 60 at% Fe, 0 to 45 at% Co, and 0 to 40 at% Ni in the $\mathbf{(Fe,Co,Ni)_{75}P_{10}C_{10}B_5}$ alloy system by copper mold casting. The ΔT_x and T_g are 55°C and 370°C, for $\mathbf{Fe_{40}Ni_{35}P_{10}C_{10}B_5}$,

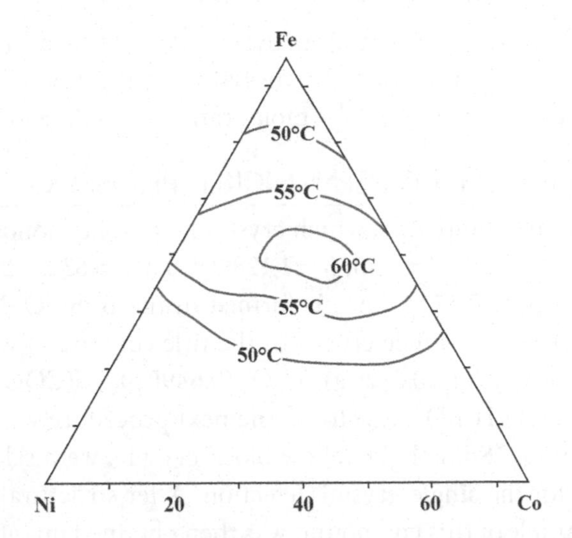

FIGURE 2.2 Composition dependence of ΔT_x for $Fe_{75-x-y}Co_xNi_yP_{10}C_{10}B_5$ (x=0–45; y= 0–45) metallic glasses. (With permission from Zhang, W., et al., *J. Alloys Compd.*, **707**, 57, 2017.)

56°C and 431°C, for $\mathbf{Fe_{40}Co_{35}P_{10}C_{10}B_5}$, and 60°C and 402°C, for $\mathbf{Fe_{40}Co_{20}Ni_{15}P_{10}C_{10}B_5}$ alloy, respectively. The last alloy is located near a eutectic point of the B–C–Fe–Co–Ni–P multinary system. Composition dependence of ΔT_x for $Fe_{75-x-y}Co_xNi_yP_{10}C_{10}B_5$ (x = 0–45; y = 0–45) metallic glasses is given in Figure 2.2 (Zhang et al. 2017).

The $(Fe,Co,Ni)_{75}P_{10}C_{10}B_5$ alloy ingots were prepared by induction melting the mixtures of pure Fe, Ni, Co, B, C, and Fe_3P precursor in a high purified Ar atmosphere. The ribbon samples with a cross section of about 0.02 mm × 1.0 mm were produced by melt spinning in an Ar atmosphere. The rod samples with diameters of 1–3 mm were prepared by injection casting into copper molds.

2.144 BORON–SILICON–IRON–COBALT–PHOSPHORUS

Thermodynamic calculations based on Miedema's extended model showed formation enthalpy changes and Gibbs free energy for amorphization of $\mathbf{Fe_{70}B_8Si_8Co_7P_7}$ alloy are –179 kJ M^{-1} and –181 kJ M^{-1}, respectively (Imani and Enayati 2017).

2.145 BORON–BISMUTH–OXYGEN–COBALT–PHOSPHORUS

The $\mathbf{BiCo_2BP_2O_{10}}$ quinary compound, which crystallizes in the monoclinic structure with the lattice parameters a = 507.82 ± 0.09, b = 1,128.89 ± 0.10, c = 640.20 ± 0.11 pm, and β = 107.827 ± 0.009° and a calculated density of 5.319 g cm^{-3}, is formed in the B–Bi–O–Co–P system (Zhang et al. 2012). It was synthesized by a solid-state reaction. A powder mixture of Cs_2CO_3 (0.0890 g), Bi_2O_3 (0.6481 g), $CoC_2O_4{\cdot}2H_2O$ (0.3831 g), $NH_4H_2PO_4$ (0.3381 g), and H_3BO_3 (0.2991 g) was first ground in an agate mortar and then transferred to a Pt crucible. The sample was gradually heated in air at 300°C for 6 h to decompose H_3BO_3 and $NH_4H_2PO_4$ and finally heated at 960°C for 24 h. The intermediate product was slowly cooled to 800°C at a rate of 2°C h^{-1}, where it was kept for 10 h and then quenched to room temperature. Some violet crystals of different shapes were mechanically removed from the solidified flux. After structural analysis, the red crystalline powder sample of this compound was then obtained quantitatively by the reaction of a mixture of Bi_2O_3, $C_4H_6CoO_4{\cdot}4H_2O$, H_3BO_3, and $NH_4H_2PO_4$ (molar ratio 1:4:2:4) at 850°C for 30 h.

2.146 BORON–BISMUTH–OXYGEN–NICKEL–PHOSPHORUS

The $\mathbf{BiNi_2BP_2O_{10}}$ quinary compound, which crystallizes in the monoclinic structure with the lattice parameters $a = 506.82 \pm 0.13$, $b = 1{,}128.2 \pm 0.3$, $c = 632.2 \pm 0.2$ pm, and $\beta = 90°$ and a calculated density of 5.370 g cm^{-3}, is formed in the B–Bi–O–Ni–P system (Zhang et al. 2012). The synthesis of a single crystal of the title compound was carried out using the powder mixture of Cs_2CO_3 (0.0829 g), Bi_2O_3 (0.6490 g), $Ni(COOH)_2{\cdot}2H_2O$ (0.3295 g), $NH_4H_2PO_4$ (0.3851 g), and H_3BO_3 (0.2946 g). The next procedure was the same as for the formation of $BiCo_2BP_2O_{10}$. Some light yellow block crystals were selected carefully from the sintered product for the single-crystal detection. After structural analysis, the yellow crystalline powder sample of this compound was then obtained quantitatively by the reaction of a mixture of Bi_2O_3, $Ni(COOH)_2{\cdot}2H_2O$, H_3BO_3, and $NH_4H_2PO_4$ (molar ratio 1:4:2:4) at 920°C for 30 h.

2.147 BORON–NIOBIUM–CHROMIUM–IRON–PHOSPHORUS

The $\mathbf{B_yNb_2Cr_1Fe_{97-x-y}P_x}$ ($x = 5–13$; $y = 7–15$) quinary phases have a high glass-forming ability with a wide range supercooled liquid region of 29°C–37°C (Matsumoto et al. 2010). They were prepared by induction melting the mixtures of prealloyed Fe–P, Fe–B, Fe–Nb, and Fe–Cr and pure Fe in an Ar atmosphere. The glassy phases were produced in a ribbon form by a single-roller melt-spinning method in an Ar atmosphere.

CHAPTER 3

Systems Based on BAs

3.1 BORON–HYDROGEN–CALCIUM–MAGNESIUM–OXYGEN–ARSENIC

$Ca_4MgB_{12}As_2O_{28}·18H_2O$ multinary compound [mineral teruggite (Rodgers et al. 2002; Aristarain and Hurlbut, Jr. 1968; Dal Negro et al. 1973)], which crystallizes in the monoclinic structure with the lattice parameters $a = 1{,}567.5 \pm 1.3$, $b = 1{,}992.0 \pm 1.4$, $c = 625.5 \pm 0.4$ pm, and $\beta = 99°20' \pm 5'$ and the calculated and experimental densities of 2.192 and 2.20 g cm^{-3}, respectively (Dal Negro et al. 1973) [$a = 1{,}568 \pm 1$, $b = 1{,}990 \pm 1$, $c = 625 \pm 1$ pm, and $\beta = 100°05' \pm 10'$ and the calculated and experimental densities of 2.139 and 2.149 ± 0.005 g cm^{-3}, respectively (Aristarain and Hurlbut, Jr. 1968)], is formed in the B–H–Ca–Mg–O–As system.

3.2 BORON–HYDROGEN–CALCIUM–OXYGEN–ARSENIC

$Ca_2BAsO_4(OH)_4$ quinary compound (mineral cahnite), which crystallizes in the tetragonal structure with the lattice parameters $a = 709.52 \pm 0.15$ and $c = 619.04 \pm 0.03$ pm (Embrey 1960) [$a = 711$ and $c = 620$ pm (Prewitt and Buerger 1961)] and an experimental density of 3.156 g cm^{-3} (Palache and Bauer 1927) [3.06 g cm^{-3} (Malinko 1966)], is formed in the B–H–Ca–O–As system.

3.3 BORON–HYDROGEN–ALUMINUM–SILICON–ANTIMONY–TANTALUM–OXYGEN–ARSENIC

$Al_6(Al,Ta)(BO_3)[(Si,Sb,As)O_4]_3(O,OH)_3$ multinary compound (mineral holtite), which crystallizes in the orthorhombic structure with the lattice parameters $a = 468.18 \pm 0.04$, 469.81 ± 0.07, 469.00 ± 0.06, and 472.64 ± 0.09, $b = 1{,}186.7 \pm 0.1$, $1{,}192.6 \pm 0.2$, $1{,}189.7 \pm 0.4$, and $1{,}198.1 \pm 0.2$, $c = 2{,}030.2 \pm 0.2$, $c = 2{,}041.3 \pm 0.3$, $c = 2{,}038.5 \pm 0.6$, and $c = 2{,}056.6 \pm 0.4$ pm and the calculated densities of 4.916, 4.848, 4.875, and 4.762 g cm^{-3} for four samples from the three localities (Groat et al. 2009) [$a = 1{,}190.5 \pm 0.5$, $b = 2{,}035.5 \pm 0.5$, and $c = 469.0 \pm 0.1$ pm and an experimental density of 3.90 ± 0.02 g cm^{-3} (Pryce 1971; Fleischer 1972); 469.14 ± 0.05, $b = 1{,}189.6 \pm 0.2$, and $c = 2{,}038.3 \pm 0.4$ pm (Hoskins et al. 1989); $a = 469.5 \pm 0.1$, $b = 1{,}190.6 \pm 0.3$, and $c = 2{,}038 \pm 3$ pm and a calculated density of 3.738 g cm^{-3} (Kazantsev et al. 2005, 2006)], is formed in the B–H–Al–Si–Sb–Ta–O–As system.

According to the data of Kazantsev et al. (2005, 2006) and Zubkova et al. (2006), there are holtite I [$a = 468.80 \pm 0.01$, $b = 1{,}188.4$, and $c = 2{,}035 \pm 3$ pm (Kazantsev et al. 2005, 2006; Locock et al. 2006b)], which is characterized by a low Sb content, and holtite II [$a = 468.75 \pm 0.01$, $b = 1{,}188.1$, and $c = 2{,}041.8 \pm 0.9$ pm (Kazantsev et al. 2005, 2006; Locock et al. 2006b); $a = 468.93 \pm 0.01$, $b = 1{,}188.1 \pm 0.1$, and $c = 2{,}039.4 \pm 0.1$ pm (Zubkova et al. 2006)] with high Sb content. Given the lack of any established structural difference between the holtite I and holtite II and that this terminology has not been approved by the *International Mineralogical Association*, its continued use seems ill advised (Locock et al. 2006b).

3.4 BORON–SODIUM–ALUMINUM–OXYGEN–ARSENIC

$Na_2AlBAs_4O_{14}$ quinary compound, which crystallizes in the monoclinic structure with the lattice parameters $a = 459.16 \pm 0.05$, $b = 2{,}070.6 \pm 0.3$, $c = 1{,}093.3 \pm 0.1$ pm, and $\beta = 90.39 \pm 0.01°$ and the calculated and experimental densities of 3.88 and 3.93 g cm^{-3}, respectively, is formed in the B–Na–Al–O–As system (Driss and Jouini 1988).

CHAPTER 4

Systems Based on AlN

4.1 ALUMINUM–HYDROGEN–POTASSIUM–PHOSPHORUS–OXYGEN–IRON–NITROGEN

The **$(NH_4,K)(Fe,Al)_2(PO_4)_2(OH)\cdot 2H_2O$** multinary compound (mineral spheniscidite), which crystallizes in the monoclinic structure with the lattice parameters a = 975 ± 1, b = 963 ± 2, c = 970 ± 1 pm, and β = 102°34′ ± 7′ and a calculated density of 2.71 g cm^{-3}, is formed in the Al–H–K–P–O–Fe–N system (Wilson and Bain 1986; Hawthorne et al. 1987).

4.2 ALUMINUM–HYDROGEN–PHOSPHORUS–OXYGEN–NITROGEN

The **$Al(NH_4)HP_3O_{10}$**, **$Al_2(NH_4)(PO_4)_2(OH)\cdot 2H_2O$**, and **$Al_2(NH_4)[(H_2P_3O_{10})(PO_3)_4$** quinary compounds are formed in the Al–H–P–O–N system. $Al(NH_4)H_3P_3O_{10}$ crystallizes in the monoclinic structure with the lattice parameters a = 1,164.3 ± 0.8, b = 491.8 ± 0.4, c = 870.5 ± 0.5 pm, and β = 119.27 ± 0.05° (Averbuch-Pouchot et al. 1977). This compound was prepared by calcinating at 350°C a mixture of Al_2O_3 (2 g) and $BaCO_3$ (1 g) in a large excess of $(NH_4)_2HPO_4$ (about 30 g) overnight. Washing with hot water to remove excess phosphoric flux allows obtaining the crystals of the title compound to be collected as squat prisms. This ammonium aluminum phosphate could also be synthesized by thermal reactions in surplus H_3PO_4 (Grunze and Grunze 1984).

$Al_2(NH_4)(PO_4)_2(OH)\cdot 2H_2O$ also crystallizes in the monoclinic structure with the lattice parameters a = 961.67 ± 0.03, b = 957.20 ± 0.04, c = 955.63 ± 0.03 pm and β = 103.589 ± 0.002° and a calculated density of 2.45 g cm^{-3} (Pluth et al. 1984) [a = 955.3 ± 0.1, b = 957.7 ± 0.1, c = 961.7 ± 0.1 pm and β = 103.56 ± 0.01° and a calculated density of 2.45 g cm^{-3} (Parise 1984).

$Al_2(NH_4)[(H_2P_3O_{10})(PO_3)_4]$ could be synthesized by thermal reactions in surplus of H_3PO_4 (Grunze and Grunze 1984).

4.3 ALUMINUM–HYDROGEN–PHOSPHORUS–OXYGEN–IRON–NITROGEN

The **$(NH_4)Al_{1.85}Fe_{1.15}(PO_4)_3(H_2O)_2$** multinary compound, which crystallizes in the monoclinic structure with the lattice parameters a = 1,327.54 ± 0.06, b = 1,016.63 ± 0.05, c = 877.93 ± 0.04 pm and β = 108.821 ± 0.001°, is formed in the Al–H–P–O–Fe–N system (Bieniok et al. 2008).

Solvothermal synthesis method was used for the preparation of this metalophosphate. The synthesis was carried out in Teflon-lined steel autoclave at 200°C and autogenous pressure. As an appropriate solvent, ethylene glycol ($C_2H_6O_2$) was used. Imidazole ($C_3H_4N_2$) was added as a cationic organic template and NH_4^+-supplier. Phosphate gel precursor with initial pH values of 5–7 was prepared from the mixture of FeO, Al_2O_3, P_2O_5, $C_3H_4N_2$, and $C_2H_6O_2$ (molar ratio 1:1:5:10:50). After homogenous stirring, the mixture was heated for 3–5 days. After the heating period, the product was cooled down quickly and washed several times with distilled water and finally acetone. Violet distorted octahedra of the title compound were obtained.

4.4 ALUMINUM–HYDROGEN–PHOSPHORUS–OXYGEN–COBALT–NITROGEN

Some multinary phases are formed in the Al–H–P–O–Co–N system. $\mathbf{Al_{1-x}Co_xPO_4[(NH_4)_x(NH_3)_{0.5-x}]}$ ($x \leq 0.4$) crystallizes in the monoclinic structure with the lattice parameters $a = 1{,}331.2 \pm 0.3$, $b = 1{,}316.0 \pm 0.3$, $c = 891.4 \pm 0.2$ pm and $\beta = 100.94 \pm 0.03°$ and a calculated density of 2.478 g cm^{-3} (Bontchev and Sevov 1997). A typical reaction for obtaining this compound was run with approximately threefold excess of aluminum and phosphate. Mixed in such a reaction were Al_2O_3 (46 mg), $Co_3(PO_4)_2{\cdot}8H_2O$ (36 mg), H_3PO_4 (130 mg), and H_2O (10 mL). The mixture was continuously stirred while aqueous solution of ammonia was added, until the pH reaches 8.5–8.8. The mixture was heated in an autoclave for 3–5 days at 200°C–205°C. Despite the excess mentioned earlier, the product usually contains the purplish $Co_7P_6O_{20}(OH)_4$ and $Co_2(PO_4)(OH)$ in addition to the major phase of the deep-blue title compound.

$\mathbf{(NH_4)_2[(OH)_{0.95}(H_2O)_{0.05}]_2[Co_{0.05}Al_{0.95}](PO_4)_2}$ crystallizes in the orthorhombic structure with the lattice parameters $a = 942.9 \pm 0.2$, $b = 959.2 \pm 0.2$, and $c = 989.7 \pm 0.2$ pm and $\mathbf{(NH_4)_2[(OH)_{0.95}(H_2O)_{0.05}]_2[Co_{0.025}Al_{0.975}](PO_4)_2{\cdot}2H_2O}$ crystallizes in the monoclinic structure with the lattice parameters $a = 955.0 \pm 0.2$, $b = 958.1 \pm 0.2$, $c = 958.3 \pm 0.2$ pm and $\beta = 103.64 \pm 0.03°$ (Bontchev and Sevov 1999). The attempts to produce the two title compounds as single-phase products were unsuccessful. They usually coexisted with yet another cobalt aluminophosphate and small amounts of various cobalt hydroxyphosphates. The composition of a typical reaction expressed as a molar ratio of oxides contained CoO/Al_2O_3/P_2O_5 = 3:3.5:1 (total of 200–500 mg) dissolved in H_2O (10 ml). Ammonia solution was added until the pH reached about 8.5. The mixture was loaded in a Teflon-lined stainless steel autoclave (23 mL) and heated at 200°C for 4 days. The product was repeatedly washed with water, ethanol, and acetone. The first compound crystallizes as light blue well-shaped octahedra. The crystals of the second compound are pale purple to pink and plate-like in shape.

$\mathbf{AlCo(NH_4)(PO_4)_2}$ crystallizes in the two various monoclinic structures with the lattice parameters $a = 1{,}340.4$, $b = 1{,}320.5$, $c = 893.5$ pm and $\beta = 103.89°$ and $a = 893.1$, $b = 1{,}318.8$, $c = 731.4$ pm and $\beta = 115.96°$, respectively (Feng et al. 1997).

$\mathbf{(NH_4)[Co(H_2O)_2Al_2(PO_4)_3]}$ crystallizes in the monoclinic structure with the lattice parameters $a = 1{,}311.0 \pm 0.2$, $b = 1{,}016.5 \pm 1.5$, $c = 874.24 \pm 0.14$ pm and $\beta = 108.830 \pm 0.013°$ and a calculated density of 2.721 g cm^{-3} (Panz et al. 1998). This compound was synthesized by the next two methods.

1. According to the first method, a mixture of the following components was prepared: 1,1′-dimethylcobalticinium chloride, 85% H_3PO_4, aluminum isopropoxide, 40% HF (40%), concentrated NH_3, and water. The molar composition (based on oxide formula) was $1.03CoO + 1.00Al_2O_3 + 2.55P_2O_5 + 1.46HF + 7.66NH_3 + 104.73H_2O$. First, a 0.65 M solution of 1,1′-dimethylcobalticinium chloride in H_2O was prepared. The aluminum isopropoxide was allowed to hydrolyze in that solution for 4 h under stirring. Then, HF and subsequently H_3PO_4 were added slowly. Finally, the pH was adjusted to 7 with ammonia. The resulting gel was transferred to a Teflon autoclave. The reaction was carried out under microwave heating at 200°C for 12 h. After that, the temperature was reduced to 180°C and held at this value for 7 days. Using this procedure, single crystals of the title compound were obtained.
2. Alternatively, it is possible to work in a common Teflon-lined steel autoclave that is heated in a convection oven. The starting mixture, prepared in the same way as described earlier, was first heated for 24 h at 180°C and then $CoCl_2{\cdot}6H_2O$ was added. Heating was continued for at least 7 days.

$(NH_4)_3[Co(NH_3)_6]_3[Al_2(PO_4)_3]_2{\cdot}2H_2O$ crystallizes in the orthorhombic structure with the lattice parameters $a = 950.2 \pm 0.2$, $b = 2{,}969.9 \pm 0.6$, and $c = 1{,}721.0 \pm 0.3$ pm and a calculated density of 1.97 g cm^{-3} (Morgan et al.1997). The reaction of AlO(OH) (0.5 g), $[Co(NH_3)_6]Cl_3$ (1 g), and 85% H_3PO_4 (1.9 g) in concentrated NH_4OH (10 mL) at 150°C for 12 h resulted in the formation of long orange needles of the title compound, large purple plates of NH_4CoPO_4, and minor amounts of unidentified small blue plates.

4.5 ALUMINUM–LITHIUM–CALCIUM–CERIUM–SILICON–NITROGEN

The **$Li_xCa_{1-2x}AlCe_xSiN_3$** multinary phase, which crystallizes in the orthorhombic structure (the lattice parameters for $Li_{0.01}Ca_{0.98}Al_{0.6}Ce_{0.01}Si_{1.3}N_3$ are the next: $a = 979.50 \pm 0.01$, $b = 564.65 \pm 0.01$, and $c = 505.90 \pm 0.01$ pm and a calculated density of 3.216 g cm^{-3}), is formed in the Al–Li–Ca–Ce–Si–N system (Li et al. 2008). The solubility limit for Ce is about $x = 0.02$. Lithium can increase the solubility of Al. With increasing Ce content, the lattice parameters first slightly increase for $x \leq 0.02$ and then decrease. This phase was prepared by a solid-state reaction. The appropriate amounts of Ca_3N_2, α-Si_3N_4, AlN, and Li_3N as well as CeN or CeO_2 were stoichiometrically mixed and ground in a Si_3N_4 mortar within a dried N_2-filled glove box. Thereafter, the powder mixture was transferred into a BN crucible and heated at 1,700°C for 4 h under high-purity N_2 atmosphere (0.48 MPa). For the prevention of lithium evaporation, a powder bed of Si_3N_4 + 10 mass% AlN mixtures was used to cover the small BN crucibles within a big BN crucible.

4.6 ALUMINUM–LITHIUM–STRONTIUM–EUROPIUM–NITROGEN

The **$LiSrAl_3N_4{:}Eu^{2+}$** quinary compound, which crystallizes in the triclinic structure with the lattice parameters $a = 584.61 \pm 0.02$, $b = 747.91 \pm 0.04$, and $c = 991.99 \pm 0.04$ pm and $\alpha = 83.598 \pm 0.002°$, $\beta = 76.766 \pm 0.005°$, and $\gamma = 79.519 \pm 0.006°$ (Park 2017) [$a = 586.631 \pm$

0.012, $b = 751.099 \pm 0.015$, and $c = 996.545 \pm 0.017$ pm and $\alpha = 83.6028 \pm 0.0012°$, $\beta = 76.7720 \pm 0.0013°$, and $\gamma = 79.5650 \pm 0.0014°$ for 0.4 at% Eu (Pust et al. 2014)], is formed in the Al–Li–Sr–Eu–N system. This compound was prepared as follows (Park 2017). AlN, Li_3N, Sr, and EuN were weighed and mixed according to the stoichiometry. They were then pulverized to prepare samples in a glove box that included oxygen and moisture less than 2 ppm. Then, 0.3 g of each sample was put into a BN container that has 18 holes with a diameter of 8.5 mm. Subsequently, the BN container was transported to a sealed tube-type heat treatment system to calcinate the samples in a (N_2/H_2 = 95:5) atmosphere at 1,000°C for 4 h.

The title compound was also synthesized by heating a stoichiometric mixture of $LiAlH_4$, AlN, SrH_2, and EuF_3 for 2 h to 1,000°C in a forming gas atmosphere (N_2/H_2 = 95:5) (Pust et al. 2014). The starting materials were thoroughly ground in a ball mill, filled in a W crucible and heated to a target temperature at a rate of 50°C min^{-1} in a radio-frequency furnace. After a short reaction period, the product was obtained in the form of a pink-colored powder, which is stable in ambient air up to 400°C. Likewise, the material is fully stable under normal air humidity conditions.

4.7 ALUMINUM–LITHIUM–SILICON–OXYGEN–NITROGEN

The $\mathbf{LiSi_9Al_3O_2N_{14}}$ and $\mathbf{LiSi_{10}Al_2ON_{15}}$ quinary compounds are formed in the Al–Li–Si–O–N system. The first of them crystallizes in the trigonal structure with the lattice parameters $a = 782.77 \pm 0.02$ and $c = 569.09 \pm 0.02$ pm (Grins et al. 2003). The title compound was synthesized from a starting powder mixture with an overall composition of $LiSi_{8.5}Al_{3.5}O_{3.41}N_{12.89}$ using the following precursor powders: AlN, Si_3N_4, Al_2O_3, and Li_2CO_3. The powder mixture was ball-mixed in water-free propanol for 24 h, dried, loaded in a cylindrical carbon die with an inner diameter of 20 mm, and compacted in vacuum in a spark plasma sintering apparatus. The sample was heated at a rate of 100°C min^{-1} under a uniaxial pressure of 50 MPa. Before reaching the final sintering temperature of 1,650°C, it was held at 1,000°C for 1 min. After holding at 1,650°C for 1 min, the spark plasma sintering unit was turned off and the sample was cooled at a rate of 400°C min^{-1}. Close to full densification of the sample was achieved.

$LiSi_{10}Al_2ON_{15}$ also crystallizes in the trigonal structure with the lattice parameters a = 783 and c = 5.67 pm and the calculated and experimental densities of 3.14 and 3.12 g cm^{-3}, respectively (Hampshire et al. 1978). It was prepared by heating appropriate mixtures of nitrides or nitrides plus oxides at 1,750°C in an atmosphere of N_2 or Ar for 15 min.

AlN–Li_2O–Si_3N_4. The isothermal section of this quasiternary system at 1,750°C is shown in Figure 4.1 (Kuang et al. 1990a; Huang and Yan 1992). It is seen that an extensive liquid phase region starts from Li_2O-rich region and expands widely towards the side of AlN–Si_3N_4. Single-phase α′-sialon ($\mathbf{Li_{2x}Si_{12-3x}Al_{3x}O_xN_{16-x}}$) exists on the line 3AlN·Li_2O_3–Si_3N_4 with the lower and upper limits at $2x = 0.25$ and 1.50, respectively. Lattice parameters of this phase increase with increasing $2x$ and the maximum data are the next: a = 783.6 and c = 568.7 pm (Kuang et al. 1990a). $Li_{2x}Si_{12-3x}Al_{3x}O_xN_{16-x}$ starts to form at 1,600°C, which rapidly increases in amount with increasing temperature. The maximum amount of α′-sialon can be obtained at around 1,700°C. O′-sialon ($\mathbf{LiAlSiON_2}$) occurs in this

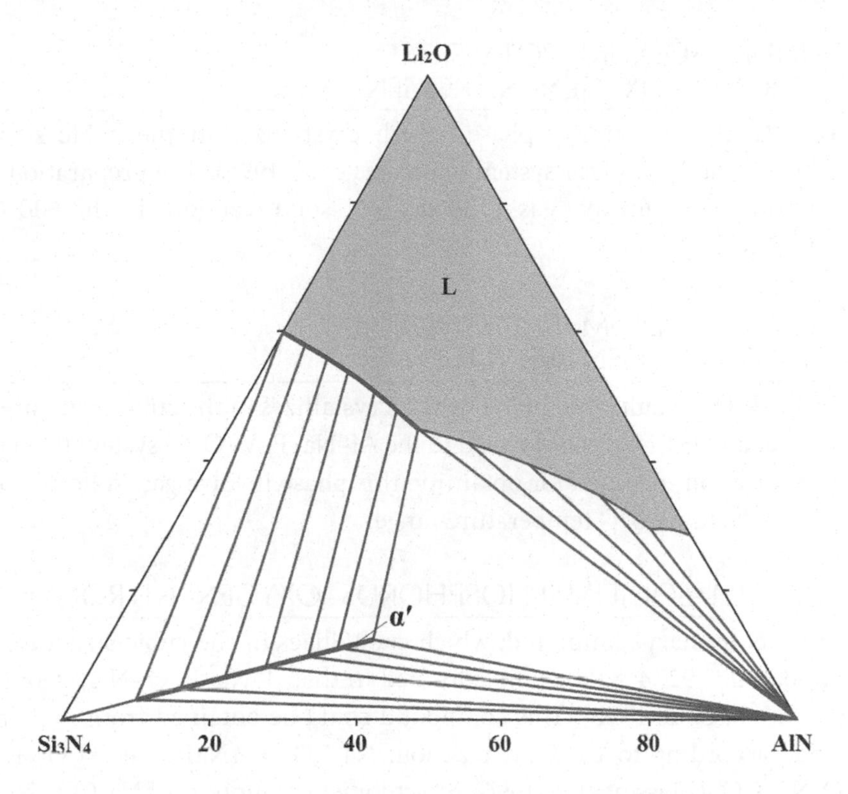

FIGURE 4.1 Isothermal sections of the $AlN–Li_2O–Si_3N_4$ quasiternary system at 1,750°C. (With permission from Kuang, S.-F., et al., *J. Mater. Sci. Lett.*, **9**(1), 72, 1990.)

system as a transitional phase before the formation of α′-sialon. O′-sialon formation starts at about 1,300°C and slightly increases with increasing temperature. The composition $AlN:10Li_2O:Si_3N_4$ has the lowest liquidus temperature (~900°C) among the composition explored.

To prepare the samples for the investigations, the starting powders used were Si_3N_4, AlN, and Li_2O (Huang and Yan 1992). Selected compositions were made by mixing the required amounts of the starting powders in an agate mortar under absolute EtOH for 1.5–2.0 h. The dried mixtures were isostatically pressed into a cylinder under 200 MPa for measuring the melting point and liquid-phase region. For the determination of phase compositions, the samples were prepared by firing or hot pressing (30 MPa) at different temperatures in BN-coated graphite dies under a mild flow of N_2. After hot pressing, the furnace was cooled at a rate of ~200°C min^{-1} in the high-temperature region.

The extent of oxynitride glass region has been determined within this quinary system (Rocherulle et al. 1989a). Oxynitride glasses are situated within the **$AlN–Li_2O–SiO_2$** quasiternary cut near the $Li_2O–SiO_2$ quasibinary system. The glass transition and crystallization temperatures increase at the increase of nitrogen content up to 2.21 mass%. It can be noted that the crystallization becomes weaker as the nitrogen content increases.

4.8 ALUMINUM–SODIUM–POTASSIUM–PHOSPHORUS–OXYGEN–NITROGEN

The $\mathbf{Na_{1.5}K_{1.5}AlP_3O_9N}$ multinary phase, which crystallizes in the cubic structure, is formed in the Al–Na–K–P–O–N system (Conanec et al. 1996). The preparation methods for obtaining this phase involve gas–solid or solid–solid reactions in the 600°C–800°C temperature range.

4.9 ALUMINUM–SODIUM–PHOSPHORUS–VANADIUM–OXYGEN–NITROGEN

The $\mathbf{Na_3Al_{0.5}V_{0.5}P_3O_9N}$ multinary phase, which crystallizes in the cubic structure with the lattice parameter $a = 933 \pm 1$ pm, is formed in the Al–Na–P–V–O–N system (Conanec et al. 1996). The preparation methods for obtaining this phase involve gas–solid or solid–solid reactions in the 600°C–800°C temperature range.

4.10 ALUMINUM–SODIUM–PHOSPHORUS–OXYGEN–NITROGEN

The $\mathbf{Na_3AlP_3O_9N}$ quinary compound, which crystallizes in the cubic structure with the lattice parameter $a = 927.4 \pm 0.01$ pm, is formed in the Al–Na–P–O–N system (Conanec et al. 1996; Massiot et al. 1996). This compound could be obtained from PON phosphorus oxynitride according to the next reaction: $Na_2CO_3 + 4NaPO_3 + Al_2O_3 + 2PON = 2Na_3AlP_3O_9N + CO_2$ (Massiot et al. 1996). Stoichiometric amounts of Na_2CO_3, $NaPO_3$, and Al_2O_3 were first heated at 800°C in a muffle furnace. In the second step, amorphous PON was finely ground with the decarbonated mixture which was then heated at 800°C under inert atmosphere. $Na_3AlP_3O_9N$ could also be prepared when the mixture of amorphous $(PON)_x$ (4 mM), $Na_5P_3O_{10}$ (32 mM), and Al_2O_3 (8 mM) were heated at 800°C for 1.5 h and then at 480°C for 2 h (Feldmann 1987a). It could also be obtained as a result of the following reactions: $6NaPO_3 + Al_2O_3 + 2NH_3 = 2Na_3AlP_3O_9N + 3H_2O$ (Conanec et al. 1996) or $3Na_2O + 2P_2O_5 + Al_2O_3 + 2PON = 2Na_3AlP_3O_9N$ (Feldmann 1987b).

Single crystals of $Na_3AlP_3O_9N$ were obtained by reacting at 800°C AlN in a molten flow of $(NaPO_3)_n$ (Conanec et al. 1996).

4.11 ALUMINUM–SODIUM–PHOSPHORUS–OXYGEN–CHROMIUM–NITROGEN

The $\mathbf{Na_3Al_{0.5}Cr_{0.5}P_3O_9N}$ multinary phase, which crystallizes in the cubic structure, is formed in the Al–Na–P–O–Cr–N system (Conanec et al. 1996). The preparation methods for obtaining this phase involve gas–solid or solid–solid reactions in the 600°C–800°C temperature range.

4.12 ALUMINUM–POTASSIUM–PHOSPHORUS–OXYGEN–NITROGEN

The $\mathbf{K_3AlP_3O_9N}$ quinary compound, which crystallizes in the cubic structure with the lattice parameter $a = 968 \pm 1$ pm, is formed in the Al–K–P–O–N system (Conanec et al. 1996). The preparation methods for obtaining of the title compound involve gas–solid or solid–solid reactions in the 600°C–800°C temperature range (Feldmann 1987b; Conanec et al. 1996).

4.13 ALUMINUM–POTASSIUM–PHOSPHORUS–OXYGEN–CHROMIUM–NITROGEN

The $\mathbf{K_3Al_{0.5}Cr_{0.5}P_3O_9N}$ multinary phase, which crystallizes in the cubic structure, is formed in the Al–K–P–O–Cr–N system (Conanec et al. 1996). The preparation methods for obtaining this phase involve gas–solid or solid–solid reactions in the 600°C–800°C temperature range.

4.14 ALUMINUM–MAGNESIUM–SILICON–OXYGEN–NITROGEN

AlN–MgO–$\mathbf{Si_3N_4}$. The isothermal section of this quasiternary system at 1,750°C is given in Figure 4.2 (Kuang et al. 1990b; Huang and Yan 1992). The liquid phase starts from the MgO–Si_3N_4 side at the composition near 4MgO·Si_3N_4 (1,630°C) and expands towards the quasiternary system with increasing temperature. At 1,750°C, the liquid phase has already covered 12*H* phase (**AlN·3MgO·$\mathbf{Si_3N_4}$**), which occurs in this system. This phase crystallizes in the hexagonal structure with the lattice parameters $a = 307$ and $c = 3{,}290$ pm, starts to form at 1,400°C, is stable up to the melting point (1,730°C), and is compatible with the three end members in the AlN–MgO–Si_3N_4 system (Kuang et al. 1990b). The polytypoid 12*H* with a formula of about $\mathbf{Mg_{2.3}Al_{0.7}Si_3O_{2.3}N_{4.7}}$ was obtained below 1,700°C (its melting point is 1,730°C) (Huang and Yan 1992).

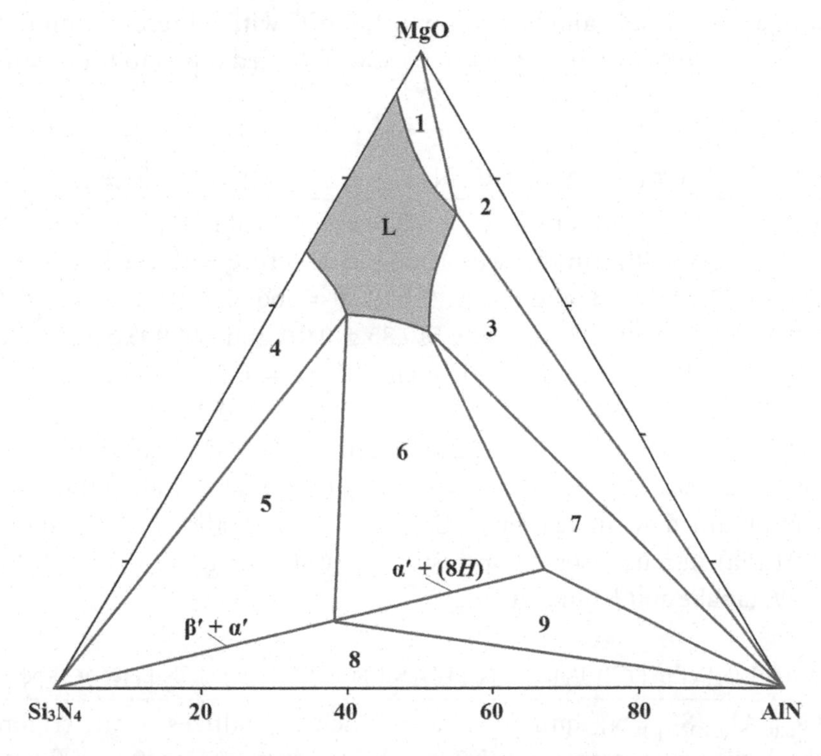

FIGURE 4.2 Isothermal sections of the AlN–MgO–Si_3N_4 quasiternary system at 1,750°C: 1, MgO + L; 2, MgO + L + AlN + (8*H*); 3, L + AlN + (8*H*); 4, L + Si_3N_4; 5, L + β′ + α′; 6, L + α′ + (8*H*); 7, L + α′ + AlN; 8, β′ + α′ + AlN; 9, α′ + 8*H* + AlN. (With permission from Kuang, S.-F., et al., *J. Mater. Sci. Lett.*, **9**(1), 69, 1990.)

To prepare the samples for the investigations, the starting powders used were Si_3N_4, AlN, and MgO (Huang and Yan 1992). Selected compositions were made by mixing the required amounts of the starting powders in an agate mortar under absolute EtOH for 1.5–2.0 h. The dried mixtures were isostatically pressed into a cylinder under 200 MPa for measuring the melting point and liquid-phase region. For the determination of phase compositions, the samples were prepared by firing or hot pressing (30 MPa) at different temperatures in BN-coated graphite dies under a mild flow of N_2. After hot pressing, the furnace was cooled at a rate of ~200°C min^{-1} in the high-temperature region.

4.15 ALUMINUM–CALCIUM–STRONTIUM–EUROPIUM–SILICON–OXYGEN–NITROGEN

The $\mathbf{(Ca_xSr_{1-x})_6Al_{6.4}Si_{25.6}O_{4.4}N_{41.6}{:}Eu^{2+}}$ multinary phases, which crystallize in the monoclinic structure, are formed in the Al–Ca–Sr–Eu–Si–O–N system (Zhao et al. 2016). It can be noted that the crystal cell became smaller as the Ca^{2+} ions saturated into the structures. These phases were prepared by a solid-state reaction method under N_2 atmosphere. Stoichiometric amounts of $CaCO_3$, $SrCO_3$, Si_3N_4, AlN, SiO_2, and Eu_2O_3 were selected as starting powders, weighed according to the composition with $x = 0$–0.1, and then ground in an agate mortar to form homogeneous mixtures. The mixed powders were placed in a BN crucible and heated up to 1,800°C with a preservation time of 10 h in a furnace. After the calcination, the samples were cooled down to room temperature in the furnace.

4.16 ALUMINUM–CALCIUM–STRONTIUM–SILICON–NITROGEN

The $\mathbf{Ca_{1-x}Sr_xAlSiN_3}$ solid solutions ($0.2 \leq x \leq 0.8$) are formed in the Al–Ca–Sr–Si–N quinary system. They crystallize in the orthorhombic structure with the lattice parameters $a = 981.3 \pm 0.3$, 982.97 ± 0.03, and 981.52 ± 0.01, $b = 566.67 \pm 0.02$, 568.79 ± 0.02, and 573.644 ± 0.007, and $c = 508.378 \pm 0.009$, 510.85 ± 0.01, and 514.905 ± 0.007, and a calculated density of 3.383, 3.576, and 3.925 g cm^{-3}, for $x = 0.2$, 0.5, and 0.8, respectively (Watanabe and Kijima 2009) [$a = 981.84 \pm 0.03$, 982.69 ± 0.03, and 981.52 ± 0.01, $b = 566.68 \pm 0.02$, 568.80 ± 0.02, and 573.65 ± 0.01, and $c = 508.39 \pm 0.01$, 510.80 ± 0.01, and 514.91 ± 0.01 for $x = 0.2$, 0.5, and 0.8, respectively (Watanabe et al. 2008a). These solid solutions were obtained by nitridation of $Ca_{1-x}Sr_x(Al_{0.5}Si_{0.5})_2$ alloy powder under high N_2 pressure (190 MPa) in a hot isostatic pressing apparatus at 1,900°C for 2 h (Watanabe et al. 2008a; Watanabe and Kijima 2009).

4.17 ALUMINUM–CALCIUM–LANTHANUM–SILICON–NITROGEN

The $\mathbf{Ca_{0.27}La_{0.03}Al_{0.62}Si_{11.38}N_{16}}$ quinary phase, which crystallizes in the trigonal structure with the lattice parameters $a = 783.8 \pm 0.1$ and $c = 570.3 \pm 0.1$, is formed in the Al–Ca–La–Si–N system (Grins et al. 2003). For its preparation, a powder mixture of La (960 mg), AlN (540 mg), Si_3N_4 (390 mg), and SiO_2 (168 mg) was heated up to 1,500°C in a N_2 atmosphere in a graphite furnace at a rate of 600°C h^{-1}, held at this temperature for

12 h, and then cooled to room temperature within 2 h. The surface of the reacted pellet was covered with hexagonal prismatic crystals of the title phase.

4.18 ALUMINUM–CALCIUM–CERIUM–SILICON–OXYGEN–NITROGEN

The $\mathbf{Ca_{1-x}AlCe_xSiO_{3x/2}N_{3-2x/3}}$ solid solution, which crystallize in the orthorhombic structure (the lattice parameters for $Ca_{0.99}Al_{0.501}Ce_{0.01}Si_{1.374}O_{0.015}N_{2.9933}$ are the next: $a = 978.78 \pm 0.02$, $b = 564.65 \pm 0.01$, and $c = 505.80 \pm 0.01$ pm and a calculated density of 3.218 g cm^{-3}) are formed in the Al–Ca–Ce–Si–O–N system (Li et al. 2008)]. These solid solutions were prepared by a solid-state reaction. The appropriate amounts of Ca_3N_2, α-Si_3N_4, and AlN, as well as CeN or CeO_2 were stoichiometrically mixed and ground in a Si_3N_4 mortar within a dried N_2-filled glove box. Thereafter, the powder mixture was transferred into a BN crucible and heated at 1,700°C for 4 h under high-purity N_2 atmosphere (0.48 MPa).

4.19 ALUMINUM–CALCIUM–SILICON–OXYGEN–NITROGEN

The $\mathbf{Ca_xAl_{3x}Si_{12-3x}O_xN_{16-x}}$ (x = 0.3–1.4) quinary solid solutions and $\mathbf{Ca_{0.913}AlSiO_{0.099}N_{2.901}}$ quinary phase are formed in the Al–Ca–Si–O–N system (Hampshire et al. 1978; Izumi et al. 1984; Kuang 1990a; Piao et al. 2007). The solid solutions crystallize in the trigonal structure with the lattice parameters a = 783.832 ± 0.007 and c = 570.327 ± 0.006 pm for $Ca_{0.67}Al_{2.0}Si_{10.0}O_{0.7}N_{15.3}$ composition (Izumi et al. 1984) [a = 782 and 786 and c = 568 and 571 pm and the calculated and experimental densities of 3.20 and 3.16 and 3.26 and 3.19 g cm^{-3} for $Ca_{0.5}Al_{1.5}Si_{10.5}O_{0.5}N_{15.5}$ and $Ca_{0.8}Al_{2.8}Si_{9.2}O_{1.2}N_{14.8}$ compositions, respectively (Hampshire et al. 1978); a = 785.1 and 791.7 and c = 570.8 and 575.6 pm for $Ca_{0.7}Al_{2.6}Si_{9.4}O_{1.2}N_{14.8}$ and $Ca_{1.2}Al_4Si_8O_{1.6}N_{14.4}$ compositions, respectively (Huang et al. 1985)]. Starting materials for the preparation of these solid solutions was a mixture of $CaCO_3$, Si_3N_4, and AlN (molar ratio 1:4:3) (Izumi et al. 1984). The mixture was hot pressed for 1 h at 1,700°C under a pressure of 19.6 MPa in a graphite die 50 mm in diameter. They could also be prepared by heating appropriate mixtures of nitrides or nitrides plus oxides at 1,750°C in an atmosphere of N_2 or Ar for 15 min (Hampshire et al. 1978).

$Ca_{0.913}AlSiO_{0.099}N_{2.901}$ crystallizes in the orthorhombic structure with the lattice parameters a = 949.9 ± 0.5, b = 565.9 ± 0.6, and c = 499.8 ± 0.2 pm (Piao et al. 2007). It was synthesized through self-propagating high-temperature synthesis method using CaAlSi as precursor at the obtaining $CaAlSiN_3$.

AlN–CaO–Si$_3$N$_4$. The isothermal section of this system at 1,700°C is given in Figure 4.3 (Huang et al. 1985; Huang and Yan 1992). A limited range of solid solutions was determined to exist extending on the tie line Si_3N_4–CaO:3AlN. The generalized formula of this solid solution can be represented as $Ca_xAl_{3x}Si_{12-3x}O_xN_{16-x}$. A metastable phase **2CaO·Si$_3$N$_4$·AlN** (τ_1) was synthesized by hot pressing at only 1,450°C in this system (Huang et al. 1985). By elevating the temperature it decomposes to two crystalline phases, $CaAlSiN_3$ (τ_2) and AlN, and with some glass phase. In fact, τ_2 does not occur on the plane of this system. The composition Si_3N_4 + 9CaO + 10AlN possesses the lowest melting temperature (~1,450°C), although the exact eutectic point has not been determined thoroughly.

The subsolidus phase diagram of this system was also constructed and is given in Figure 4.4 (Huang et al. 1985).

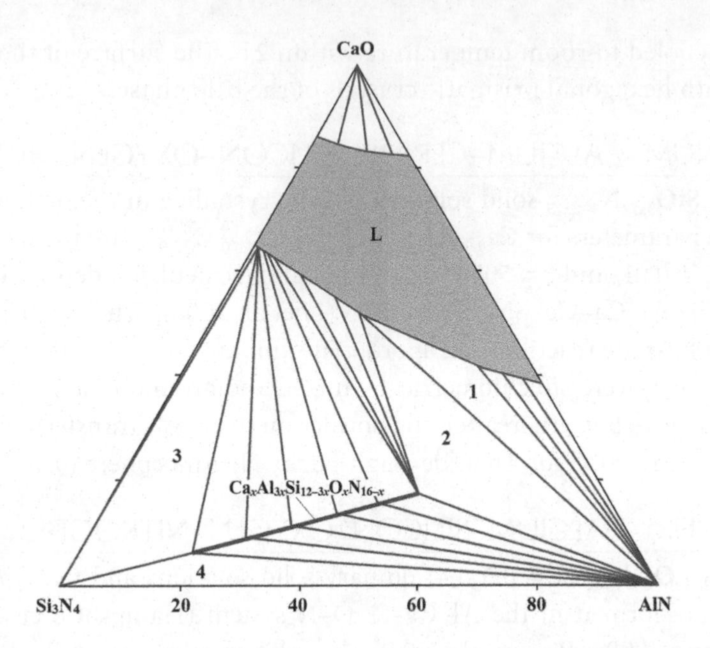

FIGURE 4.3 Isothermal sections of the AlN–CaO–Si_3N_4 system at 1,700°C: 1, L + AlN + τ_2; 2, L + AlN + $Ca_xAl_{3x}Si_{12-3x}O_xN_{16-x}$; 3, L + β-$Si_3N_4$ + $Ca_xAl_{3x}Si_{12-3x}O_xN_{16-x}$; 4, β-$Si_3N_4$ + $Ca_xAl_{3x}Si_{12-3x}O_xN_{16-x}$ + AlN. (With permission from Huang, Z.-K., et al., *J. Mater. Sci. Lett.*, **4**(3), 255, 1985.)

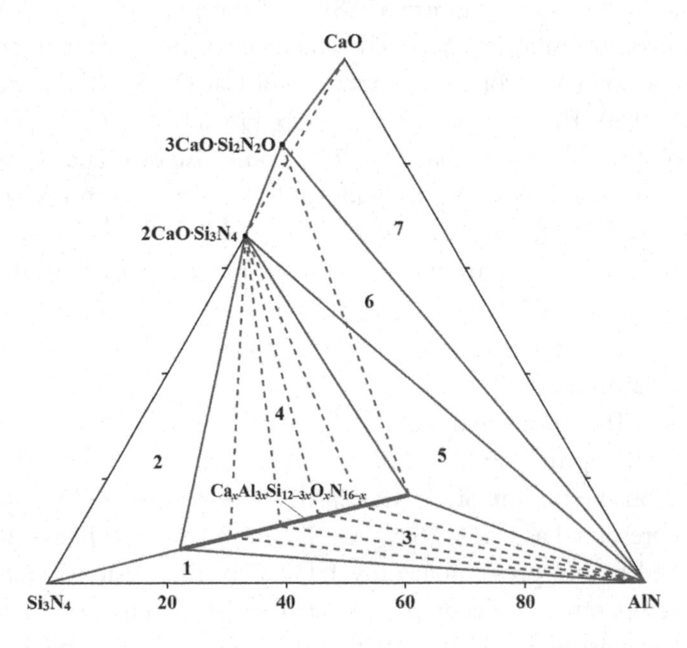

FIGURE 4.4 Subsolidus phase relations in the AlN–CaO–Si_3N_4 system: 1, β-Si_3N_4 + $Ca_xAl_{3x}Si_{12-3x}O_xN_{16-x}$+AlN; 2, β-$Si_3N_4$+$Ca_xAl_{3x}Si_{12-3x}O_xN_{16-x}$+2CaO·$Si_3N_4$; 3, $Ca_xAl_{3x}Si_{12-3x}O_xN_{16-x}$+ AlN; 4, $Ca_xAl_{3x}Si_{12-3x}O_xN_{16-x}$ + 2CaO·Si_3N_4; 5, $Ca_xAl_{3x}Si_{12-3x}O_xN_{16-x}$ + 2CaO·Si_3N_4 + AlN; 6, 2CaO·Si_3N_4 + 3CaO·Si_2N_2O + AlN; 7, CaO + 3CaO·Si_2N_2O + AlN. (With permission from Huang, Z.-K., et al., *J. Mater. Sci. Lett.*, **4**(3), 255, 1985.)

4.20 ALUMINUM–STRONTIUM–YTTRIUM–EUROPIUM–SILICON–OXYGEN–NITROGEN

The $\mathbf{SrAl_xYSi_{4-x}O_xN_{7-x}:Eu^{2+}}$ multinary solid solution ($x = 0–1$), which crystallizes in the hexagonal structure with the lattice parameters $a = 603.86 \pm 0.02$ and $c = 987{,}37 \pm 0.03$ pm for $x = 0.5$, is formed in the Al–Sr–Y–Eu–Si–O–N system (Kurushima et al. 2010). This solid solution was obtained by a solid-state reaction using stoichiometric amounts of $SrCO_3$, Y_2O_3, Si_3N_4, Al_2O_3, Eu_2O_3, and graphite; NH_4Cl (1–10 mass%) was added as a flux. The starting materials were finely ground and fired in BN crucibles to 1,600°C for 6 h in a N_2 atmosphere.

4.21 ALUMINUM–STRONTIUM–CERIUM–SILICON–OXYGEN–NITROGEN

The $\mathbf{Sr_3Al_{12}Ce_{10}Si_{18}O_{18}N_{36}}$ multinary compound, which crystallizes in the cubic structure with the lattice parameter $a = 1{,}338.2 \pm 0.2$ pm and a calculated density of 4.554 g cm^{-3}, is formed in the Al–Sr–Ce–Si–O–N system (Lauterbach et al. 2000). It was synthesized by a high-temperature reaction of $SrCO_3$ (0.62 mM), $Si(NH)_2$ (0.89 mM), AlN (1.36 mM), and Ce (1.22 mM). Under an atmosphere of pure Ar, the starting materials were placed in a W crucible, and the reaction was then performed under a pure atmosphere of N_2. The mixture was first heated up to 1,200°C with a rate of 20°C min^{-1} and then up to 1,550°C with a rate of 5.8°C min^{-1}. The sample was annealed at 1,550°C for 30 min and then cooled down to 900°C with a rate of 0.2°C min^{-1} and then down to room temperature with a rate of 100°C min^{-1}. The title compound was obtained as orange coarsely crystalline material.

4.22 ALUMINUM–STRONTIUM–PRASEODYMIUM–SILICON–OXYGEN–NITROGEN

The $\mathbf{Sr_3Al_{12}Pr_{10}Si_{18}O_{18}N_{36}}$ multinary compound, which crystallizes in the cubic structure with the lattice parameter $a = 1{,}334.54 \pm 0.06$ pm and a calculated density of 4.602 g cm^{-3} (according to the data of neutron powder diffraction $a = 1{,}335.63 \pm 0.01$ pm), is formed in the Al–Sr–Pr–Si–O–N system (Lauterbach et al. 2000). It was synthesized by a high-temperature reaction of $SrCO_3$ (0.82 mM), $Si(NH)_2$ (1.46 mM), AlN (1.71 mM), and Pr (0.85 mM). Under an atmosphere of pure Ar, the starting materials were placed in a W crucible, and the reaction was then performed under a pure atmosphere of N_2. The mixture was first heated up to 1,200°C with a rate of 20°C min^{-1} and then up to 1,650°C with a rate of 7.5°C min^{-1}. The sample was annealed at 1,650°C for 30 min and then cooled down to 1,400°C with a rate of 0.3°C min^{-1} and then down to room temperature with a rate of 100°C min^{-1}. The title compound was obtained as dark green coarsely crystalline material.

4.23 ALUMINUM–STRONTIUM–NEODYMIUM–SILICON–OXYGEN–NITROGEN

The $\mathbf{Sr_3Al_{12}Nd_{10}Si_{18}O_{18}N_{36}}$ multinary compound, which crystallizes in the cubic structure with the lattice parameter $a = 1{,}332.85 \pm 0.0.06$ pm and a calculated density of 4.666 g cm^{-3}, is formed in the Al–Sr–Nd–Si–O–N system (Lauterbach et al. 2000). It was synthesized by a high-temperature reaction of $SrCO_3$ (0.72 mM), $Si(NH)_2$ (1.30 mM), AlN (1.46 mM), and Nd (0.83 mM). Under an atmosphere of pure Ar, the starting materials were placed in a W crucible, and the reaction was then performed under a pure atmosphere of N_2. The mixture was first heated up to 1,200°C with a rate of 20°C min^{-1} and then up to 1,650°C with a rate of 7.5°C min^{-1}. The sample was annealed at 1,650°C for 30 min and then cooled down to 900°C with a rate of 0.3°C min^{-1} and then down to room temperature with a rate of 100°C min^{-1}. The title compound was obtained as coarsely crystalline blue-violet material.

4.24 ALUMINUM–STRONTIUM–SAMARIUM–SILICON–OXYGEN–NITROGEN

The $\mathbf{Sr_{10}Al_6Sm_6Si_{30}O_7N_{54}}$ multinary compound, which crystallizes in the orthorhombic structure with the lattice parameters $a = 1{,}706.94 \pm 0.09$, $b = 3{,}332.4 \pm 0.2$, and $c = 995.36 \pm 0.05$ pm and a calculated density of 4.284 g cm^{-3}, is formed in the Al–Sr–Sm–Si–O–N system (Lauterbach and Schnick 2000a). This compound has been obtained by high-temperature reaction of $SrCO_3$ (0.68 mM), $Si(NH)_2$ (1.38 mM), AlN (1.23 mM), Sr (0.57 mM), and Sm (0.67 mM). Under an atmosphere of pure Ar, the starting materials were filled into a W crucible, which was positioned in the center of the induction coil of a radio-frequency furnace. The reaction was then performed under an atmosphere of pure N_2. The mixture was heated up to 1,200°C with a rate of 20°C min^{-1} and then with 7°C min^{-1} to 1,600°C. The temperature was held constant for 30 min before the product was allowed to cool with 0.1°C min^{-1} to 900°C. Subsequently, the mixture was quenched to room temperature. The needle-shaped brownish-yellow single crystals were obtained.

4.25 ALUMINUM–STRONTIUM–EUROPIUM–SILICON–NITROGEN

The $\mathbf{Sr_{0.99}AlEu_{0.01}SiN_3}$ and $\mathbf{Sr_{0.98}AlEu_{0.02}Si_4N_7}$ quinary phases are formed in the Al–Sr–Eu–Si–N system. The first of them crystallizes in the orthorhombic structure with the lattice parameters $a = 984.3 \pm 0.3$, $b = 576.03 \pm 0.16$, and $c = 517.7 \pm 0.2$ pm and a calculated density of 4.195 g cm^{-3} (Watanabe et al. 2008b). It was synthesized by heating a mixture of binary nitrides at 1,900°C and a N_2 gas pressure of 190 MPa.

$Sr_{0.98}AlEu_{0.02}Si_4N_7$ exists in two polymorphic modifications (Yoshimura et al. 2018). The first polymorph crystallizes in the orthorhombic structure with the lattice parameters $a = 1{,}171.75 \pm 0.03$, $b = 2{,}131.03 \pm 0.06$, and $c = 494.83 \pm 0.02$ pm and a calculated density of 3.508 g cm^{-3} (Yoshimura et al. 2018) [$a = 1{,}170.46 \pm 0.02$, $b = 2{,}131.48 \pm 0.03$, and $c = 495.03 \pm 0.01$ pm and a calculated density of 1.755 g cm^{-3} (Hecht et al. 2009)]. β-$Sr_{0.98}AlEu_{0.02}Si_4N_7$ crystallizes in the monoclinic structure with the lattice parameters $a = 810.62 \pm 0.01$, $b = 909.53 \pm 0.01$, $c = 898.21 \pm 0.01$ pm, and $\beta = 111.6550 \pm 0.0005°$ and a calculated density of 3.529 g cm^{-3} (Yoshimura et al. 2018). Because the β-$Sr_{0.98}AlEu_{0.02}Si_4N_7$ has a higher calculated density, it may be either a low temperature or stable phase. Either a high-temperature or metastable α-$Sr_{0.98}AlEu_{0.02}Si_4N_7$ presumably crystallized in advance of the formation of the stable β-phase at 2,030°C.

To obtain this phase, "Sr_3N_2", Mg_3N, EuN, AlN, and α-Si_3N_4 powders were used as starting materials (Yoshimura et al. 2018). Sr_3N_2 was actually a mixture of Sr_2N, SrN_2, SrN, and SrH. The respective quantities were 155.4 mg of Sr_3N_2, 70.8 mg of Mg_3N_2, 5.4 mg of EuN, 77.1 mg of AlN, and 191.2 mg of α-Si_3N_4, and these were mixed with an alumina mortar and pestle under Ar in a glove box. The mixed powder was transferred into a BN crucible with a lid and heated in a furnace from room temperature to 800°C under 0.008 Pa of N_2 at a heating rate of 20°C min^{-1}. From 800°C to 2,030°C, the sample was heated under 0.85 MPa of N_2 at the same rate. After heating at 2,030°C and 0.85 MPa of N_2 for 4 h, the sample was cooled to 1,200°C at 20°C min^{-1} and then cooled to room temperature in the furnace by shutting off the electric power of the heater, maintaining the N_2 pressure at 0.85 MPa. The product was primarily composed of vermillion prismatic single crystals together with a white powder. Yellow thick-platelet single crystals also grew at and near the surface of the product. The vermillion single crystals of the α-phase were previously reported by Hecht et al. (2009). The powder X-ray diffraction (XRD) pattern for the white powder is similar to that for AlN. Mg_3N_2 evaporated during the heating process, and so no Mg was contained in the resulting crystals.

For the synthesis of α-$Sr_{0.98}AlEu_{0.02}Si_4N_7$, 2 mol% of Sr into $SrAlSi_4N_7$ could be substituted for Eu, using EuF_3, $EuCl_3$, or Eu as dopants (Hecht et al. 2009).

Ruan et al. (2011) prepared these phases by a solid-state reaction of "Sr_3N_2", EuN, AlN, and α-Si_3N_4. Starting materials (~2 g) with chemical composition of $Sr_{1-x}AlEu_xSi_4N_7$ (x = 0.01, 0.02, 0.03, 0.05, 0.075, and 0.10) were mixed and then filled into a BN crucible under a N_2 atmosphere in a glove box. Then, the mixtures were fired at 1,750°C for 2 h under 0.48 MPa of N_2 atmospheres in a gas pressure sintering furnace. Subsequently, the samples were mixed properly with additive AlN with contents equal to the first time and refired for 6 h under the same conditions.

4.26 ALUMINUM–STRONTIUM–EUROPIUM–SILICON–OXYGEN–NITROGEN

Some multinary phases are formed in the Al–Sr–Eu–Si–O–N system. Polycrystalline samples of $\mathbf{Sr_{1-x}AlEu_xSi_5O_2N_7}$ ($0 \le x \le 0.1$) were synthesized by high-temperature solid-state reactions (Duan et al. 2008). The starting materials were $SrCO_3$, α-Si_3N_4, AlN, and Eu_2O_3. The appropriate amounts of starting materials were weighed out separately and subsequently mixed and ground together in an agate mortar. The powder mixture was then transferred into Mo crucibles. Subsequently, those powder mixtures were fired in a horizontal tube furnace at 1,600°C for 16 h under flowing 95 vol% N_2 + 5 vol% H_2 atmosphere. After firing, the samples were gradually cooled down to room temperature in the furnace.

$\mathbf{Sr_{0.94}Al_{1.2}Eu_{0.06}Si_{2.8}O_{1.2}N_{4.8}}$ crystallizes in the orthorhombic structure with the lattice parameters $a = 580.61 \pm 0.05$, $b = 3{,}776.2 \pm 0.3$, and $c = 959.36 \pm 0.09$ pm and a calculated density of 3.650 g cm^{-3} (Yamane et al. 2012). To prepare colorless crystals of this phase, $SrCO_3$ (0.84 g), AlN (0.31 g), Si_3N_4 (0.74 g), SiO_2 (0.05 g), and Eu_2O_3 (0.06 g) were weighted and mixed in a N_2 gas filled glow box. The mixed powder was charged in a BN crucible and heated in a furnace from room temperature to 800°C in a vacuum. From 800°C–1,800°C, the sample was heated under 0.92 MPa of N_2. After heating at 1,800°C and 0.92 MPa of N_2 for 2 h, the sample was cooled down to room temperature in the furnace by shutting off the electric power of the heater.

$\mathbf{Sr_{2.94}Al_3Eu_{0.06}Si_{13}O_2N_{21}}$ crystallizes in the monoclinic structure with the lattice parameters $a = 1{,}472.50 \pm 0.3$, $b = 903.72 \pm 0.02$, $c = 746.42 \pm 0.02$ pm, and $\beta = 90.233 \pm 0.003°$

(Shioi et al. 2011). It was prepared from α-Si_3N_4 (35.25 mol%), SrO (15.62 mol%), $SrSi_2$ (16.27 mol%), AlN (32.54 mol%), and Eu_2O_3 (0.33 mol%).

$Sr_{2-y}Al_{2-x}Eu_ySi_{1+x}O_{7-x}N_x$ ($x = 0.4, y = 0.02$) crystallizes in the tetragonal structure with the lattice parameters $a = 781.59 \pm 0.01$ and $c = 525.97 \pm 0.01$ pm (Li et al. 2007). With increasing Eu concentration up to $y = 0.06$, the lattice parameters do not exhibit a significant decrease because of the similar ionic sizes of Eu^{2+} and Sr^{2+}. This phase with $x = 0–1.0$ and $y = 0–0.06$ was prepared by a solid-state reactions. Appropriate amounts of $SrCO_3$, Al_2O_3, SiO_2, Si_3N_4, and Eu_2O_3 were weighed and mixed in hexane solution using a ball mill with Si_3N_4 balls. Considering the necessity of reducing Eu_2O_3 to EuO, a small amount of additional Si_3N_4 was used. After drying, the mixture was loaded into a BN boat and fired in a horizontal tube furnace at 1,300°C–1,400°C for 5 h under 95 vol% N_2 + 5 vol% H_2 atmosphere.

$Sr_6Al_{6.4+x}Si_{25.6-x}O_{4.4+x}N_{41.6-x}$:$Eu^{2+}$ crystallizes in the monoclinic structure with the lattice parameters $a = 903.663$, $b = 1,472.27$ and $c = 746.039$ pm (β was not specified) (Ma et al. 2015). It was synthesized at 1,800°C and 0.5 MPa under N_2 atmosphere. Stoichiometric amount of $SrCO_3$, Si_3N_4, AlN, Eu_2O_3, and SiO_2 were weighed and mixed in an agate mortar. The mixed powder was placed in a W crucible and heated in a graphite resistance furnace from room temperature to 1,800°C over a period of 6 h with a heating preservation time of 10 h in the furnace. The powder after sintering was cooled down to room temperature in the furnace. A small amount of $SrAlSiN_7$ appeared as an impurity phase at $x \geq 0.30$. It seems that Si_3N_4 is excessive when $x \leq -0.15$.

$Sr_{13.72}Al_{12.32}Eu_{0.28}Si_{60.2}O_{4.2}N_{99.4}$ also crystallizes in the monoclinic structure with the lattice parameters $a = 6,778.9 \pm 0.3$, $b = 904.16 \pm 0.02$, $c = 746.86 \pm 0.01$ pm, and $\beta = 90.121 \pm 0.004°$ (Shioi et al. 2011). It was prepared from α-Si_3N_4 (33.84 mol%), SrO (9.80 mol%), $SrSi_2$ (24.50 mol%), AlN (31.51 mol%), and Eu_2O_3 (0.35 mol%). The powder mixture was filled in a *h*-BN crucible and then fired in a furnace at 1,900°C for 6 h under 1 MPa of N_2 atmosphere.

4.27 ALUMINUM–STRONTIUM–HOLMIUM–SILICON–OXYGEN–NITROGEN

The **$SrAl_{1.2}HoSi_{2.8}O_{1.2}N_{5.8}$** multinary compound, which crystallizes in the hexagonal structure with the lattice parameters $a = 606.39 \pm 0.09$ and $c = 985.8 \pm 0.2$ pm, is formed in the Al–Sr–Ho–Si–O–N system (Lieb et al. 2007). This compound was prepared as follows. Ho (0.75 mM), $Si(NH)_2$ (1.5 mM), AlN (2.0 mM), $SrCO_3$ (0.61 mM), and $SrCl_2$ (3.2 mM) were thoroughly mixed and transferred into a W crucible in a glove box (Ar atmosphere). The crucible was then positioned in a water-cooled silica glass reactor of a radio-frequency furnace, heated to 1,200°C within 1 h, kept at that temperature for 2 h, and then heated to 2,050°C within 4 h. After that, the furnace was cooled to 1,200°C within 60 h and the sample was quenched to room temperature. The crystals of the title compound could be separated from the by-product due to their differing crystal habit and color.

4.28 ALUMINUM–STRONTIUM–ERBIUM–SILICON–OXYGEN–NITROGEN

The **$SrAl_{1.5}ErSi_{2.5}O_{1.5}N_{5.5}$** and **$SrAl_3ErSiO_3N_4$** multinary compounds are formed in the Al–Sr–Er–Si–O–N system. Both compounds crystallize in the hexagonal structure with the lattice parameters $a = 606.53 \pm 0.09$ and $c = 986.0 \pm 0.2$ pm for the first compound (Lieb et al. 2007) and $a = 606.53 \pm 0.09$ and $c = 985.90 \pm 0.08$ pm and a calculated density of 5.245 g cm^{-3} for the second one (Schnick et al. 1999). Since these compounds crystallize in the same structure with almost identical lattice parameters, it is obviously the same compound.

$SrAl_{1.5}ErSi_{2.5}O_{1.5}N_{5.5}$ was prepared by the same way as $SrAl_{1.2}HoSi_{2.8}O_{1.2}N_{5.8}$ was synthesized using Er instead of Ho (Lieb et al. 2007).

$SrAl_3ErSiO_3N_4$ has a thermal stability up to 1,800°C (Schnick et al. 1999). The transparent, pink crystals of this compound were obtained according to the next procedure. Er (0.74 mM), $Si(NH)_2$ (1.49 mM), AlN (2.19 mM), and $SrCO_3$ (0.64 mM) were thoroughly mixed under an Ar atmosphere in a glove box. In a W crucible, the mixture was then transferred into the quartz reactor of the radio-frequency furnace. Under pure N_2 atmosphere, the temperature was increased to 1,200°C with a heating rate of 20°C min^{-1} and then to 1,670°C at 8°C min^{-1}. This temperature was maintained for 30 min, and then the reaction mixture was cooled to 900°C with a cooling rate of 0.2°C min^{-1}. Finally, a room temperature was reached within 10 min.

4.29 ALUMINUM–STRONTIUM–THULIUM–SILICON–OXYGEN–NITROGEN

The $\mathbf{SrAl_{1.2}TmSi_{2.8}O_{1.2}N_{5.8}}$ multinary compound, which crystallizes in the hexagonal structure with the lattice parameters $a = 605.93 \pm 0.09$ and $c = 905.93 \pm 0.09$ pm, is formed in the Al–Sr–Tm–Si–O–N system (Lieb et al. 2007). This compound was prepared by the same way as $SrAl_{1.2}HoSi_{2.8}O_{1.2}N_{5.8}$ was synthesized using Tm instead of Ho.

4.30 ALUMINUM–STRONTIUM–YTTERBIUM–SILICON–OXYGEN–NITROGEN

The $\mathbf{SrAl_{1.9}YbSi_{2.1}O_{1.9}N_{5.1}}$ multinary compound, which crystallizes in the hexagonal structure with the lattice parameters $a = 605.61 \pm 0.09$ and $c = 983.0 \pm 0.2$ pm, is formed in the Al–Sr–Yb–Si–O–N system (Lieb et al. 2007). This compound was prepared by the same way as $SrAl_{1.2}HoSi_{2.8}O_{1.2}N_{5.8}$ was synthesized using Yb instead of Ho.

4.31 ALUMINUM–STRONTIUM–SILICON–OXYGEN–NITROGEN

Some quinary compounds are formed in the Al–Sr–Si–O–N system. $\mathbf{SrAl_2SiO_3N_2}$ crystallizes in the orthorhombic structure with the lattice parameters $a = 491.98 \pm 0.06$, $b = 789.73 \pm 0.07$, and $c = 1{,}134.94 \pm 0.18$ pm and a calculated density of 3.701 g cm^{-3} (Lauterbach and Schnick 1998; Schnick et al. 1999) [according to the calculation using density functional theory $a = 491.2$, $b = 795.3$, and $c = 1{,}125.8$ pm and a bulk modulus $B = 131.9 \pm 0.3$ GPa (Winkler et al. 2001)]. It was obtained as coarsely crystalline solid by a reaction of $Si(NH)_2$ (1.70 mM), AlN (2.52 mM), and $SrCO_3$ (0.80 mM) in a W crucible (Lauterbach and Schnick 1998; Schnick et al. 1999). The mixture was heated under N_2 atmosphere to 1,200°C for 2 h (10°C min^{-1}), then the temperature was raised to 1,550°C (3°C min^{-1}) and held at this temperature for 30 min. Afterwards, the furnace was cooled slowly to 800°C (0.2°C min^{-1}) and reached after a further 30 min room temperature. The title compound has a thermal stability up to 1,800°C (Schnick et al. 1999).

$\mathbf{Sr_2Al_2Si_{10}O_4N_{14}}$ also crystallizes in the orthorhombic structure with the lattice parameters $a = 827.88 \pm 0.09$, $b = 957.57 \pm 0.09$, and $c = 491.58 \pm 0.04$ pm (Shen et al. 1999b).

It was synthesized from an appropriate mixture of AlN, Si_3N_4, and $SrCO_3$. Corrections were made for the oxygen present in the Si_3N_4 and AlN raw materials, corresponding to 2.7 and 1.9 mass% SiO_2 and Al_2O_3, respectively. The precursor powders were ball-mixed in water-free propanol for 24 h. The dried powder mixture was then hot pressed at 1,800°C for 2 h under 35 MPa pressure in a graphite resistance furnace and in a N_2 atmosphere.

$Sr_3Al_3Si_{13}O_2N_{21}$ crystallizes in the monoclinic structure with the lattice parameters $a = 1{,}475.57 \pm 0.05$, $b = 746.27 \pm 0.02$, $c = 903.48 \pm 0.03$ pm, and $\beta = 90°$ and a calculated density of 3.475 g cm^{-3} (Ishizawa et al. 2010) [in the orthorhombic structure with the lattice parameters $a = 903.7 \pm 0.6$, $b = 1{,}473.4 \pm 0.9$, and $c = 746.4 \pm 0.5$ pm (Fukuda et al 2009)]. The powder sample of this compound was prepared from α-Si_3N_4, SrO, and AlN (Ishizawa et al. 2010). The mixture was ground in the Si_3N_4 mortar and pestle. The powder sample was filled in *h*-BN crucible and fired in a furnace at 1,900°C for 24 h under 1 MPa of N_2 atmosphere. It could also be obtained by firing the mixture of $SrCO_3$, Si_3N_4, and AlN at temperatures ranging from 1,700°C to 1,900°C for several hours under 0.7 MPa of nitrogen (Fukuda et al 2009).

$Sr_5Al_5Si_{21}O_2N_{35}$ crystallizes in the orthorhombic structure with the lattice parameters $a = 2{,}361.4 \pm 0.5$, $b = 748.69 \pm 0.15$, and $c = 905.86 \pm 0.18$ pm and a calculated density of 3.495 g cm^{-3} (Oeckler et al. 2009). It is stable up to about 1,500°C under He atmosphere. Above this temperature, rapid decomposition involving the formation of gaseous products was observed. The title compound can be obtained by various synthetic routes. Large single crystals were obtained by the following procedure. AlN (1 mM), Si_3N_4 (0.7 mM), $SrCO_3$ (0.3 mM), and Sr (0.4 mM) were mixed in an agate mortar and filled into a W crucible under Ar atmosphere in a glove box. Under purified N_2, the crucible was heated to 1,750°C at a rate of about 600°C h^{-1} in the reactor of a radio-frequency furnace, and this maximum temperature was kept for 2 h. Subsequently, the crucible was cooled to 1,300°C at a rate of about 90°C h^{-1} and then quenched to room temperature by switching off the furnace. Samples synthesized by this method contain $Sr_5Al_5Si_{21}O_2N_{35}$ as comparatively large colorless, air- and water-resistant crystals in addition to compounds like $SrAl_2SiO_3N_2$, $SrAlSi_4N_7$, and $Sr_2Si_5N_8$.

4.32 ALUMINUM–BARIUM–SAMARIUM–SILICON–OXYGEN–NITROGEN

The **$BaSm_5[Al_3Si_9N_{20}]O$** multinary compound, which crystallizes in the trigonal structure with the lattice parameters $a = 952.94 \pm 0.13$, $c = 612.60 \pm 0.12$ pm and a calculated density of 5.236 g cm^{-3} at 200 ± 2 K, is formed in the Al–Ba–Sm–Si–O–N system (Lieb and Schnick 2006). To prepare the title compound, a mixture of Ba (0.7 mM), $BaCO_3$ (0.7 mM), Sm (0.6 mM), AlN (1.2 mM), and $Si(NH)_2$ (1.4 mM) was thoroughly mixed and transferred into a W crucible in a glove box (Ar atmosphere). The crucible was then positioned in a water-cooled silica glass reactor of a radio-frequency furnace. It was heated under a pure N_2 atmosphere to 1,450°C within 1 h, then to 1,850°C within 2 h, maintained at that temperature for 30 min, and subsequently cooled to 1,150° within 60 h. Finally, the product was quenched to room temperature. The reactions yielded single crystals of trigonal prismatic shape and reddish brown color, together with crystalline $Sm_2Si_2N_2O_7$, AlN, and sialon glasses as by-products. The desired crystals could be separated from the by-products due to their differing crystal habit and color.

4.33 ALUMINUM–BARIUM–EUROPIUM–SILICON–NITROGEN

The $\mathbf{Ba_{0.99}Eu_{0.01}Al_3Si_4N_9}$ quinary phase, which exists in two polymorphic modifications, is formed in the Al–Ba–Eu–Si–N system (Yamane and Yoshimura 2015). The first modification crystallizes in the orthorhombic structure with the lattice parameters $a = 1{,}002.8 \pm 0.2$, $b = 5{,}335.3 \pm 0.9$, and $c = 592.15 \pm 0.11$ pm and a calculated density of 3.831 g cm^{-3}, and the second one crystallizes in the monoclinic structure with the lattice parameters $a = 583.76 \pm 0.04$, $b = 2{,}668.95 \pm 0.12$, $c = 583.93 \pm 0.03$ pm, and $\beta = 118.8428 \pm 0.0015°$ and a calculated density of 3.808 g cm^{-3}.

To prepare this compound, the starting powders of Ca_3N_2, Ba_3N_2, EuN, Si_3N_4, and AlN were weighed with predetermined molar ratios and mixed in an Ar-filled glove box. Mixed powders of about 6 g were placed in sintered BN crucibles, and placed in a gas-pressure furnace. The powders were first heated to 800°C under a reduced condition of 0.008 Pa at a heating rate of 20°C min^{-1}. Nitrogen was introduced into the chamber of the furnace up to 0.85 MPa at 800°C and then heated to 1,900°C or 2,000°C at a rate of 20°C min^{-1}, keeping the same N_2 pressure. After heating at 1,900°C or 2,000°C for 2 h, the sample was cooled to 1,200°C at a rate of 20°C min^{-1} and then cooled in the furnace by shutting off the electric power of the heater of the furnace.

4.34 ALUMINUM–BARIUM–EUROPIUM–SILICON–OXYGEN–NITROGEN

The $\mathbf{Ba_{1-x}AlEu_xSi_5O_2N_7}$ or $\mathbf{(Ba_{0.9}Eu_{0.1})_2Al_2Si_{10}O_4N_{14}}$ multinary phase is formed in the Al–Ba–Eu–Si–O–N system. Polycrystalline samples of $Ba_{1-x}AlEu_xSi_5O_2N_7$ ($0 \leq x \leq 0.1$) were synthesized by high-temperature solid-state reactions (Duan et al. 2008). The starting materials were $BaCO_3$, α-Si_3N_4, AlN, and Eu_2O_3. The appropriate amounts of starting materials were weighed out separately and subsequently mixed and ground together in an agate mortar. The powder mixture was then transferred into Mo crucibles. Subsequently, those powder mixtures were fired in a horizontal tube furnace at 1,600°C for 16 h under flowing 95 vol% N_2 + 5 vol% H_2 atmosphere. After firing, the samples were gradually cooled down to room temperature in the furnace.

$(Ba_{0.9}Eu_{0.1})_2Al_2Si_{10}O_4N_{14}$ was prepared using Eu_2O_3, $BaCO_3$, Si_3N_4, and Al_2O_3 (Esmaeilzadeh et al. 2004). Powder mixtures were ball-mixed in water-free propanol for 24 h, and the dried powder mixtures were compacted using spark plasma sintering in 12-mm graphite dies, with the densification monitored with a dilatometer. Pressure of 50 MPa was used, and the temperature was raised at a rate of 200°C min^{-1} to 1,000°C, held there for 2 min, and then raised at the same rate to 1,600°C, where full densification was achieved. The spark plasma sintering unit was then shut down, and the sample was allowed to cool to room temperature. The compacted samples were heat-treated at 1,600°C for 15 h in a graphite furnace and N_2 atmosphere.

4.35 ALUMINUM–BARIUM–HOLMIUM–SILICON–OXYGEN–NITROGEN

The $\mathbf{BaAl_{1.5}HoSi_{2.5}O_{1.5}N_{5.5}}$ multinary compound, which crystallizes in the hexagonal structure with the lattice parameters $a = 611.52 \pm 0.09$ and $c = 1{,}000.2 \pm 0.2$ pm, is formed in the Al–Ba–Ho–Si–O–N system (Lieb et al. 2007). This compound was prepared as follows.

Ho (0.75 mM), $Si(NH)_2$ (1.5 mM), AlN (2.0 mM), $BaCO_3$ (0.61 mM), and $BaCl_2$ (2.4 mM) were thoroughly mixed and transferred into a W crucible in a glove box (Ar atmosphere). The crucible was then positioned in a water-cooled silica glass reactor of a radio-frequency furnace, heated to 1,200° within 1 h, kept at that temperature for 2 h and then heated to 2,150°C within 4 h. After that, the furnace was cooled to 1,200°C within 60 h and the sample was quenched to room temperature. The crystals of the title compound could be separated from the by-product due to their differing crystal habit and color.

4.36 ALUMINUM–BARIUM–ERBIUM–SILICON–OXYGEN–NITROGEN

The $\mathbf{BaAl_{1.5}ErSi_{2.5}O_{1.5}N_{5.5}}$ multinary compound, which crystallizes in the hexagonal structure with the lattice parameters $a = 610.29 \pm 0.09$ and $c = 994.3 \pm 0.2$ pm, is formed in the Al–Ba–Er–Si–O–N system (Lieb et al. 2007). The title compound was prepared by the same way as $BaAl_{1.5}HoSi_{2.5}O_{1.5}N_{5.5}$ was synthesized using Er instead of Ho.

4.37 ALUMINUM–BARIUM–THULIUM–SILICON–OXYGEN–NITROGEN

The $\mathbf{BaAl_{1.6}TmSi_{2.4}O_{1.6}N_{5.4}}$ multinary compound, which crystallizes in the hexagonal structure with the lattice parameters $a = 610.04 \pm 0.09$ and $c = 992.8 \pm 0.2$ pm, is formed in the Al–Ba–Tm–Si–O–N system (Lieb et al. 2007). The title compound was prepared by the same way as $BaAl_{1.5}HoSi_{2.5}O_{1.5}N_{5.5}$ was synthesized using Tm instead of Ho.

4.38 ALUMINUM–BARIUM–YTTERBIUM–SILICON–OXYGEN–NITROGEN

The $\mathbf{BaAl_{2.0}YbSi_{2.0}O_{2.0}N_{5.0}}$ multinary compound, which crystallizes in the hexagonal structure with the lattice parameters $a = 609.96 \pm 0.09$ and $c = 992.3 \pm 0.2$ pm, is formed in the Al–Ba–Yb–Si–O–N system (Lieb et al. 2007). The title compound was prepared by the same way as $BaAl_{1.5}HoSi_{2.5}O_{1.5}N_{5.5}$ was synthesized using Yb instead of Ho.

4.39 ALUMINUM–BARIUM–SILICON–OXYGEN–NITROGEN

The $\mathbf{Ba_2Al_2Si_{10}O_4N_{14}}$ quinary compound, which crystallizes in the orthorhombic structure with the lattice parameters $a = 823.0 \pm 0.1$, $b = 965.5 \pm 0.2$, and $c = 490.93 \pm 0.08$ pm, is formed in the Al–Ba–Si–O–N system (Esmaeilzadeh et al. 2004). Single crystals of the title compound were obtained by heat treating a 0.5 g mixture of Ba metal, Si_3N_4, and AlN, with nominal overall composition $Ba_{7.25}Al_{10}Si_{10}N_{23.3}$, in a graphite furnace and N_2 atmosphere. The mixture was placed in an Al_2O_3 crucible, and the sample was heated to 1,600°C during 2 h, held at 1,600°C for 8 h, and then allowed to cool to room temperature. The surface of the reacted material and parts of the Al_2O_3 crucible were found to be covered with crystals of $Ba_2Al_2Si_{10}O_4N_{14}$.

4.40 ALUMINUM–YTTRIUM–CERIUM–SILICON–OXYGEN–NITROGEN

The $\mathbf{Al_xY_{1.98}Ce_{0.02}Si_{3-x}O_{3+x}N_{4-x}}$ multinary solid solution was prepared from a mixture of CeO_2, Y_2O_3, γ-Al_2O_3, and α-Si_3N_4 (Krevel van et al. 2000). The starting powders were suspended in propan-2-ol in a planetary ball mill for 10 h. The resulting mixture was dried and fired in Mo crucibles for 2 h in a N_2/H_2 (volume ratio 95:5) atmosphere at 1,730°C. The lattice parameters of the solid solution showed a linear increase between $x = 0$ and $x = 0.6$.

The doping of the solid solution with small amount of Ce has only a limited influence on the value of the cell parameters.

4.41 ALUMINUM–YTTRIUM–SILICON–CARBON–OXYGEN–NITROGEN

$AlN–Y_2O_3–SiC$. The results of XRD and electron probe microanalysis (EPMA) show that, except for three original constituents of this system, $2Y_2O_3·Al_2O_3$ phase was also observed (Chen et al. 2013). This could be the result of impurity Al_2O_3 in the starting AlN powder reacted with Y_2O_3. After introducing $2Y_2O_3·Al_2O_3$, the phase relationships transformed this ternary system into the quasi-quaternary system $AlN–Y_2O_3–2Y_2O_3·Al_2O_3–SiC$. The coexistence of four phases brings significance to the manufacture of SiC ceramics with $AlN–Y_2O_3$ as sintering additives, because of the formation of $2Y_2O_3·Al_2O_3$ with higher melting temperatures above 2,000°C as the second phase might enhance the high-temperature properties of SiC ceramics. To investigate this system, the powders of AlN, Y_2O_3, and α-SiC were mixed and ground in an agate mortar for 2 h with alcohol. After drying up, the prepared powders were placed in a round graphite mold with 10 mm diameter and afterward hot pressed. To prevent the sample from sticking to the mold, the latter was lined with a thin layer of BN. The conditions used for hot pressing were as follows: Ar atmosphere, 30 MPa, 1450°C–1600°C, 1–2 h. To achieve the reaction equilibria, the temperature was further kept for 1–3 h after releasing the pressure.

4.42 ALUMINUM–YTTRIUM–SILICON–OXYGEN–NITROGEN

The subsolidus phase relationships in the Al–Y–Si–O–N quinary system were determined by Sun et al. (1988, 1991a, 1991b, 1992). Sixty-eight compatibility tetrahedra were established in this system. The solubility limit of α′-sialon solid solution on the $3AlN·YN–Si_3N_4$ composition join in the $AlN–YN–Si_3N_4$ system has been determined at 1,800°C. The end members of these solid solutions are $Y_{0.43}Al_{1.3}Si_{10.7}N_{16}$ and $Y_{0.8}Al_{2.4}Si_{9.6}N_{16}$. The single-phase boundary of the α′-sialon solid solution on the composition triangle $3AlN·YN–AlN·Al_2O_3$ Si_3N_4 was delineated.

The part of this quinary system bounded by the compounds AlN, Al_2O_3, Y_2O_3, Si_3N_4, and SiO_2 was also investigated by Naik and Tien (1979) and the important compatibility regions were determined. No quinary compounds were found in the part of the system investigated. It was shown that the homogeneity ranges of the phases $Y_2O_3·Si_2N_2O$, $Si_3N_4·Y_2O_3$, and $Y_4(SiO_4)_3N$ are limited. For investigation, the mixtures of AlN, Al_2O_3, Y_2O_3, and Si_3N_4 were dried in shallow glass disks over a laboratory heater and then cold pressed into pellets at 110 MPa. The pellets were placed in screw-top graphite crucibles that were lined with BN. The crucibles were placed in a furnace. The specimens were fired in nitrogen atmosphere at 1,550°C and atmospheric pressure. The heating rate was 200°C min^{-1} and the holding time at 1,550°C was 1 h. Then the specimens were cooled in the furnace at a rate of ≈200°C min^{-1} to 800°C. Further cooling to room temperature was somewhat slower.

$3AlN·YN–Al_2SiO_2N_2–Si_3N_4$. The phase relations in this system established from XRD data after sintering at 1,700°C are shown in Figure 4.5 (Stutz et al. 1986). It is necessary to note that an additional glassy phase is present in all sintered specimens. The phases 12*H* and 21*R* are silicon- and oxygen-containing AlN polytype phases with compositions

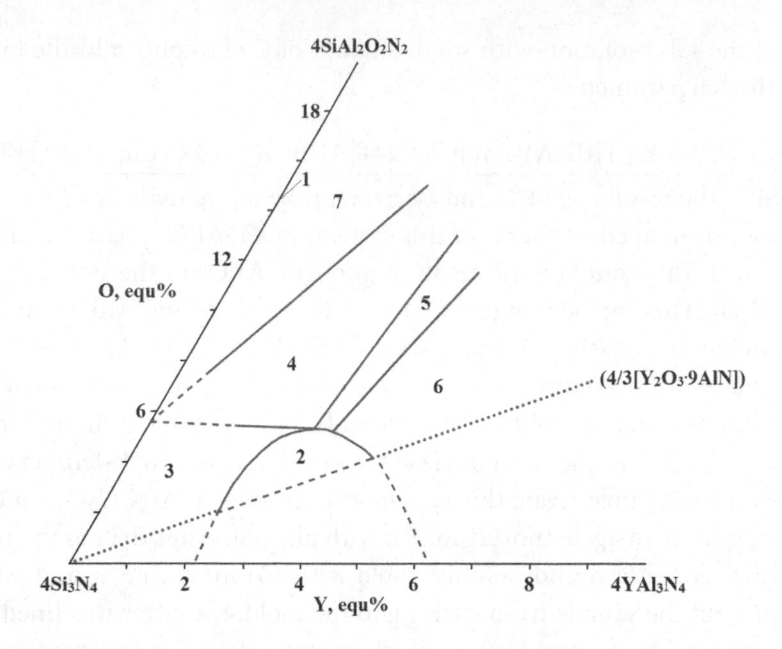

FIGURE 4.5 Phase relations in the concentration plane $Si_{12}N_{16}$–$Si_4Al_8O_8N_8$–$Y_4Al_{12}N_{16}$ (ratio of Si + Al to N + O is 3 to 4) as determined by XRD after 1 h sintering at 1,700°C: 1, (β-Si_3N_4); 2, (α-Si_3N_4); 3, (α-Si_3N_4) + (β-Si_3N_4); 4, (α-Si_3N_4) + (β-Si_3N_4) + 12*H*; 5, (α-Si_3N_4) + 12*H*; 6, (α-Si_3N_4) + 21*R*; 7, (β-Si_3N_4) + 12*H*. (With permission from Stutz, D., et al., *J. Mater. Sci. Lett.*, **5**(3), 335, 1986.)

deviating from the concentration plane given in Figure 4.5. Thus, the residual glassy phase is supposed to balance chemical composition and Figure 4.5 does not represent a quasiternary section, but a random section in the Al–Y–Si–O–N quinary system with phase equilibria comprising, theoretically, up to five crystalline phases. The phase analysis was restricted to oxygen-rich compositions lying above the concentration line Si_3N_4–4/3(Y_2O_3·9A1N). The α-Si_3N_4 solid-solution single-phase field extends parabolically into the oxygen-rich side, but it was suggested that it also continues into the N-rich side beyond the Si_3N_4–4/3(Y_2O_3·9A1N) line. The solubility limit in α-Si_3N_4 can be approximated by $0.33 \le x \le 0.67$, $1.5 \le (3x + y) \le 3$, and $0.5 \le y$ 1.24, if x and y are chosen as independent composition parameters in the general formula α-$Y_xAl_{3x+y}Si_{12-3x-y}O_yN_{16-y}$. At 1,700°C the maximum oxygen solubility of 4.3 at% ($y = 1.24$) was found in α-$Y_{0.44}Al_{2.5}Si_{9.5}O_{1.24}N_{14.76}$, which is much less than the maximum oxygen solubility of about 28 at% in β-$Al_2SiO_2N_2$ (β-sialon). The maximum solubility of Y and Al were found as 2.4 at% and 10.6 at%, respectively.

To investigate the phase relations in the 3AlN·YN–$Al_2SiO_2N_2$–Si_3N_4 system, powder mixtures of α-Si_3N_4, AlN, Al_2O_3, and Y_2O_3 were attritor milled in *n*-hexane for 1 h, dried, sieved, and cold isostatically pressed into cylinders with 630 MPa (Stutz et al. 1986). Only specimens with compositions in the single-phase α-Si_3N_4 solid solution field were milled with Si_3N_4 balls in a polyethylene mill to avoid oxygen contamination. The green compacts were sintered in a BN crucible at 1,700°C for 1–4 h in a flowing N_2 atmosphere of 0.1 MPa and cooled to 1,000°C within 2 min.

$AlN–Y_2O_3–Si_3N_4$. The subsolidus phase diagram of this system was established by Huang et al. (1983) and is presented in Figure 4.6. In this system, α-Si_3N_4 forms a solid solution with $0.1Y_2O_3·0.9AlN$. The solubility limits are represented by $Y_{0.33}Al_{1.5}Si_{10.5}O_{0.5}N_{15.5}$ and $Y_{0.67}Al_3Si_9ON_{15}$. At 1,700°C, equilibrium exists between α-Si_3N_4 and this solid solution. It is necessary to note that the $Si_2N_2O·2Y_2O_3$ compound does not lie in the plane of this system; therefore, this system is not truly quasiternary.

To investigate this system, the starting powders were AlN, Si_3N_4, and Y_2O_3 (Huang et al. 1983). Selected compositions were made by mixing the required amounts of the starting powders in agate mills with *n*-hexane. These mixtures were hot pressed in graphite dies 10 mm in diameter lined with BN. The specimens were then fired in nitrogen at 1,700°C and atmospheric pressure for 1–2 h and cooled to 650°C within 3 min. Samples that underwent <2% mass loss on firing were considered in deriving the phase relations.

The isothermal sections of the $AlN–Y_2O_3–Si_3N_4$ system at 1,450°C and 1,750°C are shown in Figures 4.7 and 4.8. At 1,450°C, two single-phase fields, three two-phase fields, and one three-phase field were observed in the Si_3N_4-rich portion of the system (Figure 4.7) (Fukuhara 1988). Above 47 mass% AlN, a new phase appears, but the boundary of this phase is not clear. With increasing Al_2O_3 content, the existence region of the α-sialon decreases remarkably and the amount of the β-sialon phase increases, and moreover, a new phase disappears; as a result, the three-phase field expands.

The liquid-phase region exists in this system at 1,750°C (Figure 4.8) (Huang and Tien 1996). The ternary composition with 25 mol% AlN, 15 mol% Si_3N_4, and 60 mol% Y_2O_3 has the lowest melting temperature of 1,650°C (ternary eutectic).

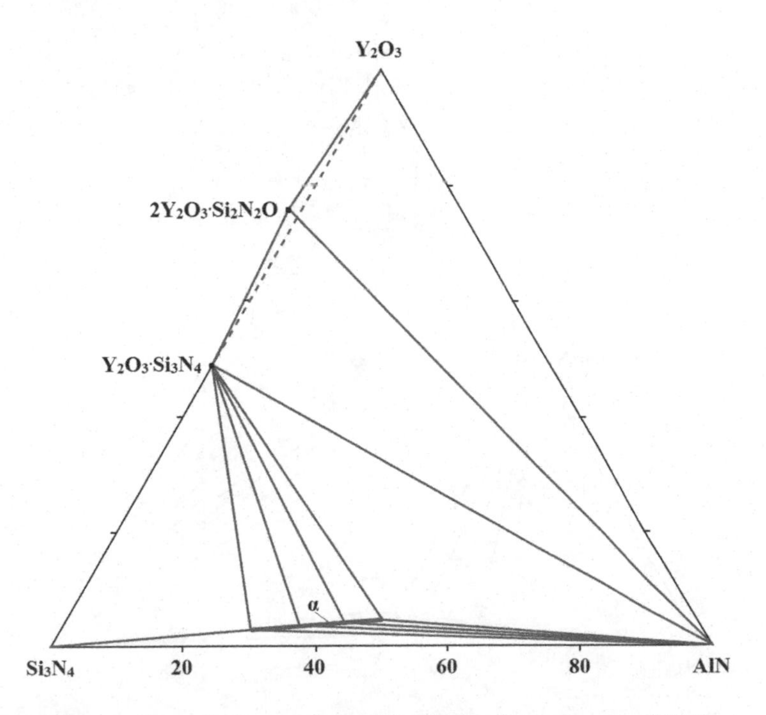

FIGURE 4.6 Subsolidus phase diagram of the $AlN–Y_2O_3–Si_3N_4$ system. (With permission from Huang, Z.-K., et al., *J. Am. Ceram. Soc.*, **66**(6), C-96, 1983.)

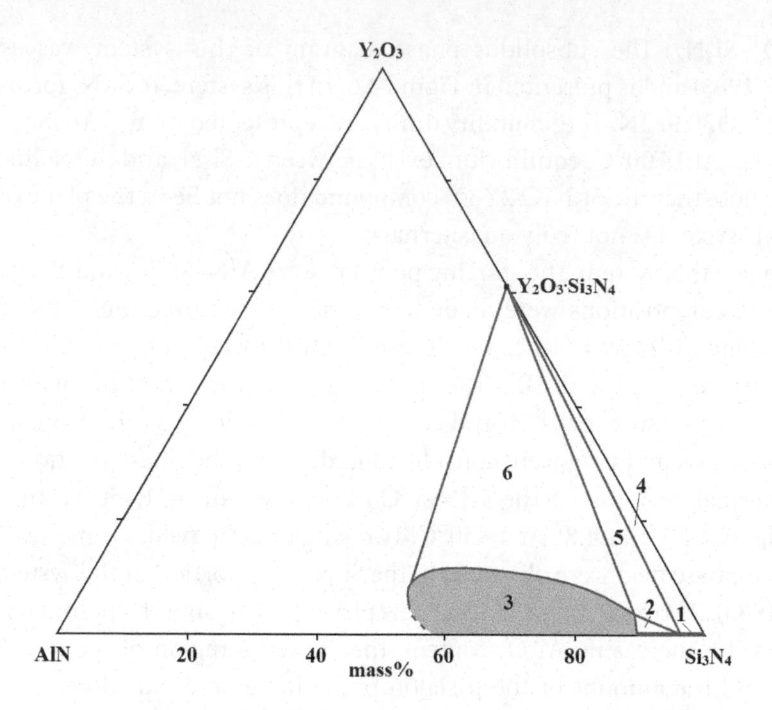

FIGURE 4.7 Isothermal sections of the AlN–Si_3N_4–Y_2O_3 system at 1,450°C: 1, (β-Si_3N_4); 2, (α-Si_3N_4) + (β-Si_3N_4); 3, (α-Si_3N_4); 4, (β-Si_3N_4) + Si_3N_4·Y_2O_3; 5, (α-Si_3N_4) + (β-Si_3N_4) + Si_3N_4·Y_2O_3; 6, (α-Si_3N_4) + Si_3N_4·Y_2O_3. (With permission from Fukuhara, M., *J. Am. Ceram. Soc.*, **71**(7), C-359, 1988.)

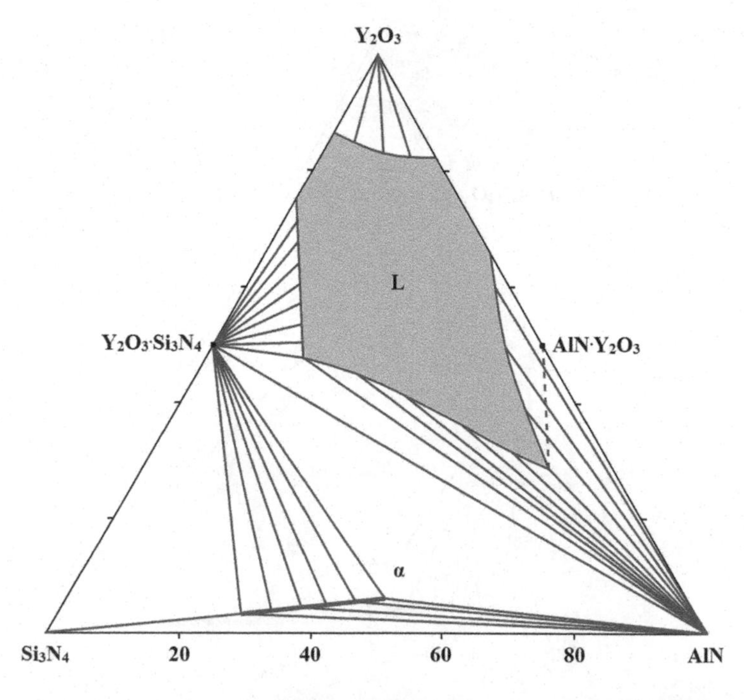

FIGURE 4.8 Isothermal sections of the AlN–Si_3N_4–Y_2O_3 system at 1,750°C under 1 MPa of N_2. (With permission from Huang, Z.-K., and Tien, T.-Y., *J. Am. Ceram. Soc.*, **79**(6), 1717, 1996.)

The $\mathbf{Y_x(Si,Al)_{12}(O,N)_{16}}$ (x = 0.3 to 0.67; yttrium α′-sialon) (Kuang et al. 1990) and $\mathbf{Y_2Al_xSi_{3-x}O_{3+x}N_{4-x}}$ ($x \leq 0{,}57$) (Wang et al. 1995) [*M*′-phase according to Metselaar (1983)] quinary solid solutions and $\mathbf{Y_4Al_{0.7}Si_{1.3}O_{7.7}N_{1.3}}$ quinary phase are formed in the Al–Y–Si–O–N system. It is possible that $\mathbf{Y_3Al_{3+x}Si_{3-x}O_{12+x}N_{2-x}}$ solid solution at x = 0–1 (*U*-phase) could also exist in this system (Metselaar 1983). The $Y_x(Si,Al)_{12}(O,N)_{16}$ solid solutions crystallize in the trigonal structure with the lattice parameters a = 782.927 ± 0.007 and c = 570.757 ± 0.006 pm for $Y_{0.5}Al_{2.7}Si_{9.3}O_{0.9}N_{15.1}$ composition (Izumi et al. 1984) [a = 781 and 783 and c = 571 and 569 pm and the calculated and experimental densities of 3.23 and 3.28 and 3.25 and 3.36 g cm^{-3} for $Y_{0.4}Al_2Si_{10}O_{0.8}N_{15.2}$ and $Y_{0.6}Al_{2.8}Si_{9.2}O_{1.1}N_{14.9}$ compositions, respectively (Hampshire et al. 1978); a = 779.4 ± 0.3, 779.4 ± 0.4, 779.6 ± 0.2, 780.1 ± 0.2, 780.3 ± 0.3 and c = 567.7 ± 1.5, 567.8 ± 0.6, 567.9 ± 4.2, 568.9 ± 1.3, 569.5 ± 1.3 pm for $Y_{0.27\pm0.02}Al_{1.60\pm0.03}Si_{10.00\pm0.14}O_{1.88\pm0.03}N_{14.12\pm0.20}$, $Y_{0.33\pm0.02}Al_{1.90\pm0.03}Si_{10.02\pm0.14}O_{1.60\pm0.02}N_{14.40\pm0.22}$, $Y_{0.10\pm0.02}Al_{2.04\pm0.03}Si_{9.53\pm0.14}O_{1.40\pm0.02}N_{14.40\pm0.22}$, $Y_{0.15\pm0.02}Al_{1.97\pm0.03}Si_{9.67\pm0.14}O_{1.33\pm0.03}N_{16.67\pm0.22}$, and $Y_{0.24\pm0.02}Al_{2.00\pm0.03}Si_{9.46\pm0.14}O_{1.44\pm0.02}N_{14.56\pm0.22}$, respectively (Grand et al. 1979); a = 782.946 ± 0.007 and c = 570.765 ± 0.007 pm for $Y_{0.5}Al_{2.25}Si_{9.75}O_{0.75}N_{15.25}$ composition (Izumi et al. 1982)]. Starting materials for the preparation of these solid solutions was a mixture of Y_2O_3, Si_3N_4, and AlN (molar ratio 0.3:3.1:2.7) (Izumi et al. 1982, 1984). The mixture was hot pressed for 1 h at 1,700°C under a pressure of 14.7 MPa in a graphite die of 15 mm diameter. They could also be prepared by heating appropriate mixtures of nitrides or nitrides plus oxides at 1,750°C in an atmosphere of N_2 or Ar for 15 min (Hampshire et al. 1978).

$Y_2Al_xSi_{3-x}O_{3+x}N_{4-x}$ crystallizes in the tetragonal structure with the lattice parameters a = 762.9 and c = 492.9 pm for x = 0.6 (Huang and Chen 1996). The maximum Al content in the solid solution is x = 0.57 (Wang et al. 1995). To prepare these solid solutions, powders of AlN, Al_2O_3, Y_2O_3, and Si_3N_4 were mixed in an agate mortar with ethanol or propan-2-ol. Then, they were dried and pressed into pellets (Wang et al. 1995; Huang and Chen 1996). Either pressureless sintering or hot pressing was used to effect reaction at 1,600°C–1,750°C in N_2 atmosphere for 1.5–2 h. After the reaction, the furnace was turned off and cooling of the sample proceeded at an estimated rate of 37°C h^{-1} from 1,650°C to 1,200°C. Annealing at 1,550°C for 24 h was also used.

$Y_4Al_{0.7}Si_{1.3}O_{7.7}N_{1.3}$ crystallizes in the monoclinic structure with the lattice parameters a = 720, b = 1,026, and c = 1,093 pm and β = 108.12° (Huang and Chen 1996). It was obtained by the same way as $Y_2Al_{0.6}Si_{2.4}O_{3.6}N_{3.4}$ was prepared.

The glasses containing up to 7 at% nitrogen were prepared in the Al–Y–Si–O–N system (Loehman 1979, 1980). The glass transition temperature increases with increasing nitrogen content. These glasses were prepared by melting mixtures of AlN, Y_2O_3, and SiO_2 in BN crucibles at 1,550°C–1,700°C for 1–5 h under a static Ar atmosphere and then turning off the furnace to quench then to <800°C in 2 or 3 min.

4.43 ALUMINUM–LANTHANUM–SILICON–OXYGEN–NITROGEN

The $\mathbf{La_2AlSi_2O_4N_3}$, $\mathbf{La_2Al_4Si_{11}O_4N_{18}}$, and $\mathbf{La_{13}Al_{12}Si_{18}O_{15}N_{39}}$ quinary compounds are formed in the Al–La–Si–O–N system. $La_2AlSi_2O_4N_3$ [*M*′-phase according to Metselaar (1983)] crystallizes in the tetragonal structure with the lattice parameters a = 785.5 and c = 512.0 pm (Huang and Chen 1996). To prepare this compound, powders of AlN, Al_2O_3,

La_2O_3, and Si_3N_4 were mixed in an agate mortar with propan-2-ol (Huang and Chen 1996). Then, they were dried and pressed into pellets. Either pressureless sintering or hot pressing was used to effect the reaction at 1,650°C in N_2 atmosphere for 2h. After the reaction, the furnace was turned off, and cooling of the sample proceeded at an estimated rate of 37°C h^{-1} from 1,650°C to 1,200°C. Annealing at 1,550°C for 24h was also used.

$La_2Al_4Si_{11}O_4N_{18}$ crystallizes in the orthorhombic structure with the lattice parameters $a = 943.03 \pm 0.07$, $b = 976.89 \pm 0.08$, and $c = 893.86 \pm 0.06$ pm and a calculated density of 3.652 g cm^{-3} (Grins et al. 1995). To synthesize this compound, samples with a fixed La/Si/Al ratio ($La_2Al_4Si_{11}$) but with different N/(O+N) contents were prepared using Si_3N_4, AlN, Al_2O_3, and La_2O_3 as starting materials. La_2O_3 was calcined at 1,000°C for 2 h before use. The starting mixtures were milled in water-free *n*-propanol for 24h in a plastic jar, using sialon-milling media. Pellets of dried powders were hot pressed in BN-coated graphite dies in the temperature range 1,650°C–1,800°C (25 MPa, 2 h) in a graphite resistance furnace under a protective nitrogen atmosphere. Selected samples were then heat treated at 1,650°C for 24h.

$La_{13}Al_{12}Si_{18}O_{15}N_{39}$ crystallizes in the cubic structure with the lattice parameter $a = 1{,}349.5 \pm 0.4$ pm and a calculated density of 4.622 g cm^{-3} (Esmaeilzadeh and Schnick 2003). It was synthesized by mixing La (1.15 mM), Si_3N_4 (0.463 mM), SiO_2 (0.466 mM), and AlN (2.20 mM) thoroughly in a glove box in Ar atmosphere. The reaction mixture was then transferred into a W crucible with lid and positioned in a radio-frequency furnace. The reaction mixture was heated to 1,750°C within 1 h and held at this temperature for 1 h. At this point, a melt had been formed, which is probably essential for the crystal growth. The reaction mixture was cooled from 1,750°C to 1,200°C for 20 h and quenched from 1,200°C to room temperature by turning off the furnace. Reddish crystals of the title compound were obtained from this synthesis procedure. The crystals were found to be stuck in the solidified melt.

4.44 ALUMINUM–CERIUM–SILICON–OXYGEN–NITROGEN

The $\mathbf{Ce_2Al_{0.5}Si_{2.5}O_{3.5}N_{3.5}}$ and $\mathbf{Ce_2Al_4Si_{11}O_4N_{18}}$ quinary compounds are formed in the Al–Ce–Si–O–N system. $Ce_2Al_{0.5}Si_{2.5}O_{3.5}N_{3.5}$ crystallizes in the tetragonal structure with the lattice parameters $a = 779.20 \pm 0.03$ and $c = 506.94 \pm 0.04$ pm and a calculated density of 5.060 g cm^{-3} (Lauterbach and Schnick 1999). It was obtained by the reaction of Ce (1.29 mM), AlN (0.51 mM), $Si(NH)_2$ (1.38 mM), and $SrCO_3$ (0.67 mM) at 1,600°C.

$Ce_2Al_4Si_{11}O_4N_{18}$ crystallizes in the orthorhombic structure with a calculated density of 3.658 g cm^{-3} (Grins et al. 1995). The title compound was synthesized by the same way as $La_2Al_4Si_{11}O_4N_{18}$ was prepared using Ce_2O_3 instead of La_2O_3.

4.45 ALUMINUM–PRASEODYMIUM–SILICON–CARBON–OXYGEN–NITROGEN

$\mathbf{AlN–Al_2O_3–Pr_2O_3–SiC}$. The subsolidus phase relations in this system at 1,630°C are given in Figure 4.9 (Pan et al. 2017). To investigate this system, the starting materials AlN, α-SiC, and Pr_6O_{11} were mixed with an agate pestle and mortar for 2 h by using anhydrous ethanol as medium. After fully mixed and dried in an oven, batches of the powder mixtures were vertically axial pressed in dies of 10 mm in diameter at 70 MPa for 30 s. The pressureless

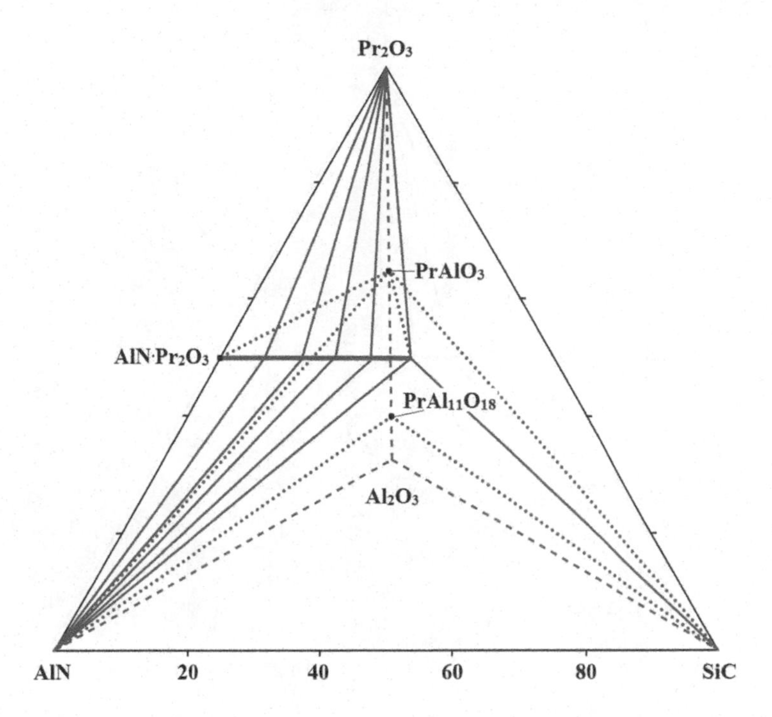

FIGURE 4.9 Subsolidus phase diagram of the AlN–Al_2O_3–Pr_2O_3–SiC system. (With permission from Pan, W., et al., *J. Phase Equilib. Dif.*, **38**(5), 676, 2017.)

sintering was carried out in a graphite furnace in an Ar atmosphere at 1,400°C–1,700°C for 2–4 h. Some samples were hot pressed under 30 MPa for comparison. The heating and cooling rate was 10 and 20°C min^{-1}, respectively.

4.46 ALUMINUM–PRASEODYMIUM–SILICON–OXYGEN–NITROGEN

The $\mathbf{Pr_2Al_{0.5}Si_{2.5}O_{3.5}N_{3.5}}$ quinary compound, which crystallizes in the tetragonal structure with the lattice parameters a = 778.26 ± 0.04 and c = 508.56 ± 0.05 pm, is formed in the Al–Pr–Si–O–N system (Lauterbach and Schnick 1999). It was obtained by the reaction of Pr (0.75 mM), AlN (1.56 mM), $Si(NH)_2$ (1.46 mM), and $SrCO_3$ (0.73 mM) at 1,625°C.

4.47 ALUMINUM–NEODYMIUM–SILICON–CARBON–OXYGEN–NITROGEN

AlN-$\mathbf{Nd_2O_3}$–SiC. The subsolidus phase diagram of this system (Figure 4.10) was constructed through the results of XRD and EPMA (Chen et al. 2013). It is shown that the solid solution with formula $Nd_2Al_{1-x}C_xSi_xO_3N_{1-x}$ (x = 0–0.5) is formed in this system. To investigate this system, the powders of AlN, Nd_2O_3, and α-SiC were mixed and ground in an agate mortar for 2 h with alcohol. After drying up, the prepared powders were placed in a round graphite mold with 10 mm diameter and afterward hot pressed. To prevent the sample from sticking to the mold, the latter was lined with a thin layer of BN. The conditions used for hot pressing were as follows: Ar atmosphere, 30 MPa, 1,450°C–1,600°C, 1–2 h. To achieve the reaction equilibria, the temperature was further kept for 1–3 h after releasing the pressure.

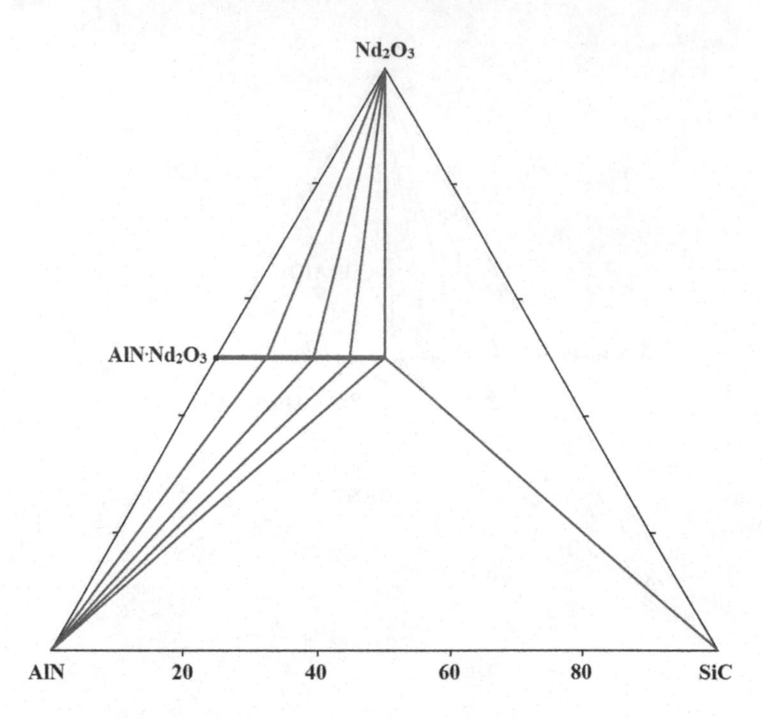

FIGURE 4.10 Subsolidus phase diagram of the AlN–Nd_2O_3–SiC system. (With permission from Chen, Y., et al., *J. Phase Equilibr. Dif.*, **34**(1), 3, 2013.)

4.48 ALUMINUM–NEODYMIUM–SILICON–OXYGEN–NITROGEN

The subsolidus phase relationships in the Al–Nd–Si–O–N quinary system were determined by Huang et al. (1986), Huang and Yan (1992), and Sun et al. (1995). Forty-four compatibility tetrahedra were established in the region **AlN–Al_2O_3–Nd_2O_3–Si_3N_4** (Sun et al. 1995). Within this region, $NdAlO_3$ and **$Nd_2Al_xSi_{3-x}O_{3+x}$**N_{4-x} (*M′* phase) are the only two important phases that have tie lines joined to β-sialon and AlN polytypoid phases. β-Sialon coexists with the M′ phase. The liquid phase crosses the tie line between $NdAlO_3$ and β-sialon at 1,600°C–1,700°C.

AlN–Nd_2O_3–Si_3N_4. The subsolidus phase diagram of this system is given in Figure 4.11 (Huang et al. 1986; Huang and Yan 1992). It should be noted that the compound $Si_2N_2O{\cdot}2Nd_2O_3$ does not lie in the plane of this system. The solubility limits of the solid solution based on α-Si_3N_4 are represented by $Nd_{0.33}Al_{1.5}Si_{10.5}O_{0.5}N_{15.5}$ and $Nd_{0.6}Al_{3.5}Si_{8.5}O_{1.7}N_{14.3}$. To investigate this system, the starting powders of AlN, Si_3N_4, and Nd_2O_3 were mixed in an agate jar under absolute EtOH on a rotating mill for 2 h. Dried mixtures were hot pressed at 1,700°C under 30 MPa in a BN-coated graphite die in a furnace under a mild flow of N_2. After hot pressing, the furnace was cooled at a rate of ~200°C min^{-1} in the high-temperature region.

The $Nd_2Al_xSi_{3-x}O_{3+x}N_{4-x}$ solid solution (*M′* phase) and **$Nd_3AlSi_5ON_{10}$**, **$Nd_3Al_3Si_3O_{12}N_2$**, and **$Nd_7[Al_3Si_8N_{20}]O$** quinary compounds are formed in the Al–Nd–Si–O–N system. $Nd_2Al_xSi_{3-x}O_{3+x}N_{4-x}$ crystallizes in the tetragonal structure with the lattice parameters a = 776.15 ± 0.04 and c = 506.7 ± 0.3 pm for x = 0.5 (Lauterbach and Schnick 1999) [a = 774.62 ±

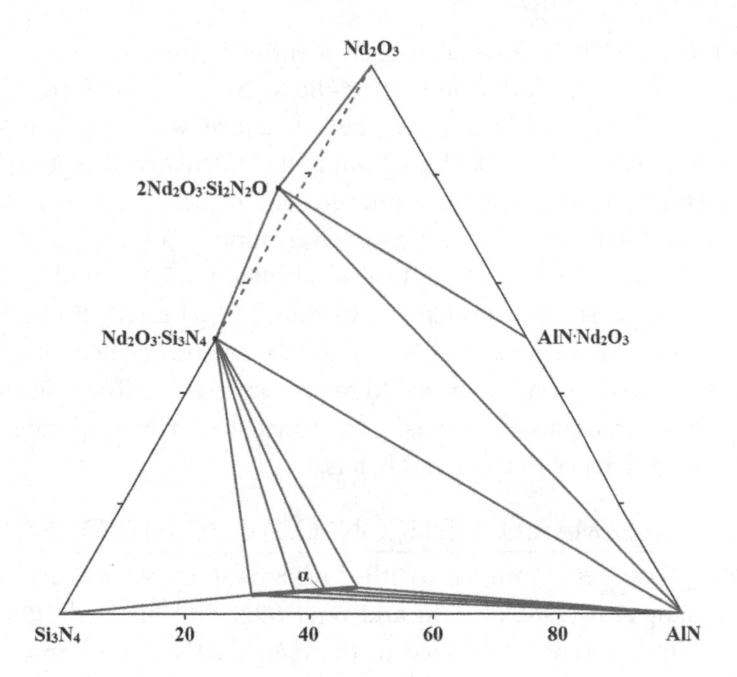

FIGURE 4.11 Subsolidus phase diagram of the AlN–Nd_2O_3–Si_3N_4 system. (With permission from Huang, Z.-K., et al., *J. Am. Ceram. Soc.*, **69**(10), C-241, 1986.)

0.05 and $c = 503.90 \pm 0.04$ pm for $x = 0.5$ (Wang 1997; Wang and Werner 1997) and $a =$ 776.6 and $c = 505.5$ pm for $x = 0.9$ (Huang and Chen 1996)]. The maximum Al content in the solid solution is $x = 0.98$ (Wang et al. 1995). To synthesize these solid solutions, powders of AlN, Al_2O_3, Nd_2O_3, and Si_3N_4 were mixed in an agate mortar with ethanol or propan-2-ol (Wang et al. 1995; Huang and Chen 1996; Wang 1997; Wang and Werner 1997). Then, they were dried and pressed into pellets. Either pressureless sintering or hot pressing was used to effect the reaction at 1,600°C–1,750°C in N_2 atmosphere for 1.5–2 h. After the reaction, the furnace was turned off and cooling of the sample was proceeded at an estimated rate of 37°C h^{-1} from 1,650°C to 1,200°C. Annealing at 1,550°C for 24 h was also used. They could also be obtained by the reaction of Nd (0.84 mM), AlN (1.08 mM), $Si(NH)_2$ (1.28 mM), and $SrCO_3$ (0.69 mM) at 1,650°C.

$Nd_3AlSi_5ON_{10}$ crystallizes in the tetragonal structure with the lattice parameters $a = 1{,}007.8 \pm 0.1$ and $c = 486.3 \pm 0.1$ pm and a calculated density of 5.086 g cm^{-3} (Lauterbach and Schnick 2000b). The title compound was synthesized by the reaction of Nd (1.11 mM), AlN (0.98 mM), Al_2O_3 (0.3 mM), and $Si(NH)_2$ (1.38 mM) in a pure N_2 atmosphere using W crucible and a radio-frequency furnace. The mixture was initially heated up to 1,200°C within 60 min and then heated up to 1,650°C within the next 60 min and held at that temperature for 30 min. Thereafter, the furnace was cooled to 900°C over a period of 42 h. Room temperature was reached for further 30 min.

$Nd_3Al_3Si_3O_{12}N_2$ crystallizes in the hexagonal structure with the lattice parameters $a = 798.7$ and $c = 487.4$ pm (Fernie et al. 1989). This compound is stable up to ~1,400°C–1450°C. The phases with approximate composition **$Nd_{3.5}Al_{2.5}Si_4ON$** and **$Nd_5Al_6Si_9ON$,** which are stable up to 1,150°C–1,250°C, have also been identified using energy-dispersive X-ray spectroscopy (EDX) analysis. Both compositions lie within the glass-forming region.

$Nd_7[Al_3Si_8N_{20}]O$ crystallizes in the trigonal structure with the lattice parameters a = 1,004.25 ± 0.09 and c = 1,095.03 ± 0.12 pm and a calculated density of 5.596 g cm^{-3} (Köllisch et al. 2001). For its synthesis, a mixture of Nd (223.2 mg), $Si(NH)_2$ (104.4 mg), $AlNH_2NH$ (46.4 mg), and SiO_2 (6.0 mg) was finely ground in a glove box and placed in a W crucible in a furnace to react. The mixture was heated for 5 min under N_2 atmosphere at 700°C, left for another 5 min at this temperature, and then heated within 2 h to 1,690°C. Thereafter, the mixture was kept 1 h at 1,690°C and subsequently heated for 42 h at 1,200°C. Finally, the reaction product was quenched to room temperature. The title compound was obtained as turquoise, transparent crystals. It was able to be further separated from the by-products due to the difference in color and habits.

4.49 ALUMINUM–SAMARIUM–SILICON–OXYGEN–NITROGEN

The subsolidus phase relationships in the Al–Sm–Si–O–N quinary system were determined by Huang et al. (1986), Huang and Yan (1992), and Sun et al. (1995). Forty-four compatibility tetrahedra were established in the region **$AlN–Al_2O_3–Sm_2O_3–Si_3N_4$** (Sun et al. 1995). Within this region, $SmAlO_3$ and **$Sm_2Al_xSi_{3-x}O_{3+x}N_{4-x}$** (*M′* phase) are the only two important phases that have tie lines joined to β-sialon and AlN polytypoid phases. β-Sialon coexists with the *M′* phase. The liquid phase crosses the tie line between $SmAlO_3$ and β-sialon at 1,600°C–1,700°C.

$AlN–Sm_2O_3–Si_3N_4$. The subsolidus phase diagram of this system was constructed by Huang et al. (1986) and Huang and Yan (1992), and it is the same as for the $AlN–Nd_2O_3–Si_3N_4$ system (Figure 4.11). It should be noted that the compound $Si_2N_2O{\cdot}2Sm_2O_3$ does not lie in the plane of this system. The solubility limits of the solid solution based on α-Si_3N_4 are represented by $Sm_{0.33}Al_{1.5}Si_{10.5}O_{0.5}N_{15.5}$ and $Sm_{0.6}Al_{2.8}Si_{9.2}ON_{15}$. To investigate this system, the starting powders of AlN, Si_3N_4, and Sm_2O_3 were mixed in an agate jar under absolute EtOH on a rotating mill for 2 h. Dried mixtures were hot pressed at 1,700°C under 30 MPa in a BN-coated graphite dies in a furnace under a mild flow of N_2. After hot pressing, the furnace was cooled at a rate of ~200°C min^{-1} in the high-temperature region.

The $Sm_2Al_xSi_{3-x}O_{3+x}N_{4-x}$ quinary solid solution (M′ phase) is formed in the Al–Sm–Si–O–N system. It crystallizes in the tetragonal structure with the lattice parameters a = 772.63 ± 0.13 and c = 502.80 ± 0.09 pm for x = 0.5 (Lauterbach and Schnick 1999) and a = 773.0 and c = 501.4 pm for x = 0.9 (Huang and Chen 1996). The maximum Al content in the solid solution is x = 0.90–0.91 (Wang et al. 1995; Cheng and Thompson 1994). To synthesize these solid solutions, powders of AlN, Al_2O_3, Sm_2O_3, and Si_3N_4 were mixed in an agate mortar with ethanol or propan-2-ol (Wang et al. 1995; Huang and Chen 1996). Then, they were dried and pressed into pellets. Either pressureless sintering or hot pressing was used to effect the reaction at 1,600°C–1,750°C in N_2 atmosphere for 1.5–2 h. After the reaction, the furnace was turned off and cooling of the sample proceeded at an estimated rate of 37°C h^{-1} from 1,650°C to 1,200°C. Annealing at 1,550°C for 24 h also was used. They could also be obtained by the reaction of Sm (0.73 mM), AlN (0.35 mM), $Si(NH)_2$ (1.05 mM), and $SrCO_3$ (0.67 mM) at 1,550°C (Lauterbach and Schnick 1999).

To study the solubility of Al in the $Sm_2Al_xSi_{3-x}O_{3+x}N_{4-x}$ solid solution, powder mixtures with $0 \le x \le 2$ were used (Cheng and Thompson 1994). Pellets weighing about 2 g were sintered at 1,600°C, 1,700°C, and 1,800°C for 2 h in a flowing N_2 atmosphere and cooled down at the maximum cooling rate (about 50°C min^{-1} between 1,800°C and 1,200°C). Heat treatment of the samples was carried out in an alumina tube furnace in nitrogen atmosphere. It was shown that the lattice parameters increases linearly with increasing Al content according to the following relationships: a(pm) = 769.0 + 4.9x and c(pm) = 498.5 + 3.7x ($0 \le x \le 0.91$) at 1,600°C. The solubility of Al increases significantly when the sintering temperature rises to 1,700°C and the terminal composition of the solid solution is $Sm_2AlSi_2O_4N_3$.

The glass-forming region and the neighboring crystalline areas in the Al–Sm–Si–O–N system at 1,700°C are represented in Figure 4.12 (Tu et al. 1995). It is seen that the

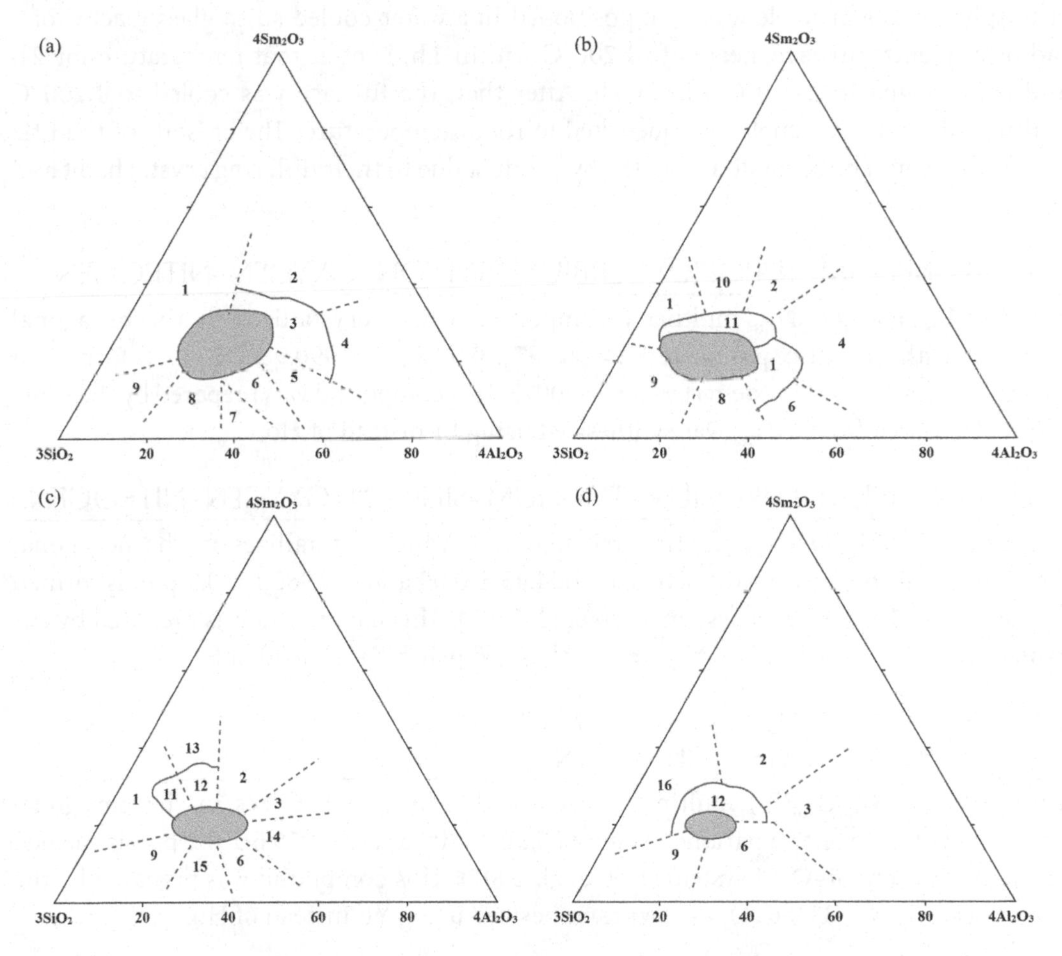

FIGURE 4.12 Glass-forming region and the neighboring crystalline areas in the Al–Sm–Si–O–N system at 1,700°C and (a) 25, (b) 30, (c) 35, and (d) 40 eq.% of nitrogen: 1, $Sm_{10}N_2(SiO_4)_6$; 2, $Sm_{10}N_2(SiO_4)_6$ + $Sm_2Al_xSi_{3-x}O_{3+x}N_{4-x}$; 3, $SmAlO_3$; 4, Al_2O_3 + $SmAlO_3$; 5, Al_2O_3 + β′-sialon; 6, β′-sialon; 7, β′-sialon + $3Al_2O_3{\cdot}2SiO_2$; 8, β′-sialon + $Al_6Si_3O_{12}N_2$; 9, O-sialon; 10, $Sm_{10}N_2(SiO_4)_6$ + $SmAlO_3$; 11, $Sm_3Si_3O_6N_3$; 12, $Sm_2Al_xSi_{3-x}O_{3+x}N_{4-x}$; 13, $Sm_{10}N_2(SiO_4)_6$ + $Sm_2Al_xSi_{3-x}O_{3+x}N_{4-x}$; 14, $Sm_3Al_3Si_3O_{12}N_2$; 15, $Al_6Si_3O_{12}N_2$; 16, $Sm_3Si_3O_6N_3$ + $Sm_2Al_xSi_{3-x}O_{3+x}N_{4-x}$. (With permission from Tu, H.Y., et al., *J. Mater. Sci. Lett.*, **14**(16), 1118, 1995.)

glass-forming region at 25 and 30 eq.% of nitrogen are relatively extensive. With increasing nitrogen content, glass formation gradually contracted towards the composition with a cation ratio of Sm/Si/Al = 20:60:20. The maximum solubility of nitrogen in this glass can be determined to be slightly above 40 eq.% of nitrogen.

4.50 ALUMINUM–EUROPIUM–HOLMIUM–SILICON–OXYGEN–NITROGEN

The $\mathbf{EuAl_{1.2}HoSi_{2.8}O_{1.2}N_{5.8}}$ multinary compound, which crystallizes in the hexagonal structure with the lattice parameters $a = 606.87 \pm 0.09$ and $c = 985.0 \pm 0.2$ pm, is formed in the Al–Eu–Ho–Si–O–N system (Lieb et al. 2007). This compound was prepared as follows. Ho (0.75 mM), $Si(NH)_2$ (1.5 mM), AlN (2.2 mM), $Eu_2(CO_3)_3$ (0.21 mM), and "$EuCl_2$" (500 mg) were thoroughly mixed and transferred into a W crucible in a glove box (Ar atmosphere). The crucible was then positioned in a water-cooled silica glass reactor of a radio-frequency furnace, heated to 1,200°C within 1 h, kept at that temperature for 2 h and then heated to 2,050°C within 4 h. After that, the furnace was cooled to 1,200°C within 60 h and the sample was quenched to room temperature. The crystals of the title compound could be separated from the by-product due to their differing crystal habit and color.

4.51 ALUMINUM–EUROPIUM–ERBIUM–SILICON–OXYGEN–NITROGEN

The $\mathbf{EuAl_{1.2}ErSi_{2.8}O_{1.2}N_{5.8}}$ multinary compound, which crystallizes in the hexagonal structure with the lattice parameters $a = 603.57 \pm 0.09$ and $c = 990.9 \pm 0.2$ pm, is formed in the Al–Eu–Er–Si–O–N system (Lieb et al. 2007). This compound was prepared by the same way as $EuAl_{1.2}HoSi_{2.8}O_{1.2}N_{5.8}$ was synthesized using Er instead of Ho.

4.52 ALUMINUM–EUROPIUM–THULIUM–SILICON–OXYGEN–NITROGEN

The $\mathbf{EuAl_{1.2}TmSi_{2.8}O_{1.2}N_{5.8}}$ multinary compound, which crystallizes in the hexagonal structure with the lattice parameters $a = 604.43 \pm 0.09$ and $c = 982.1 \pm 0.9$ pm, is formed in the Al–Eu–Tm–Si–O–N system (Lieb et al. 2007). This compound was prepared by the same way as $EuAl_{1.2}HoSi_{2.8}O_{1.2}N_{5.8}$ was synthesized using Tm instead of Ho.

4.53 ALUMINUM–EUROPIUM–YTTERBIUM–SILICON–OXYGEN–NITROGEN

The $\mathbf{EuAl_{1.6}YbSi_{2.4}O_{1.6}N_{5.4}}$ multinary compound, which crystallizes in the hexagonal structure with the lattice parameters $a = 605.30 \pm 0.09$ and $c = 979.6 \pm 0.2$ pm, is formed in the Al–Eu–Yb–Si–O–N system (Lieb et al. 2007). This compound was prepared by the same way as $EuAl_{1.2}HoSi_{2.8}O_{1.2}N_{5.8}$ was synthesized using Yb instead of Ho.

4.54 ALUMINUM–EUROPIUM–SILICON–OXYGEN–NITROGEN

The $\mathbf{Eu_{0.48}Al_{2.703}Si_{9.227}O_{1.178}N_{14.701}}$, $\mathbf{Eu_2AlSi_2O_4N_3}$, and $\mathbf{Eu_3Al_{1+x}Si_{15-x}O_xN_{23-x}}$ ($x \approx 5/3$) quinary phases are formed in the Al–Eu–Si–O–N system. The first of them crystallizes in the trigonal structure with the lattice parameters $a = 778.74$ and $c = 565.90$ pm (Shen et al. 1998). It was prepared by hot pressing a powder mixture of AlN, Eu_2O_3, and Si_3N_4 at 1,800°C for 2 h under 35 MPa in a graphite furnace and in N_2 atmosphere.

$Eu_2AlSi_2O_4N_3$ (*M*′-phase) crystallizes in the tetragonal structure with the lattice parameters $a = 760.8$ and $c = 492.0$ pm (Huang and Chen 1996). To synthesize this phase, powders of AlN, Al_2O_3, Eu_2O_3, and Si_3N_4 were mixed in an agate mortar with propan-2-ol. Then, they were dried and pressed into pellets. Either pressureless sintering or hot pressing was used to effect the reaction at 1,650°C in N_2 atmosphere for 2h. After the reaction, the furnace was turned off and cooling of the sample proceeded at an estimated rate of 37°C h^{-1} from 1,650°C to 1,200°C. Annealing at 1,550°C for 24h was also used.

Two substructures with different periodicities were found in the structure of $Eu_3Al_{1+x}Si_{15-x}O_xN_{23-x}$ ($x \approx 5/3$); therefore, this phase is considered to be a commensurate composite crystal (Michiue et al. 2009). One of the substructure ($Eu_3Al_{2.667}Si_{13.333}O_{1.667}N_{21.333}$) is monoclinic with the lattice parameters $a = 1{,}469.70 \pm 0.09$, $b = 903.6 \pm 0.2$, $c = 746.77 \pm 0.07$ pm, and $\beta = 90.224 \pm 0.001°$ and another ($Eu_2Al_{1.778}Si_{8.889}O_{1.111}N_{14.222}$) is orthorhombic with the lattice parameters $a = 489.90 \pm 0.03$, $b = 903.6 \pm 0.2$, $c = 746.77 \pm 0.07$ pm. A powder sample with the nominal composition $Eu_6Al_6Si_{27}O_6N_{42}$ was prepared from α-Si_3N_4, AlN, and Eu_2O_3. The mixed powder was placed into an *h*-BN crucible and then sintered in a graphite resistance furnace at 1,900°C for 24h under 1MPa nitrogen atmosphere. Pale yellow crystals of this phase were obtained along with polycrystalline products.

4.55 ALUMINUM–GADOLINIUM–SILICON–CARBON–OXYGEN–NITROGEN

AlN-Gd_2O_3–SiC. The results of XRD and EPMA show that, except for the three original constituents of this system, $2Gd_2O_3{\cdot}Al_2O_3$ phase was also observed (Chen et al. 2013). This could be the result of impurity Al_2O_3 in the starting AlN powder reacted with Gd_2O_3. After introducing $2Gd_2O_3{\cdot}Al_2O_3$, the phase relationships transformed this ternary system into the quasi-quaternary system AlN-Gd_2O_3–$2Gd_2O_3{\cdot}Al_2O_3$–SiC. The coexistence of four phases brings significance to the manufacture of SiC ceramics with AlN-Gd_2O_3 as sintering additives, because of the formation of $2Gd_2O_3{\cdot}Al_2O_3$ with higher melting temperatures above 2,000°C, as the second phase might enhance the high-temperature properties of SiC ceramics.

To investigate this system, the powders of AlN, Gd_2O_3, and α-SiC were mixed and ground in an agate mortar for 2h with alcohol. After drying up, the prepared powders were placed in a round graphite mold with 10mm diameter and afterward hot pressed. To prevent the sample from sticking to the mold, the latter was lined with a thin layer of BN. The conditions used for hot pressing were as follows: Ar atmosphere, 30MPa, 1,450°C–1,600°C, 1–2h. To achieve the reaction equilibria, the temperature was further kept for 1–3h after releasing the pressure.

4.56 ALUMINUM–GADOLINIUM–SILICON–OXYGEN–NITROGEN

AlN-Gd_2O_3–Si_3N_4. The subsolidus phase diagram of this system is given in Figure 4.13 (Huang et al. 1986; Huang and Yan 1992). It should be noted that the compound $Si_2N_2O{\cdot}2Gd_2O_3$ does not lie in the plane of this system. The solubility limits of the solid solution based on α-Si_3N_4 are represented by $Gd_{0.33}Al_{1.5}Si_{10.5}O_{0.5}N_{15.5}$ and $Gd_{0.6}Al_{2.8}Si_{9.2}ON_{15}$. To investigate this system, the starting powders of AlN, Si_3N_4, and Gd_2O_3 were mixed in an agate jar under absolute EtOH on a rotating mill for 2h. Dried mixtures were hot pressed

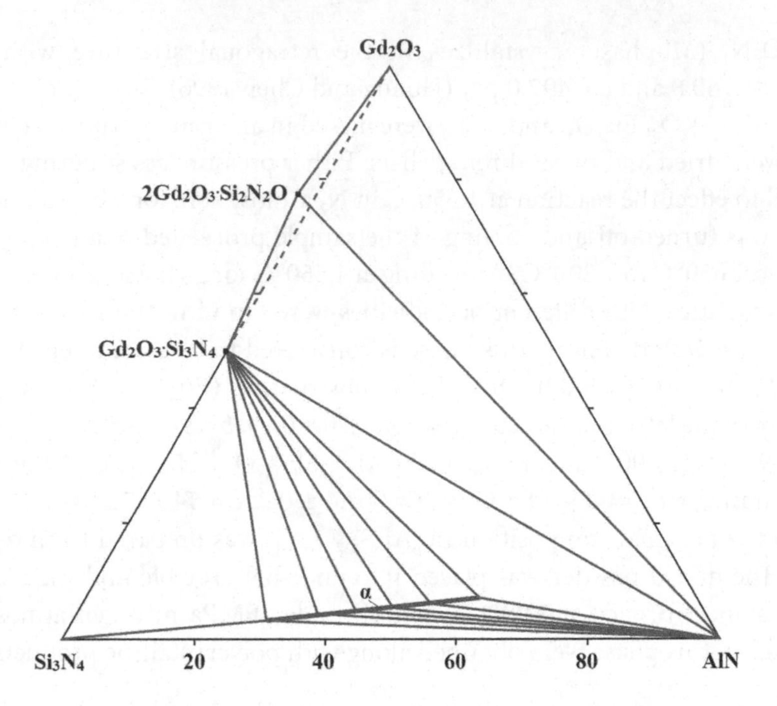

FIGURE 4.13 Subsolidus phase diagram of the AlN–Gd_2O_3–Si_3N_4 system. (With permission from Huang, Z.-K., et al., *J. Am. Ceram. Soc.*, **69**(10), C-241, 1986.)

at 1,700°C under 30 MPa in a BN-coated graphite dies in a furnace under a mild flow of N_2. After hot pressing, the furnace was cooled at a rate of ~200°C min^{-1} in the high-temperature region.

The glass-forming region in the AlN–Gd_2O_3–SiO_2 is shown in Figure 4.14 (Rocherulle et al. 1989b). The maximum AlN content in the glasses is 25 mol%.

The $\mathbf{Gd_2Al_xSi_{3-x}O_{3+x}N_{4-x}}$ quinary solid solution (*M′* phase) is formed in the Al–Gd–Si–O–N system. It crystallizes in the tetragonal structure with the lattice parameters $a = 774.15 \pm 0.05$ and $c = 504.46 \pm 0.04$ pm for $x = 0.5$ (Lauterbach and Schnick 1999) and $a = 768.9$ and $c = 499.3$ pm for $x = 0.8$ (Huang and Chen 1996). The maximum Al content in the solid solution is $x = 0.78$ (Wang et al. 1995). To synthesize these solid solutions, powders of AlN, Al_2O_3, Gd_2O_3, and Si_3N_4 were mixed in an agate mortar with ethanol or propan-2-ol (Wang et al. 1995; Huang and Chen 1996). Then, they were dried and pressed into pellets. Either pressureless sintering or hot pressing was used to effect the reaction at 1,600°C–1,750°C in N_2 atmosphere for 1.5–2 h. After the reaction, the furnace was turned off, and cooling of the sample proceeded at an estimated rate of 37°C h^{-1} from 1,650°C to 1,200°C. Annealing at 1,550°C for 24 h was also used. They could also be obtained by the reaction of Gd (0.65 mM), AlN (1.09 mM), $Si(NH)_2$ (0.68 mM), and $SrCO_3$ (0.41 mM) at 1,550°C (Lauterbach and Schnick 1999).

4.57 ALUMINUM–DYSPROSIUM–SILICON–OXYGEN–NITROGEN

The subsolidus phase relationships in the Al–Dy–Si–O–N quinary system were determined by Huang et al. (1986), Huang and Yan (1992), and Sun et al. (1996). Forty-two

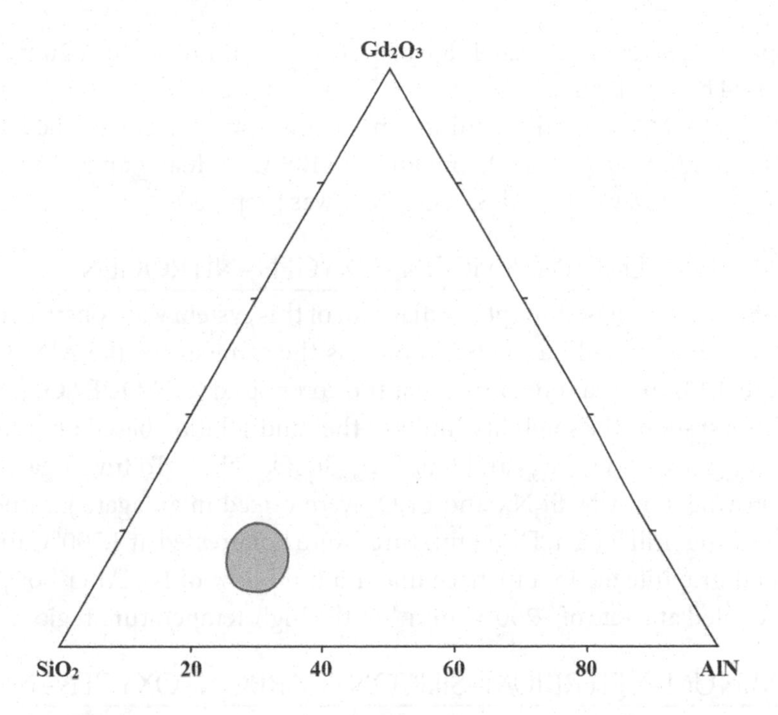

FIGURE 4.14 Glass-forming region in the $AlN–Gd_2O_3–SiO_2$ system. (With permission from Rocherulle, J., et al., *Mater. Sci. Eng. B*, **2**(4), 265, 1989.)

compatibility tetrahedra were established in the region **$AlN–Al_2O_3–Dy_2O_3–Si_3N_4$** (Sun et al. 1996). Within this region, $Dy_3Al_5O_{12}$ and **$Dy_2Al_xSi_{3-x}O_{3+x}N_{4-x}$** (*M′* phase) are the only two important phases that have tie lines joined to β-sialon and AlN polytypoid phases. α-Sialon coexists with both the phases.

$AlN–Dy_2O_3–Si_3N_4$. The subsolidus phase diagram of this system was constructed by Huang et al. (1986) and Huang and Yan (1992), and it is the same as for the $AlN–Gd_2O_3–Si_3N_4$ system (Figure 4.13). It should be noted that the compound $Si_2N_2O{\cdot}2Dy_2O_3$ does not lie in the plane of this system. The solubility limits of the solid solution based on α-Si_3N_4 are represented by $Dy_{0.33}Al_{1.5}Si_{10.5}O_{0.5}N_{15.5}$ and $Dy_{0.67}Al_3Si_9ON_{15}$. To investigate this system, the starting powders of AlN, Si_3N_4, and Dy_2O_3 were mixed in an agate jar under absolute EtOH on a rotating mill for 2 h. Dried mixtures were hot pressed at 1,700°C under 30 MPa in a BN-coated graphite dies in a furnace under a mild flow of N_2. After hot pressing, the furnace was cooled at a rate of ~200°C min^{-1} in the high-temperature region.

The $Dy_2Al_xSi_{3-x}O_{3+x}N_{4-x}$ (M′-phase) quinary solid solutions and **$Dy_4A_{10.75}Si_{1.25}O_{7.75}N_{1.25}$** quinary phase are formed in the Al–Dy–Si–O–N system. $Dy_2Al_xSi_{3-x}O_{3+x}N_{4-x}$ crystallizes in the tetragonal structure with the lattice parameters $a = 765.5$ and $c = 495.5$ pm for $x = 0.8$ (Huang and Chen 1996). The maximum Al content in the solid solution is $x = 0.71$ (Wang et al. 1995). To prepare these solid solutions, powders of AlN, Al_2O_3, Dy_2O_3, and Si_3N_4 were mixed in an agate mortar with propan-2-ol. Then, they were dried and pressed into pellets. Either pressureless sintering or hot pressing was used to effect reaction at 1,650°C in N_2 atmosphere for 2 h. After the reaction, the furnace was turned off, and cooling of the sample was proceeded at an estimated rate of 37°C h^{-1} from 1,650°C to 1,200°C. Annealing at 1,550°C for 24 h was also used.

$Dy_4Al_{0.75}Si_{1.25}O_{7.75}N_{1.25}$ crystallizes in the monoclinic structure with the lattice parameters $a = 720$, $b = 1{,}026$, and $c = 1{,}093$ pm and $\beta = 108.12°$ (Huang and Chen 1996). It was obtained by the same way as $Dy_2Al_xSi_{3-x}O_{3+x}N_{4-x}$ was prepared.

4.58 ALUMINUM–ERBIUM–SILICON–OXYGEN–NITROGEN

$AlN–Er_2O_3–Si_3N_4$. The subsolidus phase diagram of this system was constructed by Huang et al. (1986) and Huang and Yan (1992), and it is the same as for the $AlN–Gd_2O_3–Si_3N_4$ system (Figure 4.13). It should be noted that the compound $Si_2N_2O{\cdot}2Er_2O_3$ does not lie in the plane of this system. The solubility limits of the solid solution based on α-Si_3N_4 are represented by $Er_{0.33}Al_{1.5}Si_{10.5}O_{0.5}N_{15.5}$ and $Sm_{0.75}Al_{3.5}Si_{8.5}O_{1.25}N_{14.75}$. To investigate this system, the starting powders of AlN, Si_3N_4, and Er_2O_3 were mixed in an agate jar under absolute EtOH on a rotating mill for 2 h. Dried mixtures were hot pressed at 1,700°C under 30 MPa in a BN-coated graphite die in a furnace under a mild flow of N_2. After hot pressing, the furnace was cooled at a rate of ~200°C min^{-1} in the high-temperature region.

4.59 ALUMINUM–YTTERBIUM–SILICON–CARBON–OXYGEN–NITROGEN

$AlN–Yb_2O_3–SiC$. The results of XRD and EPMA show that, except for the three original constituents of this system, $2Yb_2O_3{\cdot}Al_2O_3$ phase was also observed (Chen et al. 2013). This could be the result of impurity Al_2O_3 in the starting AlN powder reacted with Yb_2O_3. After introducing $2Yb_2O_3{\cdot}Al_2O_3$, the phase relationships transformed this ternary system into the quasi-quaternary system $AlN–Yb_2O_3–2Yb_2O_3{\cdot}Al_2O_3–SiC$. The coexistence of four phases brings significance to the manufacture of SiC ceramics with $AlN–Yb_2O_3$ as sintering additives, because of the formation of $2Yb_2O_3{\cdot}Al_2O_3$ with higher melting temperatures above 2,000°C as a second phase might enhance the high-temperature properties of SiC ceramics.

To investigate this system, the powders of AlN, Yb_2O_3, and α-SiC were mixed and ground in an agate mortar for 2 h with alcohol. After drying up, the prepared powders were placed in a round graphite mold with 10 mm diameter and afterward hot pressed. To prevent the sample from sticking to the mold, the latter was lined with a thin layer of BN. The conditions used for hot pressing were as follows: Ar atmosphere, 30 MPa, 1,450°C–1,600°C, 1–2 h. To achieve the reaction equilibria, the temperature was further kept for 1–3 h after releasing the pressure.

4.60 ALUMINUM–YTTERBIUM–SILICON–OXYGEN–NITROGEN

$AlN–Yb_2O_3–Si_3N_4$. The subsolidus phase diagram of this system was constructed by Huang et al. (1986) and Huang and Yan (1992), and it is the same as for the $AlN–Gd_2O_3–Si_3N_4$ system (Figure 4.13). It should be noted that the compound $Si_2N_2O{\cdot}2Yb_2O_3$ does not lie in the plane of this system. The solubility limits of the solid solution based on α-Si_3N_4 are represented by $Yb_{0.33}Al_{1.5}Si_{10.5}O_{0.5}N_{15.5}$ and $YbAl_5Si_7O_2N_{14}$. To investigate this system, the starting powders of AlN, Si_3N_4, and Yb_2O_3 were mixed in an agate jar under absolute EtOH on a rotating mill for 2 h. Dried mixtures were hot pressed at 1,700°C under 30 MPa in a BN-coated graphite dies in a furnace under a mild flow of N_2. After hot pressing, the furnace was cooled at a rate of ~200°C min^{-1} in the high-temperature region.

The $\mathbf{Yb_4Al_xSi_{2-x}O_{7+x}N_{2-x}}$ solid solution, which crystallizes in the monoclinic structure with the lattice parameters $a = 720$, $b = 1{,}026$, and $c = 1{,}093$ pm and $\beta = 108.12°$ for $x = 0.5$, is formed in the Al–Yb–Si–O–N system (Huang and Chen 1996). To prepare these solid solutions, powders of AlN, Al_2O_3, Yb_2O_3, and Si_3N_4 were mixed in an agate mortar with propan-2-ol. Then, they were dried and pressed into pellets. Either pressureless sintering or hot pressing was used to effect reaction at 1,650°C in N_2 atmosphere for 2 h. After the reaction, the furnace was turned off, and cooling of the sample proceeded at an estimated rate of 37°C h^{-1} from 1,650°C to 1,200°C. Annealing at 1,550°C for 24 h was also used.

4.61 ALUMINUM–CARBON–SILICON–VANADIUM–CHROMIUM–MOLYBDENUM–MANGANESE–IRON–NICKEL–NITROGEN

To aid in the selection of the optimum temperature for intercritical heat treatment of an SA508 grade 3 steel, Lee et al. (1998) performed calculation of the austenite/ferrite phase equilibria in this multinary system, which was reviewed by Raghavan (2007). The calculation was done for a steel of composition (in mass%): Mn (1.24), Ni (0.88), Mo (0.47), Cr (0.21), Si (0.25), V (0.004), Al (0.008), C (0.21), N (0.005), and balance Fe. The computed equilibrium mole fractions of austenite, ferrite, and various carbides and nitrides as a function of annealing temperature were obtained.

4.62 ALUMINUM–CARBON–SILICON–OXYGEN–NITROGEN

AlN–$\mathbf{Al_2OC}$–SiC. It was shown that essentially a 2*H* solid solution can be formed by hot pressing a mixture of AlN, Al_4C_3, Al_2O_3, and SiC (Kuo et al. 1986). However, optical micrographs clearly indicated that a complete solid solution was not obtained when hot pressing was conducted at 2,000°C. At 2,050°C, a complete solid solution does not exist in this system; rather, two solid solutions, both of 2*H* structure, coexist in equilibrium. To investigate this system, AlN, Al_4C_3, Al_2O_3, and SiC were weighed in appropriate amounts in an atmosphere of argon. Subsequently, the powder mixtures were thoroughly milled in cyclohexane for up to 24 h using alumina milling media. Cyclohexane was subsequently evaporated by applying suction using a mechanical vacuum pump. Then, the powder was sieved and loaded in graphite dies in air for subsequent hot pressing in nitrogen. Hot pressing was conducted in an atmosphere of N_2 at 2,000°C ± 15°C for 0.5 h under a pressure of 30 MPa. The samples were annealed at a temperature ranging from 1,800°C to 2,050°C for up to 153 h. Principal characterization techniques employed consisted of optical microscopy, electron microscopy, XRD, and chemical analysis.

AlN–$\mathbf{Al_2O_3}$–SiC. The results of XRD, SEM and TEM showed that the polytypic content of β-SiC matrix phase after hot pressing at 2,100°C for 5 h is strongly affected by the amount of AlN and Al_2O_3 additions, with the two most predominant polytypes of α-SiC being 2*H* and 6*H* (Hilmas and Tien 1999a, 1999b). Additions of only AlN resulted in a preferred transformation from β-SiC to the 2*H*-polytype of α-SiC.

$\mathbf{(Al_{5.8}Si_{1.2})(OC_{3.5}N_{1.5})}$ and $\mathbf{(Al_{6.6}Si_{1.4})(O_{0.7}C_{4.3}N_{2.0})}$ quinary phases are formed in the Al–C–Si–O–N system (Urushihara et al. 2011). First of them crystallizes in the hexagonal structure with the lattice parameters $a = 322.508 \pm 0.004$ and $c = 3{,}171.93 \pm 0.04$ pm and a calculated density of 3.13 g cm^{-3} and the second crystallizes in the trigonal structure with the lattice parameters $a = 321.14 \pm 0.02$ and $c = 5{,}509.7 \pm 0.3$ pm and a calculated density of 3.12 g cm^{-3}.

4.63 ALUMINUM–TITANIUM–TANTALUM–OXYGEN–NITROGEN

The $\mathbf{Ti_xTa_yAl_zN_{1-\delta}O_y}$ quinary solid solution, which crystallizes in the cubic structure (the lattice parameter follows Vegard's law), is formed in the Al–Ti–Ta–O–N system (Wakabayashi et al. 2017). To prepare this solid solution, $Ti(OBu)_4$, $Ta(OEt)_5$, $Al(NO_3)_3{\cdot}9H_2O$, $Zn(NO_3)_2{\cdot}6H_2O$, EtOH, and NH_4OH were used. In a typical synthesis procedure, a stoichiometric amount (3 mM total Ti + Ta + Al precursors) of $Al(NO_3)_3{\cdot}9H_2O$ and excess $Zn(NO_3)_2{\cdot}6H_2O$ were placed in a three-necked round-bottom flask. The flask was then evacuated to 8 Pa or lower on a Schlenk line. After backfilling the flask with Ar, 20 mL of ethanol was added into the flask using a syringe, followed by the addition of stoichiometric amounts of the remaining metal alkoxide reagents, and stirred for several minutes. The solution was then transferred, using the cannula transfer technique, into approximately 30 mL of 1.5 M NH_4OH under rapid stirring, which resulted in the instantaneous formation of a white gel. The gel was then centrifuged to remove excess solvent and zinc ions, and was heated to 100°C to form a dry gel, followed by calcination at 450°C to form an amorphous mixed oxide precursor. This precursor was subsequently ball-milled to ensure that all particles were well below 1 μm in size. The mixed oxide was then heated under NH_3 flow (~6 L h^{-1}) at 800°C for up to 24 h, yielding fine black powders of the title solid solution.

CHAPTER 5

Systems Based on AlP

5.1 ALUMINUM–HYDROGEN–LITHIUM–SODIUM–CALCIUM–STRONTIUM–OXYGEN–PHOSPHORUS

The **$(Li,Na)_2(Sr,Ca)Al_4(PO_4)_4(OH)_4$** multinary phase (mineral palermoite), which crystallizes in the orthorhombic structure with the lattice parameters a = 1,155.6 ± 0.6, b = 1,584.7 ± 0.7, and c = 731.5 ± 0.4 pm and the calculated and experimental densities of 3.24 and 3.22 g cm^{-3}, respectively (Frondel and Ito 1965; Moore and Araki 1975a) [a = 731, b = 1,579, and c = 1,153 pm and the calculated and experimental densities of 3.20 and 3.22 g cm^{-3}, respectively (Mrose 1953)], is formed in the Al–H–Li–Na–Ca–Sr–O–P system.

5.2 ALUMINUM–HYDROGEN–LITHIUM–SODIUM–CALCIUM–OXYGEN–MANGANESE–FLUORINE–IRON–PHOSPHORUS

The **$Ca(Mn,Na,Li)_6FeAl_2(PO_4)_6(F,OH)_2$** multinary phase (mineral griphite), which crystallizes in the cubic structure with the lattice parameters a = 1,215.1 pm (Fransolet and Abraham 1983) [a = 1,226 pm and an experimental density of 3.40 g cm^{-3} (McConnel 1942)], is formed in the Al–H–Li–Na–Ca–O–Mn–F–Fe–P system.

5.3 ALUMINUM–HYDROGEN–LITHIUM–SODIUM–OXYGEN–PHOSPHORUS

The **$HNa_2LiAl(PO_4)_2(OH)$** multinary compound (mineral tancoite), which crystallizes in the orthorhombic structure with the lattice parameters a = 694.8 ± 0.2, b = 1,408.9 ± 0.4, and c = 1,406.5 ± 0.3 pm (Hawthorne 1983; Dunn et al. 1984b) [a = 704.1 ± 0.2, b = 1,413.0 ± 0.3, and c = 697.5 ± 0.2 pm and the calculated and experimental densities of 2.724 and 2.752 g cm^{-3}, respectively (Ramik et al. 1980; Fleischer and Cabri 1981)], is formed in the Al–H–Li–Na–O–P system.

5.4 ALUMINUM–HYDROGEN–LITHIUM–CALCIUM–OXYGEN–FLUORINE–PHOSPHORUS

The **$Li_2CaAl_4(PO_4)_4(OH,F)_4$** multinary phase (mineral bertossaite), which crystallizes in the orthorhombic structure with the lattice parameters a = 1,147.6 ± 0.1, b = 1,574.4 ± 0.1, and c = 722.8 ± 0.1 pm and a calculated density of 3.183 g cm^{-3} [a = 1,148 ± 1, b = 1,573 ± 2, and $c = 723 \pm 1$ pm and the calculated and experimental densities of 3.10 g cm^{-3} (Fleischer 1967)], is formed in the Al–H–Li–Ca–O–F–P system (Hatert et al. 2011).

5.5 ALUMINUM–HYDROGEN–LITHIUM–OXYGEN–PHOSPHORUS

The **$LiAl(H_2P_2O_7)_2$** quinary compound is formed in the Al–H–Li–O–P system (Grunze and Grunze 1984). This compound was obtained by thermal reactions in surplus H_3PO_4.

5.6 ALUMINUM–HYDROGEN–LITHIUM–OXYGEN–FLUORINE–PHOSPHORUS

The **$LiAl(PO_4)(F,OH)$** and **$LiAl(PO_4)(OH,F)$** multinary phases (minerals amblygonite and montebrasite, respectively) are formed in the Al–H–Li–O–F–P system. First of them crystallizes in the triclinic structure with the lattice parameters $a = 664.52 \pm 0.09$, $b = 773.3 \pm 0.1$, $c = 691.93 \pm 0.06$ pm and $\alpha = 90.35 \pm 0.01°$, $\beta = 117.44 \pm 0.01°$, $\gamma = 91.20 \pm 0.01°$ (Groat et al. 1990) [$a = 518$, $b = 711$, $c = 503$ pm and $\alpha = 112°02.5'$, $\beta = 97°46.5'$, $\gamma = 68°07.5'$ (Palache et al. 1943); $a = 518.4 \pm 1.0$, $b = 715.5 \pm 1.0$, $c = 504.0 \pm 1.0$ pm and $\alpha = 112°07' \pm 15'$, $\beta = 97°48' \pm 15'$, $\gamma = 67°53' \pm 15'$ and the calculated and experimental densities of 3.04 and 2.98–3.11 g cm^{-3} for different samples, respectively (Baur 1959); $a = 516.6$, 514.8, and 516, $b = 719.2$, 721.5, and 721, $c = 506.0$, 506.0, and 506 pm and $\alpha = 113.35°$, 113.97°, and 113.20°, $\beta = 98.24°$, 98.64°, and 97.90°, $\gamma = 67.60°$, 67.25°, and 67.53°, respectively, for three different samples (Moss et al. 1969)].

$LiAl(PO_4)(OH,F)$ also crystallizes in the triclinic structure with the lattice parameters $a = 671.3 \pm 0.1$, $b = 770.8 \pm 0.1$, $c = 701.94 \pm 0.07$ pm and $\alpha = 91.31 \pm 0.01°$, $\beta = 117.93 \pm 0.01°$, $\gamma = 91.77 \pm 0.01°$ (Groat et al. 1990) [$a = 518$, 518, 518.9, 517.4, and 517.5, $b = 711$, 715, 717.3, 716.4, and 717.3, $c = 503$, 504, 504.0, 504.4, and 504.7 pm, and $\alpha = 112.04°$, 112.11°, 112.50°, 112.70° and 112.89°, $\beta = 97.83°$, 97.80°, 97.90°, 98.02°, and 98.08°, $\gamma = 68.13°$, 67.89°, 67.81°, 67.78°, and 67.67°, respectively, for five different samples (Moss et al. 1969); parameters $a = 517.7 \pm 0.1$, $b = 717.1 \pm 0.2$, $c = 504.5 \pm 0.1$ pm and $\alpha = 112°42' \pm 1'$, $\beta = 98°02' \pm 1'$, $\gamma = 67°42' \pm 1'$ (Fransolet and Abraham 1983)].

The lattice parameters of amblygonite and montebrasite depend on the F/OH ratio (Černá et al. 1973).

5.7 ALUMINUM–HYDROGEN–SODIUM–POTASSIUM–BERYLLIUM–ZIRCONIUM–OXYGEN–PHOSPHORUS

The **$NaK(Be,Al)Zr_2(PO_4)_4{\cdot}2H_2O$** multinary phase (mineral selwynite), which crystallizes in the tetragonal structure with the lattice parameters $a = 657.0 \pm 0.3$ and $c = 1{,}714.2 \pm 0.6$ pm and the calculated and experimental densities of 3.08 and 2.94 g cm^{-3}, respectively, is formed in the Al–H–Na–K–Be–Zr–O–P system (Birch et al. 1995; Jambor and Roberts 1995b).

5.8 ALUMINUM–HYDROGEN–SODIUM–POTASSIUM–MAGNESIUM–CALCIUM–STRONTIUM–BARIUM–LEAD–OXYGEN–MANGANESE–IRON–PHOSPHORUS

The **$Na_3(Ba,K,Pb)(Ca,Sr)(Fe,Mg,Mn)_{14}Al(PO_4)_{12}(OH)_2$** multinary phase (mineral sigismundite), which crystallizes in the monoclinic structure with the lattice parameters $a = 1{,}639.4 \pm 0.4$, $b = 993.2 \pm 0.2$, $c = 2{,}443.7 \pm 0.7$ pm, and $\beta = 105.78 \pm 0.02°$ and a calculated density of 3.544 g cm^{-3} is formed in the Al–H–Na–K–Mg–Ca–Sr–Ba–Pb–O–Mn–Fe–P system (Demartin et al. 1996; Jambor et al. 1997a).

5.9 ALUMINUM–HYDROGEN–SODIUM–POTASSIUM–MAGNESIUM–CALCIUM–CARBON–OXYGEN–PHOSPHORUS

The **$(Na,K)(Mg,Ca)_4Al_8(PO_4)_8(CO_3)(OH)_7{\cdot}30H_2O$** multinary phase (mineral parwanite), which crystallizes in the monoclinic structure with the lattice parameters a = 2,614.8, b = 1,178.1, c = 2,049.4 pm, and β = 111.27°, is formed in the Al-H–Na–K–Mg–Ca–C–O–P system (Birch et al. 2007).

5.10 ALUMINUM–HYDROGEN–SODIUM–POTASSIUM–CALCIUM–OXYGEN–PHOSPHORUS

The **$(Na,K)CaAl_6(PO_4)_4(OH)_9{\cdot}3H_2O$** (mineral millisite) and **$Na_2K_3Ca_{10}Al_{15}(PO_4)_{21}(OH)_7{\cdot}26H_2O$** (mineral englishite) multinary phases are formed in the Al-H-Na-K-Ca-O-P system. First of them crystallizes in the hexagonal structure with the lattice parameters a = 700 and c = 1,907 pm (Owens et al. 1960). The second compound crystallizes in the monoclinic structure with the lattice parameters a = 3,843 ± 2, b = 1,186, c = 2,067 pm, and β = 111°16′ and the calculated and experimental densities of 2.67 [2.69 (Dunn et al. 1984d, 1985b)] and 2.65 g cm^{-3}, respectively (Moore 1976).

5.11 ALUMINUM–HYDROGEN–SODIUM–POTASSIUM–CALCIUM–OXYGEN–FLUORINE–MANGANESE–PHOSPHORUS

The **$Na_4KCaAlMn_{14}(PO_4)_{12}(OH,F)_2$** multinary phase [mineral dickinsonite-(KMnNa)], which crystallizes in the monoclinic structure with the lattice parameters a = 1,669.00 ± 0.09, b = 1,010.13 ± 0.05, c = 2,487.52 ± 0.13 pm, and β = 105.616 ± 0.002°, is formed in the Al-H-Na-K-Ca-O-F-Mn-P system (Cámara et al. 2006).

5.12 ALUMINUM–HYDROGEN–SODIUM–POTASSIUM–CALCIUM–OXYGEN–FLUORINE–IRON–PHOSPHORUS

The **$Na_5KCaAlFe_{13}(PO_4)_{11}(PO_3OH)(OH,F)_2$** multinary phase [mineral arrojadite-(KNa)], which crystallizes in the monoclinic structure with the lattice parameters a = 1,652.20 ± 0.11, b = 1,005.29 ± 0.07, c = 2,464.77 ± 0.16 pm, and β = 106.509 ± 0.002°, is formed in the Al-H-Na-K-Ca-O-F-Fe-P system (Cámara et al. 2006).

5.13 ALUMINUM–HYDROGEN–SODIUM–POTASSIUM–CALCIUM–OXYGEN–MANGANESE–IRON–PHOSPHORUS

The **$Na_4KCa(Mn,Fe)_{14}Al(PO_4)_{12}(OH)_2$** (mineral dickinsonite) and **$Na_4KCa(Fe,Mn)_{14}Al(PO_4)_{12}(OH)_2$** (mineral arrojadite) multinary phases are formed in the Al–H-Na-K-Ca-O-Mn-Fe-P system. First of them crystallizes in the monoclinic structure with the lattice parameters a = 2,494.0 ± 0.6, b = 1,013.1 ± 0.4, c = 1,672.2 ± 0.2 pm, and β = 105.60 ± 0.02° and the calculated and experimental densities of 3.426 and 3.41 g cm^{-3}, respectively (Moore et al. 1981) [a = 1,670, b = 995, c = 2,469 pm, and β = 104°41′ (Wolfe 1941); a = 2,489, b = 1,011, c = 1,668 pm, and β = 105°41′ and a calculated density of 3.51 g cm^{-3} (Fisher 1965)].

The second compound also crystallizes in the monoclinic structure with the lattice parameters $a = 1{,}652.6 \pm 0.4$, $b = 1{,}005.7 \pm 0.3$, $c = 2{,}473.0 \pm 0.5$ pm, and $\beta = 105.78 \pm 0.03°$ (Merlino et al. 1981) [$a = 1{,}660$, $b = 1{,}002$, $c = 2{,}399$ pm, and $\beta = 93°37' \pm 15'$ and a calculated density of 3.553 g cm^{-3} (Lindberg 1950); $a = 2{,}478$, $b = 1{,}005$, $c = 1{,}651$ pm, and $\beta = 105°41'$ and a calculated density of 3.435 g cm^{-3} (Fisher 1965); $a = 1{,}647.6 \pm 0.4$, $b = 2{,}458.1 \pm 0.5$, $c = 999.6 \pm 0.3$ pm, and $\beta = 105.79 \pm 0.02°$ (Krutik et al. 1979)].

5.14 ALUMINUM–HYDROGEN–SODIUM–POTASSIUM–OXYGEN–IRON–PHOSPHORUS

The **$(K,Na)_3(Al,Fe)_5(PO_4)_2(PO_3OH)_6{\cdot}12H_2O$** (mineral francoanellite) and **$(K,Na)_3(Al,Fe)_5(PO_4)_2(PO_3OH)_6{\cdot}18H_2O$** (mineral taranakite) multinary phases are formed in the Al–H–Na–K–O–Fe–P system. First of them crystallizes in the trigonal structure with the lattice parameters $a = 869.0 \pm 0.2$ and $c = 8{,}227 \pm 1$ pm and the calculated and experimental densities of 2.286 and 2.26 ± 0.02 g cm^{-3}, respectively (Balenzano et al. 1976; Fleischer et al. 1976a; Dick and Zeiske 1998; Jambor et al. 1999). Single crystals of synthetic francoanellite could be obtained by topochemical dehydration of taranakite crystals.

The second phase also crystallizes in the trigonal structure with the lattice parameters $a = 870.25 \pm 0.11$ and $c = 9{,}505 \pm 1$ pm and a calculated density of 2.145 g cm^{-3} (Dick et al. 1998) [in the rhombohedral structure with the lattice parameters $a = 871$ and $c = 9{,}610$ pm (in hexagonal setting) and the calculated and experimental densities of 2.11 and 2.09 g cm^{-3}, respectively (Smith and Brown 1959); in the hexagonal structure with the lattice parameters $a = 871.8 \pm 0.2$ and $c = 9{,}518 \pm 3$ pm and the calculated and experimental densities of 2.12 and 2.06 g cm^{-3}, respectively (Sakae and Sudo 1975)]. Taranakite powder was synthesized by the reaction of $Al(OH)_3$ (3.2 g) with 500 mL of aqueous potassium phosphate solution at 80°C in a glass beaker under vigorous stirring. The solution was prepared by mixing equal amounts of 1 M KH_2PO_4 and 1 M H_3PO_4 and adjusting the pH value at 4 by adding KOH. After 4 days, $Al(OH)_3$ had transformed totally into taranakite. The product was filtered off, washed phosphate-free with water, and air-dried. Large single crystals of the title phase were grown by gel crystallization.

5.15 ALUMINUM–HYDROGEN–SODIUM–MAGNESIUM–CALCIUM–OXYGEN–MANGANESE–IRON–PHOSPHORUS

The **$Na_xCa_2(Mn,Mg,Fe)_4(Fe,Mg,Al)_2(PO_4)_6{\cdot}2H_2O$** multinary phase (mineral bederite), which crystallizes in the orthorhombic structure with the lattice parameters $a = 1{,}255.9 \pm 0.2$, $b = 1{,}283.4 \pm 0.1$, and $c = 1{,}171.4 \pm 0.2$ pm and the calculated and experimental densities of 3.50 and 3.48 ± 0.01 g cm^{-3}, respectively, is formed in the Al–H–Na–Mg–Ca–O–Mn–Fe–P system (Galliski et al. 1999).

5.16 ALUMINUM–HYDROGEN–SODIUM–MAGNESIUM–CALCIUM–OXYGEN–IRON–PHOSPHORUS

The **$(Na,Ca)(Fe,Mg)Al_5(PO_4)_4(OH,O)_6{\cdot}2H_2O$** multinary phase (mineral burangaite), which crystallizes in the monoclinic structure with the lattice parameters $a = 2{,}509.9 \pm$

0.2, $b = 504.91 \pm 0.07$, $c = 1,343.8 \pm 0.1$ pm, and $\beta = 110.88 \pm 0.01°$ and a calculated density of 3.00 g cm^{-3} (Selway et al. 1997) [$a = 2,509 \pm 1$, $b = 504.8 \pm 0.3$, $c = 1,345 \pm 1$ pm, and $\beta = 110.91 \pm 0.08°$ and an experimental density of 3.05 g cm^{-3} (Knorring von et al. 1977; Fleischer et al. 1978b)], is formed in the Al–H–Na–Mg–Ca–O–Fe–P system.

5.17 ALUMINUM–HYDROGEN–SODIUM–MAGNESIUM–OXYGEN–PHOSPHORUS

The **$NaMgAl_5(PO_4)_4(OH)_6{\cdot}2H_2O$** multinary compound (mineral matioliite), which crystallizes in the monoclinic structure with the lattice parameters $a = 2,507.5 \pm 0.1$, $b = 504.70 \pm 0.03$, $c = 1,343.70 \pm 0.07$ pm, and $\beta = 110.97 \pm 0.03°$ and a calculated density of 2.948 g cm^{-3}, is formed in the Al–H–Na–Mg–O–P system (Atencio et al. 2006).

5.18 ALUMINUM–HYDROGEN–SODIUM–CALCIUM–STRONTIUM–OXYGEN–FLUORINE–IRON–PHOSPHORUS

The **$Na_2CaSrAlFe_{13}(PO_4)_{11}(PO_3OH)(OH,F)_2$** multinary phase [mineral arrojadite-(SrFe)], which crystallizes in the monoclinic structure with the lattice parameters $a = 1,639.92 \pm 0.07$, $b = 994.00 \pm 0.04$, $c = 2,444.34 \pm 0.11$ pm, and $\beta = 105.489 \pm 0.001°$, is formed in the Al–H–Na–Ca–Sr–O–F–Fe–P system (Cámara et al. 2006).

5.19 ALUMINUM–HYDROGEN–SODIUM–CALCIUM–BARIUM–OXYGEN–IRON–PHOSPHORUS

The **$BaNa_3(Na,Ca)Fe_{13}Al(PO_4)_{11}(PO_3OH)(OH)$** multinary phase [mineral arrojadite-(BaNa)], which crystallizes in the monoclinic structure with the lattice parameters $a = 1,649.84 \pm 0.06$, $b = 1,002.28 \pm 0.03$, $c = 2,464.8 \pm 0.1$ pm, and $\beta = 105.850 \pm 0.004°$, is formed in the Al–H–Na–Ca–Ba–O–Fe–P system (Vignola et al. 2015).

5.20 ALUMINUM–HYDROGEN–SODIUM–CALCIUM–OXYGEN–FLUORINE–PHOSPHORUS

The **$NaCa_2Al_2(PO_4)_2(F,OH)_5{\cdot}2H_2O$** multinary phase (mineral morinite), which crystallizes in the monoclinic structure with the lattice parameters $a = 945.4 \pm 0.3$, $b = 1,069.2 \pm 0.4$, $c = 544.4 \pm 0.2$ pm, and $\beta = 105.46 \pm 0.02°$ (Hawthorne 1979) [$a = 945.6$, $b = 1,069.0$, $c = 544.5$ pm, and $\beta = 105°28'$ and a calculated density of 2.96 g cm^{-3} (Fisher and Runner 1958)], is formed in the Al–H–Na–Ca–O–F–P system.

5.21 ALUMINUM–HYDROGEN–SODIUM–CALCIUM–OXYGEN–FLUORINE–MANGANESE–PHOSPHORUS

The **$Na(Ca,Mn)Al(PO_4)(F,OH)_5$** multinary phase (mineral viitaniemiite), which crystallizes in the monoclinic structure with the lattice parameters $a = 545.7 \pm 0.2$, $b = 715.1 \pm 0.2$, $c = 683.6 \pm 0.2$ pm, and $\beta = 109.36 \pm 0.03°$ and a calculated density of 3.242 g cm^{-3} (Pajunen and Lahti 1984) [$a = 683.2 \pm 0.3$, $b = 714.3 \pm 0.3$, $c = 544.7 \pm 0.3$ pm, and $\beta = 109°22' \pm 5'$ and the calculated and experimental densities of 3.240 and 3.245 g cm^{-3}, respectively (Cabri et al. 1981); $a = 686.5 \pm 0.2$, $b = 722.5 \pm 0.2$, $c = 552.2 \pm 0.1$ pm, and $\beta = 109°00' \pm 2'$ and an experimental density of 3.06 ± 0.01 g cm^{-3} (Ramik et al. 1983)], is formed in the Al–H–Na–Ca–O–F–Mn–P system.

5.22 ALUMINUM–HYDROGEN–SODIUM–STRONTIUM–OXYGEN–PHOSPHORUS

The **$Na_2SrAl_4(PO_4)_4(OH)_4$** multinary compound (mineral natropalermoite), which crystallizes in the orthorhombic structure with the lattice parameters a = 1,148.49 ± 0.06, b = 1,624.90 ± 0.07, and c = 729.27 ± 0.04 pm and a calculated density of 3.502 g cm^{-3}, is formed in the Al–H–Na–Sr–O–P system (Schumer et al. 2014, 2017).

5.23 ALUMINUM–HYDROGEN–SODIUM–ZIRCONIUM–OXYGEN–PHOSPHORUS

The **$NaAlZr(PO_4)_2(OH)_2 \cdot H_2O$** multinary compound (mineral wycheproofite), which crystallizes in the triclinic structure with the lattice parameters a = 526.3 ± 0.1, b = 925.1 ± 0.2, c = 948.0 ± 0.2 pm and α = 109.49 ± 0.03°, β = 98.57 ± 0.03°, γ = 90.09 ± 0.03° and a calculated density of 2.93 g cm^{-3} (Kolitsch 2003; Jambor and Roberts 2004) [a = 1,092.6 ± 0.5, b = 1,098.6 ± 0.5, c = 1,247.9 ± 0.9 pm and α = 71.37 ± 0.04°, β = 77.39 ± 0.04°, γ = 87.54 ± 0.03° and the calculated and experimental densities of 2.81 and 2.83 g cm^{-3}, respectively (Birch et al. 1994; Jambor et al. 1995)], is formed in the Al–H–Na–Zr–O–P system.

5.24 ALUMINUM–HYDROGEN–SODIUM–OXYGEN–PHOSPHORUS

The **$NaAl(H_2P_2O_7)_2$**, **$NaAl_3(PO_4)_2(OH)_4$** (mineral brazilianite), and **$NaAl_3(PO_4)_2(OH)_4 \cdot 2H_2O$** (mineral wardite) quinary compounds are formed in the Al–H–Na–O–P system. First of them was obtained by thermal reactions in surplus of H_3PO_4 (Grunze and Grunze 1984). It is formed in the temperature range of 170°C–250°C. The modification I of this compound crystallizes at Na/Al molar ratio of 2.5, while higher Na/Al ratios (5–10) cause the formation of modification II. With increase in temperature $NaAl(H_2P_2O_7)_2$ decomposes, and at 300°C–350°C $Na_2H_2(P_4O_{12})$ is formed. At higher temperature, thermally stable end product is $Al_4(P_4O_{12})_3$.

$NaAl_3(PO_4)_2(OH)_4$ crystallizes in the monoclinic structure with the lattice parameters a = 1,124.48 ± 0.05, b = 1,015.39 ± 0.06, c = 710.31 ± 0.03 pm, and β = 97.351 ± 0.004° and a calculated density of 2.989 g cm^{-3} (Gatta et al. 2013b) [a:b:c = 1.1056:1:0.6992 and β = 97°22′ and an experimental density of 2.980 ± 0.005 g cm^{-3} (Pough and Henderson 1945; Frondel and Lindberg 1948); a = 1,119, b = 1,008, c = 706 pm, and β = 97°22′ and the calculated and experimental densities 3.025 and 2.975–2.977 g cm^{-3}, respectively (Hurlbut and Weichel 1946); a = 1,123.3 ± 0.6, b = 1,014.2 ± 0.5, c = 709.7 ± 0.4 pm, and β = 97.37 ± 0.02° and the calculated and experimental densities of 2.998 ± 0.004 and 2.98 ± 0.01 g cm^{-3}, respectively (Gatehouse and Miskin 1974)].

$NaAl_3(PO_4)_2(OH)_4 \cdot 2H_2O$ crystallizes in the tetragonal structure with the lattice parameters a = 703 ± 1 and c = 1,904 ± 1 pm and the calculated and experimental densities of 2.805 and 2.81 g cm^{-3}, respectively (Fanfani et al. 1970) [a = 704 ± 2 and c = 1,888 ± 2 pm and the calculated and experimental densities of 2.81 g cm^{-3} (Hurlbut, 1952)].

5.25 ALUMINUM–HYDROGEN–SODIUM–OXYGEN–SULFUR–PHOSPHORUS

The $\mathbf{Na_3Al_{16}(PO_4)_{10}(SO_4)_2(OH)_{17}\cdot 20H_2O}$ multinary compound (mineral peisleyite), which crystallizes in the triclinic structure with the lattice parameters a = 928.0 ± 1.9, b = 1,197.6 ± 1.9, c = 1,325.0 ± 1.8 pm and α = 91.3 ± 0.1°, β = 75.6 ± 0.1°, γ = 67.67 ± 0.01° and the calculated and experimental densities of 2.23 and 2.20 ± 0.05 g cm^{-3}, respectively (Mills et al. 2011b; Gatta et al. 2012) [in the monoclinic structure with the lattice parameters a = 1,331.0 ± 0.6, b = 1,262.0 ± 0.6, c = 2,315 ± 1 pm, and β = 110.00 ± 0.03° and the calculated and experimental densities of 2.11 and 2.12 g cm^{-3}, respectively (Pilkington et al. 1982; Dunn et al. 1983)], is formed in the Al–H–Na–O–S–P system.

5.26 ALUMINUM–HYDROGEN–SODIUM–OXYGEN–MOLYBDENUM–IRON–PHOSPHORUS

The $\mathbf{NaAl_4Fe_7(PO_4)_5(PMo_{12}O_{40})(OH)_{16}\cdot 56H_2O}$ multinary compound (mineral paramendozavilite), which crystallizes in the monoclinic or triclinic structure with an experimental density of 3.35 g cm^{-3}, is formed in the Al–H–Na–O–Mo–Fe–P system (Williams 1986; Hawthorne et al. 1988).

5.27 ALUMINUM–HYDROGEN–SODIUM–OXYGEN–FLUORINE–PHOSPHORUS

The $\mathbf{NaAl_3(PO_4)_2F_2(OH)_2(H_2O)_2}$ multinary compound (mineral fluorowardite), which crystallizes in the tetragonal structure with the lattice parameters a = 707.7 ± 0.2 and c = 1,922.7 ± 0.3 pm and a calculated density of 2.760 g cm^{-3}, is formed in the Al–H–Na–O–F–P system (Kampf et al. 2012a, 2014a).

5.28 ALUMINUM–HYDROGEN–POTASSIUM–MAGNESIUM–TITANIUM–OXYGEN–PHOSPHORUS

The $\mathbf{KMg_2Al_2Ti(PO_4)_4(OH)_3\cdot 15H_2O}$ multinary compound (mineral mantienneite), which crystallizes in the orthorhombic structure with the lattice parameters a = 1,040.9 ± 0.2, b = 2,033.0 ± 0.4, and c = 1,231.2 ± 0.2 pm and the calculated and experimental densities of 2.25 and 2.31 g cm^{-3}, respectively, is formed in the Al–H–K–Mg–Ti–O–P system (Fransolet et al. 1984; Dunn et al. 1985b).

5.29 ALUMINUM–HYDROGEN–POTASSIUM–MAGNESIUM–TITANIUM–OXYGEN–MANGANESE–IRON–PHOSPHORUS

The $\mathbf{K(Mg,Mn)_2(Fe,Al)_2Ti(PO_4)_4(OH)_3\cdot 15H_2O}$ multinary compound (mineral paulkerrite), which crystallizes in the orthorhombic structure with the lattice parameters a = 1,049 ± 7, b = 2,075 ± 13, and c = 1,244 ± 2 pm and the calculated and experimental densities of 2.36 and 2.36 ± 0.04 g cm^{-3}, respectively, is formed in the Al–H–K–Mg–Ti–O–Mn–Fe–P system (Peacor et al. 1984b; Dunn et al. 1985c).

5.30 ALUMINUM–HYDROGEN–POTASSIUM–OXYGEN–PHOSPHORUS

The $\mathbf{KAl(HP_3O_{10})}$, $\mathbf{KAl(H_2P_2O_7)_2}$, $\mathbf{KAl_2[(H_2P_3O_{10})(P_4O_{12})]}$, and $\mathbf{KAl_2(PO_4)_2(OH)\ 2H_2O}$ (mineral tinsleyite) quinary compounds are formed in the Al–H–K–O–P system. $KAl(HP_3O_{10})$

crystallizes in the monoclinic structure with the lattice parameters $a = 1{,}161.9 \pm 0.8$, $b = 490.5 \pm 0.5$, $c = 861.4 \pm 0.5$ pm, and $\beta = 120.84 \pm 0.05°$ (Averbuch-Pouchot et al. 1977). This compound was prepared by calcination at 350°C for 10 h of the mixture by introducing 0.5 g of alumina and 5 g K_2CO_3 into 15 mL H_3PO_4. The product obtained was polycrystalline (Averbuch-Pouchot et al. 1977; Grunze and Grunze 1984).

$KAl(H_2P_2O_7)_2$ and $KAl_2[(H_2P_3O_{10})(P_4O_{12})]$ were prepared by thermal reactions in surplus of H_3PO_4 (Grunze and Grunze 1984). Second compound crystallizes in the monoclinic structure.

$KAl_2(PO_4)_2(OH){\cdot}2H_2O$ crystallizes in the monoclinic structure with the lattice parameters $a = 960.2 \pm 0.8$, $b = 953.2 \pm 0.6$, $c = 954.3 \pm 1.1$ pm, and $\beta = 103.16 \pm 0.06°$ and the calculated and experimental densities of 2.62 and 2.69 ± 0.05 g cm^{-3}, respectively (Dunn et al. 1984c).

5.31 ALUMINUM–HYDROGEN–POTASSIUM–OXYGEN–FLUORINE–PHOSPHORUS

The **$KAl_2(PO_4)_2(OH,F)$ $4H_2O$** multinary phase (mineral minyulite), which crystallizes in the orthorhombic structure with the lattice parameters $a = 933.7 \pm 0.5$, $b = 974.0 \pm 0.5$, and $c = 552.2 \pm 0.3$ pm and a calculated density of 2.47 g cm^{-3} (Kampf 1977) [$a = 935 \pm 2$, $b = 974 \pm 2$, and $c = 552 \pm 2$ pm and an experimental density of 2.46 g cm^{-3} (Spencer et al. 1943)], is formed in the Al–H–K–O–F–P system.

5.32 ALUMINUM–HYDROGEN–POTASSIUM–OXYGEN–NICKEL–PHOSPHORUS

The **$K[Ni(H_2O_2)_2Al_2(PO_4)_3]$** multinary compound, which crystallizes in the monoclinic structure with the lattice parameters $a = 1{,}307.5 \pm 0.1$, $b = 1{,}011.4 \pm 0.1$, $c = 872.0 \pm 0.1$ pm, and $\beta = 108.158 \pm 0.010°$ and a calculated density of 2.865 g cm^{-3}, is formed in the Al–H–K–O–Ni–P system (Meyer and Haushalter 1994). The hydrothermal reaction of $NiCl_2{\cdot}6H_2O$, $AlCl_3{\cdot}6H_2O$, KCl, Me_4NOH, 85% H_3PO_4, and H_2O (molar ratio 3:0.20:6:6:10:400) at 150°C for 100 h yields yellow crystals of the title compound.

5.33 ALUMINUM–HYDROGEN–RUBIDIUM–OXYGEN–PHOSPHORUS

The **$RbAl(HP_3O_{10})$** and **$RbAl_2[(H_2P_3O_{10})(P_4O_{12})]$** quinary compounds are formed in the Al–H–Rb–O–P system (Grunze and Grunze 1984). Both compounds were prepared by thermal reactions in surplus of H_3PO_4.

5.34 ALUMINUM–HYDROGEN–CESIUM–OXYGEN–PHOSPHORUS

Three quinary compounds are formed in the Al–H–Cs–O–P system. **$CsAl(HP_3O_{10})$** and **$CsAl_2[(H_2P_3O_{10})(P_4O_{12})]$** were prepared by thermal reactions in surplus of H_3PO_4 (Grunze and Grunze 1984). **$Cs_2AlH_3(P_2O_7)$** is formed by the heating of a mixture P_2O_5, Cs_2O, and Al_2O_3 [molar ratio 15:7.5:1] at 150°C–170°C (Chudinova et al. 1987a).

5.35 ALUMINUM–HYDROGEN–COPPER–CADMIUM–OXYGEN–FLUORINE–PHOSPHORUS

The **$CuCd_2Al_3(PO_4)_4F_2{\cdot}12H_2O$** multinary compound (mineral goldquarryite), which crystallizes in the triclinic structure with the lattice parameters $a = 677.7 \pm 0.3$, $b = 908.1 \pm 0.4$,

$c = 1{,}010.4 \pm 0.5$ pm and $\alpha = 101.40 \pm 0.04°$, $\beta = 104.24 \pm 0.04°$, $\gamma = 102.56 \pm 0.04°$ and the calculated and experimental densities of 2.81 and 2.78 ± 0.01 g cm^{-3}, respectively, is formed in the Al–H–Cu–Cd–O–F–P system (Roberts et al. 2003; Jambor and Roberts 2003; Cooper and Hawthorne 2004).

5.36 ALUMINUM–HYDROGEN–COPPER–VANADIUM–OXYGEN–FLUORINE–PHOSPHORUS

The **$Cu_{0.56}(VO)_{0.44}Al_2(PO_4)_2(F,OH)_2{\cdot}5H_2O$** (mineral cloncurryite) and **$(Cu^{2+},\square,Al,V)_6[Al_8(PO_4)_8F_8](OH)_2(H_2O)_{22}$** (mineral nevadaite) multinary compounds are formed in the Al–H–Cu–V–O–F–P system. First of them crystallizes in the monoclinic structure with the lattice parameters $a = 495.73 \pm 0.02$, $b = 1{,}218.24 \pm 0.04$, $c = 1{,}897.49 \pm 0.08$ pm, and $\beta = 90.933 \pm 0.006°$ and a calculated density of 2.525 g cm^{-3} (Colchester et al. 2007).

Second compound crystallizes in the orthorhombic structure with the lattice parameters $a = 1{,}212.3 \pm 0.2$, $b = 1{,}899.9 \pm 0.2$, and $c = 496.13 \pm 0.05$ pm and the calculated and experimental densities of 2.55 and 2.54 g cm^{-3}, respectively (Cooper et al. 2004; Jambor et al. 2005).

5.37 ALUMINUM–HYDROGEN–COPPER–OXYGEN–PHOSPHORUS

The **$CuAl_6(PO_4)_4(OH)_8{\cdot}4H_2O$** (mineral turquoise), **$Cu_3Al_4(PO_4)_2(OH)_{12}{\cdot}2H_2O$** (mineral sieleckiite) and **$Cu_3Al_4(PO_4)_3(OH)_9{\cdot}4H_2O$** (mineral zapatalite) quinary compounds are formed in the Al–H–Cu–O–P system. First of them crystallizes in the triclinic structure with the lattice parameters $a = 741.0 \pm 0.1$, $b = 763.3 \pm 0.1$, $c = 990.4 \pm 0.1$ pm and $\alpha = 68.42 \pm 0.01°$, $\beta = 69.65 \pm 0.01°$, $\gamma = 65.05 \pm 0.01°$ and calculated density of 2.935 g cm^{-3} (Kolitsch and Giester 2000) [a:b:c = 0.7910:1:0.6051 and $\alpha = 92°58'$, $\beta = 93°30'$, $\gamma = 107°41'$ and an experimental density of 2.84 g cm^{-3} (Schaller 1912); $a = 742.4 \pm 0.4$, $b = 762.9 \pm 0.3$, $c = 991.0 \pm 0.4$ pm and $\alpha = 68.61 \pm 0.03°$, $\beta = 69.71 \pm 0.04°$, $\gamma = 65.08 \pm 0.03°$ and the calculated and experimental densities of 2.90 and 2.84 g cm^{-3}, respectively (Cid-Dresdner 1964, 1965; Cid-Dresdner and Villarroel 1972)]. Complete solid solution exists between turquoise and **$Al_6(PO_4)_2(PO_3OH)_2(OH)_8{\cdot}4H_2O$** (mineral planerite) (Foord and Taggart, 1998).

$Cu_3Al_4(PO_4)_2(OH)_{12}{\cdot}2H_2O$ crystallizes in the monoclinic structure with the lattice parameters $a = 1{,}171.1 \pm 0.2$, $b = 692.33 \pm 0.14$, $c = 982.8 \pm 0.2$ pm, and $\beta = 92.88 \pm 0.03°$ and a calculated density of 3.041 g cm^{-3} (Elliott 2017) [in the triclinic structure with the lattice parameters $a = 941 \pm 8$, $b = 756 \pm 5$, $c = 595 \pm 6$ pm and $\alpha = 90.25 \pm 0.12°$, $\beta = 91.27 \pm 0.12°$, $\gamma = 104.02 \pm 0.07°$ and the calculated and experimental densities of 2.94 and 3.02 ± 0.02 g cm^{-3}, respectively (Birch and Pring 1988; Jambor and Burke 1989)].

$Cu_3Al_4(PO_4)_3(OH)_9{\cdot}4H_2O$ crystallizes in the tetragonal structure with the lattice parameters $a = 1{,}522.3$ and $c = 1{,}151.8$ pm and the calculated and experimental densities of 3.017 and 3.016 ± 0.026 g cm^{-3}, respectively (Williams 1972).

5.38 ALUMINUM–HYDROGEN–COPPER–OXYGEN–IRON–PHOSPHORUS

The **$Cu(Al,Fe)_6(PO_4)_4(OH)_8{\cdot}5H_2O$** multinary phase (mineral rashleighite), which crystallizes in the triclinic structure with the lattice parameters $a = 749$, $b = 768$, $c = 1{,}000$ pm and $\alpha = 68°36' \pm 11'$, $\beta = 69°48' \pm 20'$, $\gamma = 69°15' \pm 15'$ and the calculated and experimental densities of 3.07 and 3.02 [3.00–3.02 (Russel 1948)] g cm^{-3}, respectively, is formed in the Al–H–Cu–O–Fe–P system (Cid-Dresdner and Villarroel 1972).

5.39 ALUMINUM–HYDROGEN–BERYLLIUM–OXYGEN–PHOSPHORUS

The $\mathbf{BeAl_2(PO_4)_2(OH)_2(H_2O)_5}$ quinary compound (mineral brandãoite), which crystallizes in the triclinic structure with the lattice parameters $a = 610.0 \pm 0.4$, $b = 861.6 \pm 0.4$, $c = 1{,}026.1 \pm 0.5$ pm and $\alpha = 93.19 \pm 0.1°$, $\beta = 95.12 \pm 0.1°$, $\gamma = 96.86 \pm 0.1°$, is formed in the Al–H–Be–O–P system (Menezes et al. 2017a, 2017b).

5.40 ALUMINUM–HYDROGEN–BERYLLIUM–OXYGEN–IRON–PHOSPHORUS

The $\mathbf{BeAl_2Fe_2(PO_4)_2(OH)_6}$ multinary compound (mineral lefontite), which crystallizes in the orthorhombic structure with the lattice parameters $a = 700.87 \pm 0.03$, $b = 1{,}050.82 \pm 0.04$, and $c = 1{,}311.79 \pm 0.05$ pm, is formed in the Al–H–Be–O–Fe–P system (Yang et al. 2015).

5.41 ALUMINUM–HYDROGEN–MAGNESIUM–CALCIUM–OXYGEN–PHOSPHORUS

The $\mathbf{MgCaAl(PO_4)_2(OH){\cdot}4H_2O}$ (mineral overite), $\mathbf{MgCaAl_2(PO_4)_2(OH)_4{\cdot}7H_2O}$ (mineral angastonite) and $\mathbf{MgCa_4Al_4(PO_4)_6(OH)_4{\cdot}12H_2O}$ (mineral montgomeryite) multinary compounds are formed in the Al–H–Mg–Ca–O–P system. $MgCaAl(PO_4)_2(OH){\cdot}4H_2O$ crystallizes in the orthorhombic structure with the lattice parameters $a = 1{,}472.3 \pm 1.4$, $b = 1{,}874.6 \pm 1.6$, and $c = 710.7 \pm 0.4$ pm and the calculated and experimental densities of 2.51 and 2.53 g cm^{-3}, respectively (Moore and Araki 1977a) [$a = 1{,}478$, $b = 1{,}878$, and $c = 714$ pm and the calculated and experimental densities of 2.53 and 2.48 g cm^{-3}, respectively (Moore and Ito 1974); $a = 1{,}475 \pm 2$, $b = 1{,}874 \pm 2$, and $c = 712 \pm 2$ pm and the calculated and experimental densities of 2.47 and 2.53 g cm^{-3}, respectively (Larsen 1940)].

$MgCaAl_2(PO_4)_2(OH)_4{\cdot}7H_2O$ crystallizes in the triclinic structure with the lattice parameters $a = 1{,}330.3 \pm 0.1$, $b = 2{,}702.0 \pm 0.2$, $c = 610.70 \pm 0.07$ pm and $\alpha = 89.64 \pm 0.1°$, $\beta = 83.44 \pm 0.1°$, $\gamma = 80.444 \pm 0.008°$ and the calculated and experimental densities of 2.332 and 2.47 g cm^{-3}, respectively, at room temperature and $a = 1{,}325.9 \pm 0.2$, $b = 2{,}698.0 \pm 0.4$, $c = 612.68 \pm 0.09$ pm and $\alpha = 89.17 \pm 0.01°$, $\beta = 83.13 \pm 0.02°$, $\gamma = 80.490 \pm 0.009°$ at –186°C (Mills et al. 2008).

$MgCa_4Al_4(PO_4)_6(OH)_4{\cdot}12H_2O$ crystallizes in the monoclinic structure with the lattice parameters $a = 1{,}000.4$, $b = 2{,}408.3$, $c = 623.5$ pm, and $\beta = 92°36'$ (Fanfani et al. 1976) [$a = 1{,}002.3 \pm 0.1$, $b = 2{,}412.1 \pm 0.3$, $c = 624.3 \pm 0.1$ pm, and $\beta = 91.55 \pm 0.01°$ (Moore and Araki 1974b); $a = 999 \pm 2$, $b = 2{,}410 \pm 2$, $c = 625 \pm 5$ pm, and $\beta = 91°28'$ and the calculated and experimental densities of 2.52 and 2.53 ± 0.05 g cm^{-3}, respectively (Larsen 1940)].

5.42 ALUMINUM–HYDROGEN–MAGNESIUM–CALCIUM–OXYGEN–MANGANESE–PHOSPHORUS

The $\mathbf{Mg_2CaAl_2Mn(PO_4)_4(OH)_2{\cdot}8H_2O}$ multinary compound [mineral whiteite-(CaMnMg)], which crystallizes in the monoclinic structure with the lattice parameters $a = 1{,}484.2 \pm 0.9$, $b = 697.6 \pm 0.1$, $c = 1{,}010.9 \pm 0.4$ pm, and $\beta = 112.59 \pm 0.05°$ and the calculated and experimental densities of 2.64 and 2.63 ± 0.02 g cm^{-3}, respectively, is formed in the Al–H–Mg–Ca–O–Mn–P system (Grice et al. 1989; Jambor and Grew 1990b).

5.43 ALUMINUM–HYDROGEN–MAGNESIUM–CALCIUM–OXYGEN–MANGANESE–IRON–PHOSPHORUS

The **(Mn,Ca)(Mg,Fe,Mn)Al$(PO_4)_2$(OH)·$4H_2O$** (mineral lun'okite), **Ca(Fe,Mn)Mg_2Al_2 $(PO_4)_4(OH)_2$·$8H_2O$** [mineral whiteite-(CaFeMg)], and **(Mn,Ca)Mn$(Fe,Mn,Mg)_2(Al,Fe)_2$ $(PO_4)_4(OH)_2$·$8H_2O$** (mineral rittmannite) multinary phases are formed in the Al–H–Mg–Ca–O–Mn–Fe–P system. First of them crystallizes in the orthorhombic structure with the lattice parameters a = 1,495 ± 5, b = 1,871 ± 2, and c = 696 ± 3 pm and the calculated and experimental densities of 2.69 and 2.66 g cm^{-3}, respectively (Voloshin et al. 1983; Dunn et al. 1984b).

Ca(Fe,Mn)$Mg_2Al_2(PO_4)_4(OH)_2$·$8H_2O$ crystallizes in the monoclinic structure with the lattice parameters a = 1,487.00 ± 0.15, b = 697.85 ± 0.05, c = 992.68 ± 0.10 pm, and β = 110.110 ± 0.001° and a calculated density of 2.597 g cm^{-3} (Capitelli et al. 2011) [a = 1,490 ± 4, b = 698 ± 2, c = 1,013 ± 2 pm, and β = 113°07′ ± 10′ and an experimental density of 2.58 g cm^{-3} (Moore and Ito 1978b; Fleischer et al. 1979c)].

(Mn,Ca)Mn$(Fe,Mn,Mg)_2(Al,Fe)_2(PO_4)_4(OH)_2$·$8H_2O$ also crystallizes in the monoclinic structure with the lattice parameters a = 1,501 ± 4, b = 689 ± 3, c = 1,016 ± 3 pm, and β = 112.82 ± 0.25° and the calculated and experimental densities of 2.83 and 2.81 ± 0.01 g cm^{-3}, respectively (Marzoni Fecia di Cossato et al. 1989; Jambor and Grew 1990b).

5.44 ALUMINUM–HYDROGEN–MAGNESIUM–STRONTIUM–BARIUM–OXYGEN–MANGANESE–IRON–PHOSPHORUS

The **(Ba,Sr)$(Mn,Fe,Mg)_2Al_2(PO_4)_3(OH)_3$** multinary phase (mineral bjarebyite), which crystallizes in the monoclinic structure with the lattice parameters a = 893.0 ± 1.4, b = 1,207.3 ± 2.4, c = 491.7 ± 0.9 pm, and β = 100.15 ± 0.13° and the calculated and experimental densities of 4.02 and 3.95 g cm^{-3}, respectively, is formed in the Al–H–Mg–Sr–Ba–O–Mn–Fe–P system (Moore et al. 1973; Fleischer 1974b; Moore and Araki 1974a).

5.45 ALUMINUM–HYDROGEN–MAGNESIUM–BARIUM–CARBON–OXYGEN–PHOSPHORUS

The **Ba(Al,Mg)$(PO_4,CO_3)(OH)_2$·H_2O** multinary phase (mineral krasnovite), which crystallizes in the orthorhombic structure with the lattice parameters a = 893.9 ± 0.2, b = 566.9 ± 0.3, and c = 1,107.3 ± 0.3 pm and the calculated and experimental densities of 3.691 and 3.70 ± 0.05 g cm^{-3}, respectively, is formed in the Al–H–Mg–Ba–C–O–P system (Britvin et al. 1996; Jambor et al. 1997c).

5.46 ALUMINUM–HYDROGEN–MAGNESIUM–BARIUM–OXYGEN–MANGANESE–IRON–PHOSPHORUS

The **Ba$(Fe,Mn,Mg)_2Al_2(PO_4)_3(OH)_3$** multinary phase (mineral kulanite), which crystallizes in the monoclinic structure with the lattice parameters a = 901.4 ± 0.1, b = 1,207.4 ± 0.1, c = 492.6 ± 0.1 pm, and β = 100.48 ± 0.01° (Cooper and Hawthorne 1994a; Jambor and

Roberts 1995a) [$a = 902.4 \pm 0.1$, $b = 1{,}207.9 \pm 0.4$, $c = 492.4 \pm 0.1$ pm, and $\beta = 100.462°$ (Yang et al. 1986; Jambor and Grew 1990a); in the triclinic structure with the lattice parameters $a = 903.2$, $b = 1{,}211.9$, $c = 493.6$ pm, and $\alpha \approx 90°$, $\beta = 100°23'$, $\gamma \approx 90°$ and the calculated and experimental densities of 3.92 and 3.91 ± 0.03 g cm^{-3}, respectively (Mandarino and Sturman 1976; Fleischer and Mandarino 1977)], is formed in the Al–H–Mg–Ba–O–Mn–Fe–P system.

5.47 ALUMINUM–HYDROGEN–MAGNESIUM–BARIUM–OXYGEN–IRON–PHOSPHORUS

The **$Ba(Mg,Fe)_2Al_2(PO_4)_3(OH)_3$** multinary compound (mineral penikisite), which crystallizes in the triclinic structure with the lattice parameters $a = 899.9$, $b = 1{,}206.9$, $c = 492.1$ pm, and $\alpha \approx 90°$, $\beta = 100°31'$, $\gamma \approx 90°$ and the calculated and experimental densities of 3.82 and 3.79 ± 0.02 g cm^{-3}, respectively, is formed in the Al–H–Mg–Ba–O–Fe–P system (Mandarino et al. 1977; Fleischer et al. 1979b).

5.48 ALUMINUM–HYDROGEN–MAGNESIUM–ZINC–OXYGEN–MANGANESE–IRON–PHOSPHORUS

The **$MgMn_4(Fe_{0.5}Al_{0.5})_4Zn_4(PO_4)_8(OH)_4(H_2O)_{20}$** multinary compound (mineral ferraioloite), which crystallizes in the monoclinic structure with the lattice parameters $a = 2{,}532.0 \pm 0.6$, $b = 634.5 \pm 0.6$, $c = 1{,}526.7 \pm 0.6$ pm, and $\beta = 91.031 \pm 0.005°$ at room temperature and a calculated density of 2.59 g cm^{-3}, and $a = 2{,}533.33 \pm 0.03$, $b = 629.9 \pm 0.1$, $c = 1{,}516.1 \pm 0.3$ pm, and $\beta = 90.93 \pm 0.03°$ at 100 K, is formed in the Al–H–Mg–Zn–O–Mn–Fe–P system (Mills et al. 2015, 2016b; Cámara et al. 2016).

5.49 ALUMINUM–HYDROGEN–MAGNESIUM–SILICON–OXYGEN–MANGANESE–PHOSPHORUS

The **$Mn_8[Al_{10}(Mn,Mg)][Si_{11}P]O_{44}(OH)_{12}$** multinary phase (mineral lavoisierite), which crystallizes in the orthorhombic structure with the lattice parameters $a = 868.91 \pm 0.10$, $b = 577.55 \pm 0.03$, and $c = 3{,}695.04 \pm 0.20$ pm and a calculated density of 3.576 g cm^{-3}, is formed in the Al–H–Mg–Si–O–Mn–P system (Orlandi et al. 2012, 2013; Cámara et al. 2014b).

5.50 ALUMINUM–HYDROGEN–MAGNESIUM–OXYGEN–PHOSPHORUS

The **$MgAl_2(PO_4)_2(OH)_2$** (mineral lazulite), **$MgAl_2(PO_4)_2(OH)_2{\cdot}8H_2O$** (mineral gordonite), and **$Mg_5Al_{12}(PO_4)_8(OH)_{22}{\cdot}32H_2O$** (mineral aldermanite) quinary compounds are formed in the Al–H–Mg–O–P system. $MgAl_2(PO_4)_2(OH)_2$ crystallizes in the monoclinic structure with the lattice parameters $a = 715.1 \pm 0.1$, $b = 727.1 \pm 0.1$, $c = 723.5 \pm 0.1$ pm, and $\beta = 120.55 \pm 0.01°$ (Blanchard and Abernathy 1980) [$a = 712$, 714, and 716, $b = 724$, 727, and 725, $c = 710$, 716, and 714 pm and $\beta = 118°55'$, $119°18'$, and $118°47'$ for three various samples and the calculated and experimental densities of 3.14 and 3.12 g cm^{-3}, respectively (Berry 1948); $a = 716 \pm 2$, $b = 726 \pm 2$, $c = 724 \pm 2$ pm, and $\beta = 120°40' \pm 5'$ and the calculated and experimental densities of 3.14 and 3.12 g cm^{-3}, respectively (Lindberg and Christ 1959); $a = 715$, $b = 728$, $c = 725$ pm, and $\beta = 120°35'$ (Campbell 1962); $a = 714.6 \pm 0.3$, $b = 725.7 \pm 0.3$, $c = 723.1 \pm 0.1$ pm, and $\beta = 120°28' \pm 3'$ and a calculated density of 3.105 g cm^{-3} (Wise and Loh 1976)]. The heat capacity of synthetic lazulite has been measured between 35 and

298 K by means of low-temperature adiabatic calorimetry (Brunet et al. 2004). A lazulite third-law entropy (S^0_{298}) value of 204 ± 3 J K^{-1}mol^{-1} was retrieved after extrapolation of the lazulite heat capacity by a cubic temperature function, down to absolute temperature.

$MgAl_2(PO_4)_2(OH)_2{\cdot}8H_2O$ crystallizes in the triclinic structure with the lattice parameters a = 524.6 ± 0.2, b = 1,053.2 ± 0.5, c = 697.5 ± 0.3 pm, and α = 107.51 ± 0.03°, β = 111.03 ± 0.03°, γ = 72.21 ± 0.03° and the calculated and experimental densities of 2.22 and 2.23 g cm^{-3}, respectively (Leavens and Rheingold 1988).

$Mg_5Al_{12}(PO_4)_8(OH)_{22}{\cdot}32H_2O$ crystallizes in the orthorhombic structure with the lattice parameters a = 1,500.0 ± 0.7, b = 833.0 ± 0.6, and c = 2,660 ± 1 pm and a calculated density of 2.15 g cm^{-3} (Cabri et al. 1981; Harrowfield et al. 1981).

Phase relations in the system **Al_2O_3–$AlPO_4$–$Mg_3(PO_4)_2$–H_2O** were studied experimentally between 0.01 and 0.31 GPa at temperatures between 487°C and 704°C (Cemič and Schmid-Beurmann 1995). Two univariant reactions, which define the upper thermal stability of pure lazulite and of lazulite in the presence of $Mg_3(PO_4)_2$ and Al_2O_3, were determined by bracketing experiments. A combination of the obtained thermodynamic data with the tabulated standard enthalpies of formation and third-law entropies of Al_2O_3, $AlPO_4$, $Mg_3(PO_4)_2$, and H_2O yield ΔH^0_f = −4,532 ± 7 kJ·mol^{-1} and S^0_f = 139 ± 7 J K^{-1}mol^{-1} as the standard enthalpy of formation and third-law entropy of synthetic lazulite. Extrapolation of the experimentally determined univariant equilibria to higher pressures and temperatures predicts an invariant point at 0.36 GPa and 710°C, where lazulite, Al_2O_3, $MgAlPO_5$, $AlPO_4$, $Mg_3(PO_4)_2$, and H_2O coexist.

5.51 ALUMINUM–HYDROGEN–MAGNESIUM–OXYGEN–IRON–PHOSPHORUS

The **$(Fe,Mg)Al_2(PO_4)_2(OH)_2$** (mineral scorzalite) and **$(Mg,Fe)_3(Al,Fe)_4(PO_4)_4(OH)_6{\cdot}2H_2O$** (mineral souzalite) multinary phases are formed in the Al–H–Mg–O–Fe–P system. First of them crystallizes in the monoclinic structure with the lattice parameters a = 715.6 + 0.1, b = 730.6 ± 0.1, c = 724.9 ± 0.1 pm, and β = 120.59 ± 0.01° and a calculated density of 3.33 g cm^{-3} (Blanchard and Abernathy 1980) [a = 715, b = 732, c = 714 pm and β = 119°00′ and a calculated density of 3.39 g cm^{-3} (Berry 1948); a = 715 ± 2, b = 731 ± 2, c = 725 ± 2 pm, and β = 120°35′ ± 5′ and the calculated and experimental densities of 3.32 and 3.33 g cm^{-3}, respectively (Lindberg and Christ 1959); a = 716.0 ± 0.3, b = 729.8 ± 0.6, c = 726.4 ± 0.5 pm, and β = 120°33′ ± 3′ and a calculated density of 3.390 g cm^{-3} (Wise and Loh 1976)].

The second phase crystallizes in the triclinic structure with the lattice parameters a = 722.23 ± 0.01, b = 1,178.01 ± 0.01, c = 511.69 ± 0.01 pm, and α = 90.158 ± 0.001°, β = 109.938 ± 0.001°, γ = 81.330 ± 0.001° (Le Bail et al. 2003) [a = 1,174 ± 1, b = 511 ± 1, c = 1,358 ± 1 pm, and α = 90°55′ ± 5′, β = 99°05′ ± 5′, γ = 90°20′ ± 5′ and a calculated density of 3.07 g cm^{-3} (Sturman et al. 1981)].

5.52 ALUMINUM–HYDROGEN–CALCIUM–STRONTIUM–BARIUM–ARSENIC–OXYGEN–FLUORINE–PHOSPHORUS

The **$(Sr,Ca,Ba)Al_3(AsO_4,PO_4)_2(OH,F)_5{\cdot}H_2O$** multinary phase (mineral arsenogoyazite), which crystallizes in the rhombohedral structure with the lattice parameters a = 704 pm and $\alpha = 60.60°$ or $a = 710$ and $c = 1{,}716$ pm (in hexagonal setting) and the calculated and experimental densities of 3.33 and 3.35 ± 0.05 g cm^{-3}, respectively, is formed in the Al–H–Ca–Sr–Ba–As–O–F–P system (Walenta and Dun 1984; Hawthorne et al. 1986).

5.53 ALUMINUM–HYDROGEN–CALCIUM–STRONTIUM–SILICON–OXYGEN–MANGANESE–IRON–PHOSPHORUS

The **$(Ca,Sr)Mn(Al,Fe)_4[H(Si,P)O_4](PO_4)_3(OH)_4$** multinary phase (mineral attakolite), which crystallizes in the monoclinic structure with the lattice parameters $a = 1{,}718.8 \pm 0.4$, $b = 1{,}147.7 \pm 0.8$, $c = 732.2 \pm 0.5$ pm, and $\beta = 113.83 \pm 0.04°$ (Grice and Dunn 1992) [in the orthorhombic structure with the lattice parameters $a = 1{,}138$, $b = 1{,}322$, $c = 1{,}408$ pm and a calculated density of 3.229 g cm^{-3} (Fleischer 1966)], is formed in the Al–H–Ca–Sr–Si–O–Mn–Fe–P system.

5.54 ALUMINUM–HYDROGEN–CALCIUM–STRONTIUM–ARSENIC–OXYGEN–PHOSPHORUS

The **$(Ca,Sr)Al_3[(As,P)O_4]_2(OH)_5{\cdot}H_2O$** multinary phase (mineral arsenocrandallite), which crystallizes in the rhombohedral structure with the lattice parameters $a = 704$–706 pm and $\alpha = 60.18$–$60.19°$ or $a = 706$–708 and $c = 1{,}722$–$1{,}727$ pm (in hexagonal setting) and the calculated and experimental densities of 3.30 and 3.25 ± 0.10 g cm^{-3}, respectively, is formed in the Al–H–Ca–Sr–As–O–P system (Walenta 1981; Fleischer et al. 1982b).

5.55 ALUMINUM–HYDROGEN–CALCIUM–STRONTIUM–OXYGEN–PHOSPHORUS

The **$(Sr,Ca)_2Al(PO_4)_2(OH)$** multinary phase (mineral goedkenite), which crystallizes in the monoclinic structure with the lattice parameters $a = 845 \pm 2$, $b = 574 \pm 2$, $c = 726 \pm 2$ pm, and $\beta = 113.7 \pm 0.1°$ and a calculated density of 3.83 g cm^{-3}, is formed in the Al–H–Ca–Sr–O–P system (Moore et al. 1975b).

5.56 ALUMINUM–HYDROGEN–CALCIUM–BARIUM–OXYGEN–MANGANESE–IRON–PHOSPHORUS

The **$(Ca,Ba)Ca_8(Fe,Mn)_4Al_2(PO_4)_{10}(OH)$** multinary phase (mineral samuelsonite), which crystallizes in the monoclinic structure with the lattice parameters $a = 1{,}862.1 \pm 0.3$, $b = 684.2 \pm 0.1$, $c = 1{,}406.6 \pm 0.2$ pm, and $\beta = 112°30' \pm 1'$ and the calculated and experimental densities of 3.22 and 3.24 ± 0.05 g cm^{-3}, respectively (Fransolet et al. 1992) [$a = 1{,}849.5 \pm 1.0$, $b = 680.5 \pm 0.4$, $c = 1{,}400.0 \pm 0.8$ pm, and $\beta = 112.75 \pm 0.06°$ and the calculated and experimental densities of 3.355 and 3.353 g cm^{-3}, respectively (Moore and Araki 1975b, 1977b)], is formed in the Al–H–Ca–Ba–O–Mn–Fe–P system.

5.57 ALUMINUM–HYDROGEN–CALCIUM–GALLIUM–LEAD–OXYGEN–PHOSPHORUS

The **$(Pb_{0.87}Ca_{0.13})H(Al_{1.95}Ga_{1.05})(PO_4)_2(OH)$** multinary phase (Ga-rich plumbogummite), which crystallizes in the trigonal structure with the lattice parameters $a = 707.52 \pm 0.19$ and $c = 1{,}681.8 \pm 0.4$ pm, is formed in the Al–H–Ca–Ga–Pb–O–P system (Mills et al. 2009).

5.58 ALUMINUM–HYDROGEN–CALCIUM–YTTRIUM–THORIUM–SILICON–OXYGEN–SULFUR–PHOSPHORUS

The $\mathbf{Ca(Y,Th)Al_5(SiO_4)_2(PO_4,SO_4)_2(OH)_7{\cdot}6H_2O}$ multinary phase [mineral saryarkite-(Y)], which crystallizes in the tetragonal structure with the lattice parameters $a = 821.3 \pm 0.2$ and $c = 655 \pm 1$ pm and the calculated and experimental densities of 3.35 and 3.07–3.15 g cm^{-3}, respectively, is formed in the Al–H–Ca–Y–Th–Si–O–S–P system (Fleischer 1964; Krol' et al. 1964).

5.59 ALUMINUM–HYDROGEN–CALCIUM–SILICON–OXYGEN–PHOSPHORUS

The $\mathbf{Ca_3Al_7(SiO_4)_3(PO_4)_4(OH)_3{\cdot}16.5H_2O}$ multinary compound (mineral perhamite), which crystallizes in the trigonal structure with the lattice parameters a = 702.1 ± 0.1 and c = 2,021.8 ± 0.1 pm and a calculated density of 2.49 g cm^{-3} (Locock et al. 2006a; Mills et al. 2006) [in the hexagonal structure with the lattice parameters a = 702.2 ± 0.1 and c = 2,018.2 ± 0.5 pm and the calculated and experimental densities of 2.53 and 2.64 g cm^{-3}, respectively (Dunn and Appleman 1977; Fleischer et al. 1978b)], is formed in the Al–H–Ca–Si–O–P system.

5.60 ALUMINUM–HYDROGEN–CALCIUM–SILICON–OXYGEN–FLUORINE–PHOSPHORUS

The $\mathbf{Ca_3Al_{7.7}Si_3P_4O_{23.5}(OH)_{12.1}F_2{\cdot}8H_2O}$ (mineral krásnoite) and $\mathbf{Ca_{10}Al_{24}(SiO_4)_6(PO_4)_{14}(O,F)_{13}{\cdot}72H_2O}$ (mineral viséite) multinary phases are formed in the Al–H–Ca–Si–O–F–P system. First of them crystallizes in the trigonal structure with the lattice parameters a = 699.56 ± 0.04 and c = 2,020.0 ± 0.2 pm and the calculated and experimental densities of 2.476 and 2.48 ± 0.04 g cm^{-3}, respectively (Mills et al. 2011d, 2012d).

Second compound crystallizes in the cubic or pseudocubic structure with the lattice parameter a = 1,365 pm and the calculated and experimental densities of 2.17 and 2.2 g cm^{-3}, respectively (McConnell 1952).

5.61 ALUMINUM–HYDROGEN–CALCIUM–OXYGEN–PHOSPHORUS

The $\mathbf{Ca_2Al(PO_4)_2OH}$ (mineral bearthite), $\mathbf{CaAl(PO_4)(OH)_2{\cdot}H_2O}$ (mineral foggite), $\mathbf{CaAl_2(PO_4)_2(OH)_2{\cdot}H_2O}$ (mineral gatumbaite), and $\mathbf{CaAl_3(PO_4)(PO_3OH)(OH)_6}$ (mineral crandallite) quinary compounds are formed in the Al–H–Ca–O–P system. $Ca_2Al(PO_4)_2OH$ crystallizes in the monoclinic structure with the lattice parameters a = 722.8 ± 0.3, b = 567.7 ± 0.2, c = 824.0 ± 0.3 pm, and β = 112.30 ± 0.03° pm (Brunet and Chopin 1995) [a = 723.1 ± 0.3, b = 573.4 ± 0.2, c = 826.3 ± 0.4 pm, and β = 112.57 ± 0.08° pm and a calculated density of 3.25 g cm^{-3} (Chopin et al. 1993; Jambor and Vanko 1993)]. This compound has been synthesized from a stoichiometric mixture of γ-Al_2O_3 and $CaHPO_4{\cdot}2H_2O$ between about 585°C at 0.4 GPa and up to 815°C at 2.45 GPa (Brunet and Chopin 1995). The enthalpy and entropy of formation of the title compound are equal to −4,327.25 kJ mol^{-1} and 214.5 J K^{-1}mol^{-1}.

$CaAl(PO_4)(OH)_2{\cdot}H_2O$ crystallizes in the orthorhombic structure with the lattice parameters a = 926.4 ± 0.1, b = 2,133.4 ± 0.8, and c = 519.7 ± 0.7 pm (Onac et al. 2005) [a = 927.0 ± 0.2, b = 2,132.4 ± 0.7, and c = 519.0 ± 0.2 pm and the calculated and experimental densities of 2.771 and 2.78 ± 0.01 g cm^{-3}, respectively (Moore and Araki 1975b, 1975c)].

$CaAl_2(PO_4)_2(OH)_2{\cdot}H_2O$ crystallizes in the monoclinic structure with the lattice parameters $a = 690.7$, $b = 509.5$, $c = 1{,}076.4$ pm, and $\beta = 91°03'$ and the calculated and experimental densities of 2.95 and 2.92 g cm^{-3}, respectively (Fleischer et al. 1978b; Knorring von and Fransolet 1977).

$CaAl_3(PO_4)(PO_3OH)(OH)_6$ is stable up to 200°C (Yuh and Rockett 1981) and crystallizes in the hexagonal structure with the lattice parameters $a = 701.7 \pm 0.1$ and 701.3 ± 0.1, $c = 1{,}625.2 \pm 0.4$ and $1{,}619.6 \pm 0.3$ pm for two samples (Blanchard 1972) [$a = 700.5 \pm 1.5$ and $c = 1{,}619.2 \pm 3.2$ pm (Blount 1974); $a = 700.7$ and $c = 1{,}621.6$ pm (Gilkes and Palmer 1983)]. This compound was prepared by interaction of $NaAlO_2$, $CaCl_2$, H_2O, and H_3PO_4 taken in stoichiometric ratios (Gilkes and Palmer 1983). The resultant gel was crystallized in a stainless steel pressure vessel with a glass insert for 2 weeks at 200°C and 1.5 MPa.

5.62 ALUMINUM–HYDROGEN–CALCIUM–OXYGEN–SULFUR–PHOSPHORUS

The $\mathbf{CaAl_3(PO_4)(SO_4)(OH)_6}$ multinary compound (mineral woodhouseite), which crystallizes in the rhombohedral structure with the lattice parameters $a = 675$ and $\alpha = 62°04'$ or $a = 696.1$ and $c = 1{,}627$ pm in the hexagonal setting and the calculated and experimental densities of 3.003 g cm^{-3}, is formed in the Al–H–Ca–O–S–P system (Pabst 1947).

5.63 ALUMINUM–HYDROGEN–CALCIUM–OXYGEN–FLUORINE–PHOSPHORUS

The $\mathbf{Ca_2Al_7(PO_4)_2(PO_3OH)_2(OH,F)_{15}{\cdot}8H_2O}$ (mineral iangreyite) and $\mathbf{Ca_4Al_2(PO_4)_2F_8{\cdot}5H_2O}$ (mineral galliskiite) multinary phases are formed in the Al–H–Ca–O–F–P system. First of them crystallizes in the trigonal structure with the lattice parameters $a = 698.8 \pm 0.1$ and 698.9 ± 0.1, $c = 1{,}670.7 \pm 0.3$ and $1{,}678.2 \pm 0.4$ pm for two samples and the calculated and experimental densities of 2.451 and 2.46 ± 0.03 g cm^{-3}, respectively (Mills et al. 2011a; Welch et al. 2013).

$Ca_4Al_2(PO_4)_2F_8{\cdot}5H_2O$ crystallizes in the triclinic structure with the lattice parameters $a = 619.33 \pm 0.07$, $b = 987.1 \pm 0.1$, $c = 1{,}358.0 \pm 0.2$ pm and $\alpha = 89.716 \pm 0.003°$, $\beta = 75.303 \pm 0.004°$, $\gamma = 88.683 \pm 0.004°$ and the calculated and experimental densities of 2.674 and 2.67 ± 0.03 g cm^{-3}, respectively (Kampf et al. 2010).

5.64 ALUMINUM–HYDROGEN–CALCIUM–OXYGEN–MANGANESE–PHOSPHORUS

The $\mathbf{CaMn^{II}Mn_2Al_2(PO_4)_4(OH)_2{\cdot}8H_2O}$ multinary compound [mineral whiteite-(CaMnMn)], which crystallizes in the monoclinic structure with the lattice parameters $a = 1{,}502.0 \pm 0.5$, $b = 695.9 \pm 0.2$, $c = 1{,}013.7 \pm 0.3$ pm, and $\beta = 111.740 \pm 0.004°$ and the calculated and experimental densities of 2.768 and 2.70 ± 0.03 g cm^{-3}, respectively, is formed in the Al–H–Ca–O–Mn–P system (Yakovenchuk et al. 2012; Cámara et al. 2015).

5.65 ALUMINUM–HYDROGEN–CALCIUM–OXYGEN–MANGANESE–IRON–PHOSPHORUS

The $\mathbf{(Ca,Mn)_4(Fe,Mn)Al_4(PO_4)_6(OH)_4{\cdot}12H_2O}$ multinary phase (mineral kingsmountite), which crystallizes in the monoclinic structure with the lattice parameters $a = 1{,}002.9 \pm 0.6$,

$b = 2,446 \pm 1$, $c = 625.8 \pm 0.4$ pm, and $\beta = 91.16 \pm 0.07°$ and the calculated and experimental densities of 2.58 and 2.51 ± 0.03 g cm^{-3}, respectively, is formed in the Al–H–Ca–O–Mn–Fe–P system (Dunn et al. 1979; Fleischer and Cabri 1981).

5.66 ALUMINUM–HYDROGEN–STRONTIUM–LEAD–OXYGEN–SULFUR–PHOSPHORUS

The **$(Pb,Sr)Al_3(PO_4)(SO_4)(OH)_6$** multinary phase (mineral hinsdalite) (Baker 1963; Stanley 1987; Kolitsch et al. 1999), which crystallizes in the rhombohedral structure with the lattice parameters $a = 702.9 \pm 0.4$ and $c = 1,678.9 \pm 0.4$ pm (in hexagonal setting) (Kolitsch et al. 1999) [$a = 690$ pm, $\alpha = 61°08'$ and a calculated density of 4.06 g cm^{-3} and $a = 688$ pm, $\alpha = 61°08'$ and a calculated density of 3.92 g cm^{-3} for two samples, respectively (Baker 1963)], is formed in the Al–H–Sr–Pb–O–S–P system.

5.67 ALUMINUM–HYDROGEN–STRONTIUM–OXYGEN–PHOSPHORUS

The **$SrAl_3(PO_4)(PO_3OH)(OH)_6$** quinary compound (mineral goyazite), which crystallizes in the hexagonal structure with the lattice parameters $a = 702.4 \pm 0.1$ and $c = 1,661.1 \pm 0.3$ pm (Schwab et al. 1990) [$a = 698$, $c = 1,654$ pm and a calculated density of 3.15 g cm^{-3} (Mrose 1953); $a = 701.3$ and $c = 1,665.0$ pm (Gilkes and Palmer 1983); $a = 701.5 \pm 0.3$ and $c = 1,655.8 \pm 0.6$ pm (Kato 1987); in the rhombohedral structure with the lattice parameters $a = 698.2 \pm 0.1$ and $c = 1,654 \pm 2$ pm in hexagonal setting and the calculated and experimental densities of 3.392 and 3.386 g cm^{-3}, respectively (McKie 1962)], is formed in the Al–H–Sr–O–P system. This compound was prepared by the interaction of $NaAlO_2$, $SrCl_2$, H_2O, and H_3PO_4 taken in stoichiometric ratios (Gilkes and Palmer 1983). The resultant gel was crystallized in a stainless steel pressure vessel with a glass insert for 2 weeks at 200°C and 1.5 MPa. Goyazite could also be synthesized under hydrothermal conditions at 200°C and 1.5 MPa starting from H_3PO_4 and oxides or hydroxides of the other elements (Schwab et al. 1990).

5.68 ALUMINUM–HYDROGEN–STRONTIUM–OXYGEN–SULFUR–PHOSPHORUS

The **$SrAl_3(PO_4)(SO_4)(OH)_6$** multinary compound (mineral svanbergite), which crystallizes in the trigonal structure with the lattice parameters $a = 699.2 \pm 0.2$ and $c = 1,656.7 \pm 0.5$ pm (Kato and Miúra 1977) [in the rhombohedral structure with the lattice parameters $a = 689$ and $\alpha = 60°38'$ and the calculated and experimental densities of 3.18 and 3.20 g cm^{-3}, respectively, or $a = 696 \pm 3$ and $c = 1,680$ pm in hexagonal setting (Pabst 1947)], is formed in the Al–H–Sr–O–S–P system.

5.69 ALUMINUM–HYDROGEN–BARIUM–TITANIUM–OXYGEN–FLUORINE–PHOSPHORUS

The **$Ba(Al,Ti)(PO_4)(OH,O)F$** multinary phase (mineral curetonite), which crystallizes in the monoclinic structure with the lattice parameters $a = 697.7 \pm 0.2$, $b = 1,256.4 \pm 0.4$, $c = 522.3 \pm 0.1$ pm, and $\beta = 102.15 \pm 0.02°$ (Cooper and Hawthorne 1994b) [$a = 695.7$, $b = 1,255$, $c = 522$ pm, and $\beta = 102°0.2'$ and the calculated and experimental densities of 4.31 and

4.42 ± 0.05 g cm^{-3}, respectively (Williams 1979; Fleischer et al. 1980d)], is formed in the Al–H–Ba–Ti–O–F–P system.

5.70 ALUMINUM–HYDROGEN–BARIUM–OXYGEN–PHOSPHORUS

The $\mathbf{BaAl_2(PO_4)_2(OH)_2}$ (mineral jagowerite) and $\mathbf{BaAl_3(PO_4)(PO_3OH)(OH)_6}$ (mineral gorceixite) quinary compounds are formed in the Al–H–Ba–O–P system. $BaAl_2(PO_4)_2(OH)_2$ crystallizes in the triclinic structure with the lattice parameters $a = 604.9 \pm 0.2$, $b = 696.4 \pm 0.3$, $c = 497.1 \pm 0.2$ pm and $\alpha = 116.51 \pm 0.04°$, $\beta = 86.06 \pm 0.04°$, $\gamma = 112.59 \pm 0.03°$ and the calculated and experimental densities of 4.05 and 4.01 g cm^{-3}, respectively (Meagher et al. 1973, 1974; Fleischer et al. 1976b).

$BaAl_3(PO_4)(PO_3OH)(OH)_6$ crystallizes in the trigonal structure with the lattice parameters $a = 705.38 \pm 0.03$ and $c = 1{,}727.46 \pm 0.06$ pm (Dzikowski et al. 2006; Locock et al. 2006a) [in the hexagonal structure with the lattice parameters $a = 705.9 \pm 0.1$ and $c = 1{,}727.5 \pm 0.2$ pm (Schwab et al. 1990)]. This compound was synthesized under hydrothermal conditions at 200°C and 1.5 MPa starting from H_3PO_4 and oxides or hydroxides of the other elements (Schwab et al. 1990).

5.71 ALUMINUM–HYDROGEN–ZINC–OXYGEN–PHOSPHORUS

The $\mathbf{ZnAl_2(PO_4)_2(OH)_2{\cdot}3H_2O}$ (mineral kleemanite) and $\mathbf{ZnAl_6(PO_4)_4(OH)_8{\cdot}4H_2O}$ (mineral faustite) quinary compounds are formed in the Al–H–Zn–O–P system. First of them crystallizes in the monoclinic structure with the lattice parameters $a = 729.0 \pm 0.6$, $b = 719.4 \pm 0.6$, $c = 976.2 \pm 0.9$ pm, and $\beta = 110.20 \pm 0.04°$ and a calculated density of 2.76 g cm^{-3} (Fleischer et al. 1979a; Pilkington et al. 1979).

$ZnAl_6(PO_4)_4(OH)_8{\cdot}4H_2O$ crystallizes in the triclinic structure with the lattice parameters $a = 741.9 \pm 0.2$, $b = 762.9 \pm 0.3$, $c = 990.5 \pm 0.3$ pm and $\alpha = 69.17 \pm 0.02°$, $\beta = 69.88 \pm 0.02°$, $\gamma = 64.98 \pm 0.02°$ and the calculated and experimental densities of 2.929 and 2.92 (Erd et al. 1953) g cm^{-3}, respectively (Kolitsch and Giester 2000).

5.72 ALUMINUM–HYDROGEN–ZINC–OXYGEN–IRON–PHOSPHORUS

The $\mathbf{(Fe^{II},Zn)Al_6(PO_4)_4(OH)_8{\cdot}4H_2O}$ (mineral aheylite), which crystallizes in the triclinic structure with the lattice parameters $a = 740.0 \pm 0.1$, $b = 989.6 \pm 0.1$, $c = 762.7 \pm 0.1$ pm and $\alpha = 110.87°$, $\beta = 115.00°$, $\gamma = 69.96°$ and the calculated and experimental densities of 2.90 and 2.84 ± 0.02 g cm^{-3}, respectively, is formed in the Al–H–Zn–O–Fe–P system (Foord and Taggart, 1998).

5.73 ALUMINUM–HYDROGEN–GALLIUM–INDIUM–CARBON–ARSENIC–ANTIMONY–PHOSPHORUS

It was shown by Zhang et al. (2001) that with the help of Thermo-Calc software and based on thermodynamic database of this multinary system, the compositional spaces under conditions of different temperatures and pressures can be calculated and analyzed in the cases of stable and metastable equilibria, or irreversible reactions. Such information is useful to compute-assisted design for metal-organic vapor-phase epitaxy (MOVPE) processing of III-V semiconductors.

5.74 ALUMINUM–HYDROGEN–GALLIUM–GERMANIUM–LEAD–OXYGEN–PHOSPHORUS

The **$Pb(Ga,Al,Ge)_3(PO_4)_2(OH)_6$** multinary compound (mineral galloplumbogummite), which crystallizes in the trigonal structure with the lattice parameters a = 708.3 ± 0.5 and c = 1,674.2 ± 0.3 pm, is formed in the Al–H–Ga–Ge–Pb–O–P system (Schlüter and Malcherek 2011; Belakovskiy et al. 2016a).

5.75 ALUMINUM–HYDROGEN–LANTHANUM–CERIUM–ARSENIC–OXYGEN–PHOSPHORUS

The **$(Ce,La)Al_3(AsO_4,PO_4)_2(OH)_6$** multinary phase [mineral arsenoflorencite-(Ce)], which crystallizes in the trigonal structure with the lattice parameters a = 702.9 and c = 1,651.7 pm and the calculated and experimental densities of 4.091 and 4.096 g cm^{-3}, respectively, is formed in the Al–H–La–Ce–As–O–P system (Nickel and Temperly 1987; Jambor et al. 1988).

5.76 ALUMINUM–HYDROGEN–LANTHANUM–CERIUM–OXYGEN–PHOSPHORUS

The **$(La,Ce)Al_3(PO_4)_2(OH)_6$** [mineral florencite-(La)] and **$(Ce,La)Al_3(PO_4)_2(OH)_6$** [mineral florencite-(Ce)] multinary phases are formed in the Al–H–La–Ce–O–P system. First of them crystallizes in the rhombohedral structure with the lattice parameters a = 698.7 ± 0.2 and 697.9 ± 0.4, c = 1,624.8 ± 0.6 and 1,625 ± 1 pm in a hexagonal setting for two samples and an experimental density of 3.52 ± 0.2 g cm^{-3} (Lefebvre and Gasparrini 1980).

$(Ce,La)Al_3(PO_4)_2(OH)_6$ also crystallizes in the rhombohedral structure with the lattice parameters a = 699.1 ± 0.1 and c = 1,625 ± 1 pm in a hexagonal setting and the calculated and experimental densities of 3.69 and 3.54 g cm^{-3}, respectively (Pouliot and Hofmann 1981) [a = 697.1 ± 0.4 and c = 1,642 ± 13 pm in a hexagonal setting and the calculated and experimental densities of 3.471 and 3.457 g cm^{-3}, respectively (McKie 1962); in the trigonal structure with the lattice parameters a = 696.0 ± 0.4 and c = 1,633 ± 3 pm (Frank-Kamenetskiy et al. 1953)]. This phase remains stable in the presence of H_2O up to 535 ± 10°C at $p(H_2O)$ = 20 MPa and up to 565 ± 10°C at $p(H_2O)$ = 0.25 GPa (McKie 1962).

5.77 ALUMINUM–HYDROGEN–CERIUM–NEODYMIUM–OXYGEN–PHOSPHORUS

The **$(Nd,Ce)Al_3(PO_4)_2(OH)_6$** multinary phase [mineral florencite-(Nd)], which crystallizes in the trigonal structure with the lattice parameters a = 699.2 ± 0.1 and c = 1,645.4 ± 0.7 pm and a calculated density of 3.70 g cm^{-3}, is formed in the Al–H–Ce–Nd–O–P system (Fitzpatrick 1986).

5.78 ALUMINUM–HYDROGEN–NEODYMIUM–SAMARIUM–OXYGEN–PHOSPHORUS

The **$(Sm,Nd)Al_3(PO_4)_2(OH)_6$** multinary phase [mineral florencite-(Sm)], which crystallizes in the trigonal structure with the lattice parameters a = 697.2 ± 0.4 and c = 1,618.2 ± 0.7 pm and the calculated and experimental densities of 3.666 and 3.753 (for two mineral compositions) and 3.60 ± 0.01 g cm^{-3}, respectively, is formed in the Al–H–Nd–Sm–O–P system (Repina et al. 2011; Belakovskiy et al. 2012).

5.79 ALUMINUM–HYDROGEN–THORIUM–URANIUM–OXYGEN–PHOSPHORUS

The $\mathbf{ThAl(UO_2)_7O_2(PO_4)_4(OH)_5{\cdot}15H_2O}$ multinary phase (mineral althupite), which crystallizes in the triclinic structure with the lattice parameters a = 1,093.5 ± 0.3, b = 1,856.7 ± 0.4, c = 1,350.4 ± 0.3 pm and α = 72.64 ± 0.02°, β = 68.20 ± 0.02°, γ = 84.21 ± 0.02° and the calculated and experimental densities of 3.98 and 3.9 ± 0.1 g cm^{-3}, respectively, is formed in the Al–H–Th–U–O–P system (Piret and Deliens 1987; Hawthorne et al. 1988).

5.80 ALUMINUM–HYDROGEN–THORIUM–SILICON–LEAD–OXYGEN–PHOSPHORUS

The $\mathbf{(Th,Pb)_{1-x}Al_3(PO_4,SiO_4)_2(OH)_6}$ multinary phase (mineral eylettersite), which crystallizes in the rhombohedral structure with the lattice parameters a = 699 ± 4 and 698 ± 3 and c = 1,670 ± 10 and 1,666 ± 8 pm in hexagonal setting and the calculated and experimental densities of 3.44 ± 0.05 and 3.50 ± 0.06 and 3.38 ± 0.01 and 3.44 ± 0.01 g cm^{-3}, respectively, for two samples, is formed in the Al–H–Th–Si–Pb–O–P system (Van Wambeke 1972; Fleischer 1974a).

5.81 ALUMINUM–HYDROGEN–URANIUM–LEAD–ARSENIC–OXYGEN–PHOSPHORUS

The $\mathbf{PbAl(UO_2)_5(PO_4,AsO_4)_2(OH)_9{\cdot}9.5H_2O}$ multinary phase (mineral kamitugaite), which crystallizes in the triclinic structure with the lattice parameters a = 1,098 ± 2, b = 1,596 ± 2, c = 906.8 ± 0.6 pm and α = 95.1 ± 0.2°, β = 96.1 ± 0.2°, γ = 89.0 ± 0.2° and the calculated and experimental densities of 4.47 and 4.03 g cm^{-3}, respectively, is formed in the Al–H–U–Pb–As–O–P system (Deliens and Piret 1984; Dunn et al. 1985d).

5.82 ALUMINUM–HYDROGEN–URANIUM–OXYGEN–PHOSPHORUS

Some quinary compounds are formed in the Al–H–U–O–P system. $\mathbf{HAl(UO_2)(PO_4)(OH)_3{\cdot}4H_2O}$ (mineral ranunculite) crystallizes in the monoclinic (pseudoorthorhombic) structure with the lattice parameters a = 1,110, b = 1,770, c = 1,800 pm, and $\beta \approx 90°$ and the calculated and experimental densities of 3.39 and 3.4 g cm^{-3}, respectively (Deliens and Piret 1979c; Fleischer et al. 1980c).

$\mathbf{HAl(UO_2)_4(PO_4)_4{\cdot}16H_2O}$ (mineral sabugalite) also crystallizes in the monoclinic structure with the lattice parameters a = 1,942.6, b = 984.3, c = 985.0 pm, and β = 96.161° (Vochten and Pelsmaekers 1983) [in the tetragonal structure with the lattice parameters a = 696 and c = 1,930 pm and a calculated density of 3.20 g cm^{-3} (Frondel 1951); a = 696 and c = 1,922 pm (Magin, 1959)]. The title compound was synthesized by diluting a mixture of 0.02 M $UO_2SO_4{\cdot}3.5H_2O$ (50 mL), 0.02 M $Al_2(SO_4)_3{\cdot}18H_2O$ (50 mL), and 0.02 M H_3PO_4 (50 mL) to 1 L (Vochten and Pelsmaekers 1983). After 3 weeks, lemon-yellow quadratic crystals were obtained. It could be also prepared, if the stoichiometric amount of $UO_2(CH_3COO)_2{\cdot}2H_2O$ (200.0 mg) was added to 27.5 mg of $AlCl_3{\cdot}6H_2O$ in a large volume of H_2O (600 mL) (Magin,

1959)]. A threefold excess of H_3PO_4 (1.4 mL of a 100 g L^{-1} solution of H_3PO_4) was added. A yellow fine-grained precipitate was formed slowly.

$Al_{0.67}\square_{0.33}[(UO_2)(PO_4)]_2(H_2O)_{15.5}$ crystallizes in the triclinic structure with the lattice parameters a = 700.20 ± 0.06, b = 1,371.20 ± 0.11, c = 1,402.43 ± 0.11 pm and α = 78.418 ± 0.002°, β = 89.676 ± 0.002°, γ = 81.863 ± 0.002° and a calculated density of 2.61 g cm^{-3} (Locock et al. 2005, 2006b). The crystals of this compound were grown at room temperature over 4 months by slow diffusion of aqueous 0.1 M H_3PO_4 and aqueous 0.1 M $UO_2(NO_3)\cdot6H_2O$ into Al-bearing silica gel. The gel was formed by the hydrolysis of a mixture of $(CH_3O)_4Si$ and aqueous 0.1 M $AlCl_3\cdot6H_2O$.

$HAl(UO_2)_4(PO_4)_4\cdot32H_2O$ crystallizes in the tetragonal structure with the lattice parameters a = 697 and c = 2,643 pm and a calculated density of 2.67 g cm^{-3} (Walenta 1978).

$HAl(UO_2)_4(PO_4)_4\cdot40H_2O$ (mineral uranospathite) crystallizes in the orthorhombic structure with the lattice parameters a = 3,002.0 ± 0.4, b = 700.84 ± 0.09, and c = 704.92 ± 0.09 pm and a calculated density of 2.54 g cm^{-3} (Locock et al. 2005, 2006b) [in the tetragonal structure with the lattice parameters a = 700 and c = 3,002 pm and a calculated density of 2.49 g cm^{-3} (Walenta 1978; Fleischer et al. 1979c)]. Uranospathite is unstable unless preserved at low temperature (Walenta 1978; Fleischer et al. 1979c).

$Al(UO_2)_3(PO_4)_2(OH)_3\cdot5.5H_2O$ (mineral mundite) also crystallizes in the orthorhombic structure with the lattice parameters a = 1,708, b = 3,098, and c = 1,376 pm and a calculated density of 4.295 g cm^{-3} (Deliens and Piret 1981; Fleischer et al. 1982a).

$Al(UO_2)_3O(PO_4)_2(OH)\cdot7H_2O$ (mineral upalite) crystallizes in the monoclinic structure with the lattice parameters a = 1,370.4 ± 0.4, b = 1,682 ± 1, c = 933.2 ± 0.2 pm, and β = 111.5 ± 0.1° and the calculated and experimental densities of 3.94 and 3.9 ± 0.1 g cm^{-3}, respectively (Piret and Declercq 1983) [in the orthorhombic structure with the lattice parameters a = 3,468, b = 1,681, and c = 1,372 pm and the calculated and experimental densities of 3.58 and 3.5 g cm^{-3}, respectively, for the composition **$Al(UO_2)_3(PO_4)_2(OH)_3$** (Deliens and Piret 1979a; Fleischer et al. 1980d)].

$Al(UO_2)_2(PO_4)_2(OH)\cdot8H_2O$ (mineral threadgoldite) also crystallizes in the monoclinic structure with the lattice parameters a = 2,016.8 ± 0.8, b = 984.7 ± 0.2, c = 1,971.9 ± 0.4 pm, and β = 110.71 ± 0.02° and the calculated and experimental densities of 3.33 and 3.4 g cm^{-3}, respectively (Piret et al. 1979a) [a = 2,025, b = 985, c = 1,975 pm, and β = 111.4° and the calculated and experimental densities of 3.32 and 3.4 g cm^{-3}, respectively (Deliens and Piret 1979b; Fleischer et al. 1980d)].

$Al_2(UO_2)(PO_4)_2(OH)_2\cdot8H_2O$ (mineral furongite) crystallizes in the triclinic structure with the lattice parameters a = 1,927.1 ± 0.6, b = 1,417.3 ± 0.4, c = 1,213.6 ± 0.7 pm and α = 67.62 ± 0.03°, β = 115.45 ± 0.03°, γ = 94.58 ± 0.03° and the calculated and experimental densities of 2.75 and 3.0 g cm^{-3}, respectively (Deliens and Piret 1985a; Hawthorne et al. 1988) [a = 1,787, b = 1,418, c = 1,218 pm and α = 67.8°, β = 77.5°, γ = 79.9° and the calculated and experimental densities of 2.848 and 2.82–2.90 g cm^{-3}, respectively (Fleischer et al. 1978a; Shen and Peng 1981)].

$Al_2(UO_2)_3(PO_4)_2(OH)_6\cdot10H_2O$ (mineral phuralumite) crystallizes in the monoclinic structure with the lattice parameters a = 1,383.6 ± 0.6, b = 2,091.8 ± 0.6, c = 942.8 ± 0.3 pm, and β = 112.44 ± 0.03° and the calculated and experimental densities of 3.52 and 3.5 g cm^{-3}, respectively (Piret et al. 1979b) [a = 1,387, b = 2,079, c = 938 pm, and β = 112° and the calculated and experimental densities of 3.54 and 3.5 g cm^{-3}, respectively (Deliens and Piret 1979a; Fleischer et al. 1980d)].

$Al_3(UO_2)_4(PO_4)_4(OH)_5 \cdot 5H_2O$ (mineral triangulite) crystallizes in the triclinic structure with the lattice parameters a = 1,039, b = 1,056, c = 1,060 pm and α = 116.4°, β = 107.8°, γ = 113.4° and the calculated and experimental densities of 3.68 and 3.7 g cm^{-3}, respectively (Deliens and Piret 1982; Dunn et al. 1984b).

$Al_3(UO_2)(PO_4)_3(OH)_2 \cdot 13H_2O$ (mineral moreauite) crystallizes in the monoclinic structure with the lattice parameters a = 2,341 ± 6, b = 2,144 ± 4, c = 1,834 ± 3 pm, and β = 92.0 ± 0.1° and the calculated and experimental densities of 2.61 and 2.64 ± 0.05 g cm^{-3}, respectively (Deliens and Piret 1985b; Dunn et al. 1985b).

5.83 ALUMINUM–HYDROGEN–URANIUM–OXYGEN–SULFUR–IRON–PHOSPHORUS

$(Fe,Al)(UO_2)_4(PO_4)_2(SO_4)_2(OH)\ 22H_2O$ (mineral xiangjiangite) and **$Fe_2Al_2(UO_2)_2(PO_4)_4(SO_4)(OH)_2 \cdot 20H_2O$** (mineral coconinoite) multinary phases are formed in the Al–H–U–O–S–Fe–P system. The first of them crystallizes in the tetragonal structure with the lattice parameters a = 717 and c = 2,222 pm and the calculated and experimental densities of 2.87 and 2.9–3.1 g cm^{-3}, respectively (Fleischer et al. 1979c).

The second phase crystallizes in the monoclinic structure with the lattice parameters a = 1,245 ± 6, b = 1,296 ± 3, c = 1,722 ± 5 pm, and β = 105.7° and the calculated and experimental densities of 2.68 or 2.70 and 2.70 g cm^{-3}, respectively (Young et al. 1966; Belova et al. 1993).

5.84 ALUMINUM–HYDROGEN–LEAD–VANADIUM–OXYGEN–PHOSPHORUS

The **$AlPb_2(PO_4)(VO_4)(OH)$** multinary compound (mineral bushmakinite), which crystallizes in the monoclinic structure with the lattice parameters a = 773.4 ± 0.9, b = 581.4 ± 0.6, c = 869 ± 1 pm, and β = 112.1 ± 0.1° and a calculated density of 6.21 g cm^{-3}, is formed in the Al–H–Pb–V–O–P system (Pekov et al. 2002).

5.85 ALUMINUM–HYDROGEN–LEAD–OXYGEN–PHOSPHORUS

The **$PbAl_3(PO_4)_2(OH)_5 \cdot H_2O$** quinary compound (mineral plumbogummite), which crystallizes in the rhombohedral structure with the lattice parameters a = 703.9 ± 0.5 and c = 1,676.1 ± 0.3 pm in hexagonal setting (Kolitsch et al. 1999) [a = 690.7 ± 0.2 and α = 61°03′ ± 2′ or a = 701.8 ± 0.2 and c = 1,678.4 ± 0.3 pm in hexagonal setting (Förtsch 1,967); a = 703.3 ± 0.1 and c = 1,678.9 ± 0.2 pm in hexagonal setting (Schwab et al. 1990)], is formed in the Al–H–Pb–O–P system. This compound was synthesized under hydrothermal conditions at 200°C and 1.5 MPa starting from H_3PO_4 and oxides or hydroxides of the other elements (Schwab et al. 1990).

5.86 ALUMINUM–HYDROGEN–LEAD–OXYGEN–SULFUR–PHOSPHORUS

The **$H_6Pb_{10}Al_{20}(PO_4)_{12}(SO_4)_5(OH)_{40} \cdot 11H_2O$** multinary compound (mineral orpheite), which melts at about 1,200°C and crystallizes in the trigonal structure with the lattice parameters a = 700 ± 2 and c = 1,672 ± 2 pm and an experimental density of 3.75 ± 0.01 g cm^{-3}, is formed in the Al–H–Pb–O–S–P system (Fleischer et al. 1976b). This mineral needs further study. It is probable that it is a variety of hinsdalite [$(Pb,Sr)Al_3(PO_4)(SO_4)(OH)_6$].

5.87 ALUMINUM–HYDROGEN–LEAD–OXYGEN–IRON–PHOSPHORUS

The $\mathbf{HPb_2(Fe,Al)(PO_4)_2(OH)_2}$ multinary phase (mineral drugmanite), which crystallizes in the monoclinic structure with the lattice parameters $a = 1{,}111.1 \pm 0.5$, $b = 798.6 \pm 0.5$, $c = 464.3 \pm 0.3$ pm, and $\beta = 90.41 \pm 0.03°$ (King and Sengier-Roberts 1988) [$a = 1{,}110.0 \pm 0.6$, $b = 797.6 \pm 0.4$, $c = 464.4 \pm 0.3$ pm, and $\beta = 90°18' \pm 2'$ and a calculated density of 5.55 g cm^{-3} (Tassel van et al. 1979; Fleischer et al. 1980a)], is formed in the Al–H–Pb–O–Fe–P system.

5.88 ALUMINUM–HYDROGEN–BISMUTH–OXYGEN–PHOSPHORUS

The $\mathbf{BiAl_3(PO_4)_2(OH)_6}$ quinary compound (mineral waylandite), which crystallizes in the trigonal structure with the lattice parameters $a = 700.59 \pm 0.07$ and $c = 1{,}634.31 \pm 0.12$ pm (Mills et al. 2010a) [in the hexagonal structure with the lattice parameters $a = 698.34 \pm 0.03$ and $c = 1{,}617.5 \pm 0.1$ pm and a calculated density of 4.08 g cm^{-3} (Clark et al. 1986; Hawthorne et al. 1988); in the rhombohedral structure with the lattice parameters $a = 696.49 \pm 0.08$ and $c = 1{,}625.6 \pm 0.1$ pm (Fleischer 1963) or $a = 697.44 \pm 0.13$ and $c = 1{,}629.3 \pm 0.4$ pm in hexagonal setting (Bayliss 1986)], is formed in the Al–H–Bi–O–P system.

5.89 ALUMINUM–HYDROGEN–BISMUTH–OXYGEN–IRON–PHOSPHORUS

The $\mathbf{Bi(Fe,Al)_3(PO_4)_2(OH)_6}$ multinary phase (mineral zairite), which crystallizes in the rhombohedral structure with the lattice parameters $a = 701.5 \pm 0.5$ and $c = 1{,}636.5 \pm 1.5$ pm (in hexagonal setting) and the calculated and experimental densities of 4.42 and 4.37 ± 0.05 g cm^{-3}, respectively, is formed in the Al–H–Bi–O–Fe–P system (Fleischer and Mandarino 1977).

5.90 ALUMINUM–HYDROGEN–VANADIUM–OXYGEN–PHOSPHORUS

The $\mathbf{Al_2(PO_4)(VO_4){\cdot}8H_2O}$ (mineral schoderite) and $\mathbf{Al_2(PO_4)(VO_4){\cdot}6H_2O}$ (mineral metaschoderite) quinary compounds are formed in the Al–H–V–O–P system. First of them crystallizes in the monoclinic structure with the lattice parameters $a = 1{,}626 \pm 1$, $b = 3{,}060 \pm 4$, $c = 1{,}255 \pm 1$ pm, and $\beta = 91.77 \pm 0.08°$ and the calculated and experimental densities of 1.931 and 1.92 ± 0.02 g cm^{-3}, respectively (Pabst 1979) [$a = 1{,}140$, $b = 1{,}580$, $c = 920$ pm, and $\beta = 79°$ and an experimental density of 1.88 g cm^{-3} (Hausen 1962)]. Schoderite loses H_2O at room temperature in a dry atmosphere and converts to metaschoderite, which may rehydrate to schoderite by contact with H_2O (Hausen 1962). The dehydration is accompanied by shrinkage of the unit cell along *b*-axis to 1,490 pm. The β-angle remains unchanged for both hydration states, since *a* and *c* are unaffected by state of hydration. According to the data of Pabst (1979), at 21°C and ambient humidity conditions, no evidence of schoderite dehydration was observed for several months.

5.91 ALUMINUM–HYDROGEN–VANADIUM–OXYGEN–IRON–PHOSPHORUS

The $\mathbf{(Fe,Al)_5(VO_4,PO_4)_2(OH)_9{\cdot}3H_2O}$ multinary phase (mineral rusakovite) with an experimental density of 2.73–2,080 g cm^{-3} is formed in the Al–H–V–O–Fe–P system (Ankinovich 1960; Fleischer 1960).

5.92 ALUMINUM–HYDROGEN–OXYGEN–SULFUR–PHOSPHORUS

Some quinary compounds are formed in the Al–H–O–S–P system. **$Al_2(PO_4)(SO_4)(OH)$ $9H_2O$** (mineral sanjuanite) crystallizes in the monoclinic structure with the lattice parameters a = 1,391.63 ± 0.05, b = 1,724.22 ± 0.05, c = 611.25 ± 0.03 pm, and β = 98.255 ± 0.004° and the calculated and experimental densities of 1.95 and 1.94 g cm^{-3}, respectively (Colombo et al. 2011) [in the triclinic structure with the lattice parameters a = 1,131.4 ± 1.1, b = 901.8 ± 0.9, c = 737.6 ± 0.7 pm and α = 93°4.17′ ± 6.01′, β = 95°46.49′ ± 4.17′, γ = 105°39.37′ ± 4.42′ and the calculated and experimental densities of 1.96 and 1.94 g cm^{-3}, respectively (Abeledo de et al. 1968; Bruiyn de et al. 1989)].

$Al_3(PO_4)(SO_4)_2(OH)_2{\cdot}14H_2O$ (mineral rossiantonite) crystallizes in the triclinic structure with the lattice parameters a = 1,034.10 ± 0.05, b = 1,096.00 ± 0.05, c = 1,114.46 ± 0.05 pm and α = 86.985 ± 0.002°, β = 65.727 ± 0.002°, γ = 75.064 ± 0.002° and a calculated density of 1.958 g cm^{-3} (Galli et al. 2013a) [a = 1,034.15 ± 0.03, b = 1,095.80 ± 0.03, c = 1,114.45 ± 0.03 pm and α = 86.968 ± 0.004°, β = 65.757 ± 0.003°, γ = 75.055 ± 0.003° (Galli et al. 2013b)].

$Al_5(PO_4)(SO_4)(OH)_{10}{\cdot}8H_2O$ (mineral hotsonite) also crystallizes in the triclinic structure with the lattice parameters a = 1,128.8 ± 5.9, b = 1,165.8 ± 6.0, c = 1,055.5 ± 6.7 pm and α = 112°32.22′ ± 2.93′, β = 107°31.33′ ± 2.84′, γ = 64°27.06′ ± 2.90′ and the calculated and experimental densities of 2.03 and 2.060 ± 0.005 g cm^{-3}, respectively (Beukes et al. 1984; Bruiyn de et al. 1989) [a = 1,130 ± 6, b = 1,166 ± 6, c = 1,055 ± 6 pm and α = 112°32′ ± 3′, β = 107°31′ ± 3′, γ = 64°27′ ± 3′ and an experimental density of 2.129 g cm^{-3} (Ivanov et al. 1990)].

$Al_5(PO_4)_3(SO_4)(OH)_4{\cdot}4H_2O$ (mineral kribergite) also crystallizes in the triclinic structure with the lattice parameters a = 1,812.6 ± 2.5, b = 1,351.9 ± 22.5, c = 750.0 ± 1.3 pm and α = 70°29.99′ ± 7.17′, β = 117°52.20′ ± 7.11′, γ = 136°34.60′ ± 6.67′ and the calculated and experimental densities of 1.95 and 1.92 g cm^{-3}, respectively (Bruiyn de et al. 1989).

$(Al,Fe^{III})_6(PO_4,SO_4)_5(OH)_3{\cdot}35{-}36H_2O$ (mineral sasaite) crystallizes in the orthorhombic structure with the lattice parameters a = 2,150, b = 3,004, and c = 9,206 pm and the calculated and experimental densities of 1.747 and 1.75 g cm^{-3}, respectively (Martini 1978; Fleischer et al. 1979c) [a = 1,075, b = 1,502, and c = 4,603 pm and a calculated density of 1.780 g cm^{-3} (Johan et al. 1983)]. It is possible that this mineral crystallizes in the monoclinic or triclinic structure (Martini 1978; Fleischer et al. 1979c).

5.93 ALUMINUM–HYDROGEN–OXYGEN–SULFUR–FLUORINE–PHOSPHORUS

The **$Al_2(SO_4)(PO_4)F{\cdot}7.5H_2O$** (mineral arangasite) and **$Al_{11}(PO_4,SO_3OH)_{10}F_3{\cdot}30H_2O$** (mineral mitryaevaite) multinary phases are formed in the Al–H–O–S–F–P system. The first of them crystallizes in the monoclinic structure with the lattice parameters a = 974.0 ± 0.5, b = 1,931 ± 1, c = 1,068.8 ± 0.5 pm, and β = 98.65 ± 0.08° and the calculated and experimental densities of 2.001 and 2.01 ± 0.01 g cm^{-3}, respectively (Gamyanin et al. 2012, 2014; Cámara et al. 2014b) [a = 707.31 ± 0.13, b = 963.4 ± 0.2, c = 1,082.7 ± 0.2 pm, and β = 100.403 ± 0.005° [β = 79.60 ± 0.01° (Yakubovich et al. 2014b)] at 100 ± 2 K and the calculated and experimental densities of 1.95 and 2.01, respectively (Yakubovich et al. 2014a, 2014b)].

The second phase crystallizes in the triclinic structure with the lattice parameters a = 691.8 ± 0.1, b = 1,012.7 ± 0.2, c = 1,029.6 ± 0.2 pm and α = 77.036 ± 0.003°, β = 73.989 ±

0.004°, γ = 76.272 ± 0.004° and a calculated density of 2.057 g cm^{-3} (Cahill et al. 2001) [a = 692 ± 1, b = 1,009 ± 1, c = 2,246 ± 6 pm and α = 92.42 ± 0.04°, β = 96.43 ± 0.07°, γ = 104.3 ± 0.2° and the calculated and experimental densities of 2.033 and 2.02 g cm^{-3}, respectively (Ankinovich et al. 1997; Jambor and Roberts 1999a)].

5.94 ALUMINUM–HYDROGEN–OXYGEN–FLUORINE–PHOSPHORUS

Some quinary compounds are formed in the Al–H–O–F–P system. **$Al_2(PO_4)F_2(OH)\cdot7H_2O$** (mineral fluellite) crystallizes in the orthorhombic structure with the lattice parameters a = 854.6 ± 0.8, b = 1,122.2 ± 0.5, and c = 2,115.8 ± 0.5 pm and the calculated and experimental densities of 2.16 and 2.18 ± 0.01 g cm^{-3}, respectively (Guy and Jeffrey 1966).

$Al_3(PO_4)_2(OH)_2F\cdot5H_2O$ (mineral fluorwavellite) also crystallizes in the orthorhombic structure with the lattice parameters a = 963.11 ± 0.04, b = 1,737.31 ± 0.12, and c = 699.46 ± 0.03 pm and the calculated and experimental densities of 2.353 and 2.30 ± 0.01 g cm^{-3}, respectively (Kampf et al. 2015a, 2017a).

$Al_3(PO_4)_2(OH,F)_3\cdot5H_2O$ (mineral wavellite) also crystallizes in the orthorhombic structure with the lattice parameters a = 964.22 ± 0.07, b = 1,741.46 ± 0.15, and c = 700.94 ± 0.02 pm and a calculated density of 2.459 g cm^{-3} (Capitelli et al. 2014a) [a = 960, b = 1,731, and c = 698 pm and the calculated and experimental densities of 2.365 and 2.36 g cm^{-3}, respectively (Gordon 1950); a = 962.1 ± 0.2, b = 1,736.3 ± 0.4, and c = 699.4 ± 0.3 pm (Araki and Zoltai 1968); a = 960.4, b = 1,735.6, and c = 700.9 pm (Blanchard 1974)].

$Al_3(PO_4)_2(OH,F)_3\cdot9H_2O$ (mineral kingite) crystallizes in the triclinic structure with the lattice parameters a = 915 ± 1, b = 1,000 ± 1, c = 724 ± 2 pm and α = 98.6 ± 0.1°, β = 93.6 ± 0.1°, γ = 93.2 ± 0.1° and a calculated density of 2.465 g cm^{-3} (Kato 1970) [a = 1,000, b = 915, c = 724 pm and α = 93.6°, β = 98.6°, γ = 86.8° and the calculated and experimental densities of 2.23 and 2.2–2.3 g cm^{-3}, respectively (McConnell 1974); and experimental densities of 2.21–2.30 g cm^{-3} (Fleischer 1957b; Norrish et al. 1957)].

5.95 ALUMINUM–HYDROGEN–OXYGEN–FLUORINE–MANGANESE–PHOSPHORUS

The **$MnAl_2(PO_4)_2(F,OH)_2\cdot5H_2O$** multinary phase (mineral nordgauite), which crystallizes in the triclinic structure with the lattice parameters a = 992.0 ± 0.4, b = 993.3 ± 0.3, c = 608.7 ± 0.2 pm and α = 92.19 ± 0.03°, β = 100.04 ± 0.03°, γ = 97.61 ± 0.03° and the calculated and experimental densities of 2.46 and 2.35 g cm^{-3}, respectively, is formed in the Al–H–O–F–Mn–P system (Birch et al. 2011; Welch et al. 2013).

5.96 ALUMINUM–HYDROGEN–OXYGEN–FLUORINE–IRON–PHOSPHORUS

The **$HFe_3Al_2(PO_4)_4F\cdot18H_2O$** multinary phase (mineral mcauslanite), which crystallizes in the triclinic structure with the lattice parameters a = 1,005.5 ± 0.5, b = 1,156.8 ± 0.5, c = 688.8 ± 0.5 pm and α = 105.84 ± 0.06°, β = 93.66 ± 0.06°, γ = 106.47 ± 0.05° and the calculated and experimental densities of 2.17 and 2.22 ± 0.02 g cm^{-3}, respectively, is formed in the Al–H–O–F–Fe–P system (Richardson et al. 1988; Jambor and Vanko 1990).

5.97 ALUMINUM–HYDROGEN–OXYGEN–MANGANESE–PHOSPHORUS

$Mn^{II}Al(PO_3OH)_2(OH)\cdot 6H_2O$ (mineral sinkankasite) and **$MnAl_2(PO_4)_2(OH)_2\cdot 6H_2O$** (mineral kayrobertsonite) quinary compounds are formed in the Al–H–O–Mn–P system. The first of them crystallizes in the triclinic structure with the lattice parameters $a = 959.0 \pm 0.2$, $b = 981.8 \pm 0.2$, $c = 686.0 \pm 0.1$ pm and $\alpha = 108.04 \pm 0.03°$, $\beta = 99.63 \pm 0.03°$, $\gamma = 98.87 \pm 0.03°$ and a calculated density of 2.243 g cm^{-3} (Burns and Hawthorne 1995) [$a = 958 \pm 4$, $b = 979 \pm 5$, $c = 688 \pm 4$ pm and $\alpha = 108.1 \pm 0.2°$, $\beta = 99.6 \pm 0.3°$, $\gamma = 98.7 \pm 0.3°$ and the calculated and experimental densities of 2.25 and 2.27 g cm^{-3}, respectively (Peacor et al. 1984a)].

The second compound also crystallizes in the triclinic structure with the lattice parameters $a = 1{,}006.0 \pm 0.3$, $b = 1{,}016.7 \pm 0.2$, $c = 610.8 \pm 0.1$ pm and $\alpha = 91.79 \pm 0.03°$, $\beta = 99.99 \pm 0.03°$, $\gamma = 98.48 \pm 0.02°$ at room temperature and the calculated and experimental densities of 2.41 and 2.29 ± 0.03 g cm^{-3}, respectively, and $a = 1{,}004.9 \pm 0.2$, $b = 1{,}020.5 \pm 0.2$, $c = 608.3 \pm 0.1$ pm and $\alpha = 91.79 \pm 0.03°$, $\beta = 99.70 \pm 0.03°$, $\gamma = 98.02 \pm 0.02°$ at 100 K (Cámara et al. 2016; Mills et al. 2016).

5.98 ALUMINUM–HYDROGEN–OXYGEN–MANGANESE–IRON–PHOSPHORUS

Some multinary compounds are formed in the Al–H–O–Mn–Fe–P system. **$(Mn,Fe)Al(PO_4)(OH)_2\cdot H_2O$** (mineral eosphorite) crystallizes in the orthorhombic structure with the lattice parameters $a = 692.8 \pm 0.1$, $b = 1{,}044.5 \pm 0.1$, and $c = 1{,}350.1 \pm 0.2$ pm and a calculated density of 3.113 g cm^{-3} (Hoyos et al. 1993) [$a = 1{,}045$, $b = 1{,}349$, and $c = 693$ pm and the calculated and experimental densities of 3.112 and 3.10 g cm^{-3}, respectively (Hurlbut, 1950); $a = 1{,}052 \pm 4$, $b = 1{,}360 \pm 5$, and $c = 697 \pm 3$ pm and the calculated and experimental densities of 3.04 and 3.07 g cm^{-3}, respectively (Hanson 1960); $a = 692.63 \pm 0.04$, $b = 1{,}043.56 \pm 0.08$, and $c = 1{,}352.34 \pm 0.10$ pm at 20 K (Gatta et al. 2013a)].

$(Mn,Fe)Al(PO_4)(OH,O)_2\cdot H_2O$ (mineral ernstite) crystallizes in the monoclinic structure with the lattice parameters $a = 1{,}332 \pm 1$, $b = 1{,}049.7 \pm 0.5$, $c = 696.9 \pm 0.4$ pm, and $\beta = 90°22' \pm 20'$ and the calculated and experimental densities of 3.086 and 3.07 g cm^{-3}, respectively (Fleischer 1971).

$MnFeAl(PO_4)_2(OH)_2\cdot 8H_2O$ (mineral kummerite) crystallizes in the triclinic structure with the lattice parameters $a = 531.6 \pm 0.1$, $b = 1{,}062.0 \pm 0.3$, $c = 711.8 \pm 0.1$ pm and $\alpha = 107.33 \pm 0.03°$, $\beta = 111.33 \pm 0.03°$, $\gamma = 72.22 \pm 0.02°$ and a calculated density of 2.337 g cm^{-3} (Grey et al. 2016b; Belakovskiy et al. 2017).

$(Mn,Fe)Al_2(PO_4)_2(OH)_2\cdot 8H_2O$ (mineral kastningite) also crystallizes in the triclinic structure with the lattice parameters $a = 701.02 \pm 0.03$, $b = 1{,}020.50 \pm 0.07$, $c = 1{,}050.40 \pm 0.07$ pm and $\alpha = 71.82 \pm 0.01°$, $\beta = 89.62 \pm 0.01°$, $\gamma = 69.90 \pm 0.01°$ and the calculated and experimental densities of 2.379 and 2.35 g cm^{-3}, respectively (Jambor and Roberts 1999b; Schlüter et al. 1999) [$a = 1{,}020.5 \pm 0.1$, $b = 1{,}050.4 \pm 0.1$, $c = 701.0 \pm 0.1$ pm and $\alpha = 90.38 \pm 0.01°$, $\beta = 110.10 \pm 0.01°$, $\gamma = 71.82 \pm 0.01°$ (Adiwidjaja et al. 1999)].

$(Mn,Fe)Al_2(PO_4)_2(OH)_2\cdot 8H_2O$ (mineral mangangordonite) also crystallizes in the triclinic structure with the lattice parameters $a = 525.7 \pm 0.3$, $b = 1{,}036.3 \pm 0.4$, $c = 704.0 \pm 0.3$ pm and $\alpha = 105.44 \pm 0.03°$, $\beta = 113.07 \pm 0.03°$, $\gamma = 78.69 \pm 0.04°$ and the calculated and experimental densities of 2.319 and 2.36 ± 0.03 g cm^{-3}, respectively (Jambor and Burke 1991; Leavens et al. 1991).

$FeAl(PO_4)(OH)_2 \cdot H_2O$ (mineral childrenite) crystallizes in the orthorhombic structure with the lattice parameters $a = 1{,}038$, $b = 1{,}336$, and $c = 691.1$ pm and the calculated and experimental densities of 3.186 and 3.18–3.24 g cm^{-3}, respectively (Barnes 1949; Hurlbut, 1950).

5.99 ALUMINUM–HYDROGEN–OXYGEN–IRON–PHOSPHORUS

Some quinary compounds are formed in the Al–H–O–Fe–P system. **$AlFe_{24}O_6(PO_4)_{17}(OH)_{12} \cdot 75H_2O$** (mineral cacoxenite) crystallizes in the hexagonal structure with the lattice parameters $a = 2{,}755.9 \pm 0.1$ and $c = 1{,}055.0 \pm 0.1$ pm (Moore and Shen 1983; Dunn et al. 1985a) [$a = 2{,}760$ and $c = 1{,}040$ pm and the calculated and experimental densities of 2.25 and 2.26 g cm^{-3}, respectively (Gordon 1950); $a = 2{,}766.9$ and $c = 1{,}065.5$ pm (Fisher 1966)].

$[Fe^{II}(H_2O)_2Fe^{III}{}_{0.8}Al_{1.2}(PO_4)_3] \cdot H_3O$ crystallizes in the monoclinic structure with the lattice parameters $a = 1{,}332.00 \pm 0.14$, $b = 1{,}021.04 \pm 0.11$, $c = 884.12 \pm 0.09$ pm, and $\beta = 108.590 \pm 0.002°$ and a calculated density of 2.756 g cm^{-3} (Peng et al. 2005). The synthesis of this compound was carried out using solvothermal method in a gel with molar composition 1.0 $(Pr^iO)_3Al$/0.5 $FeCl_2$/4 H_2O/4.1 H_3PO_4/4.8 imidazole/12.0 triethylene glycol. The typical synthesis procedure was as follows: 2.0 g of $(Pr^iO)_3Al$ was first dispersed into 16 mL of triethylene glycol with stirring, 3.2 g of imidazole and 2.72 mL of 85% H_3PO_4 was then added, and the mixture was stirred until it was homogeneous. Finally, 0.97 g of $FeCl_2 \cdot 4H_2O$ was added to the earlier reaction mixture. The reaction mixture was further stirred for 1 h, and was then sealed in Teflon-lined stainless steel autoclaves and heated in an oven at 180°C for 5 days under static conditions. Deep purple-red rhombus crystals of the title compound together with light khaki byproducts were obtained. A batch of crystals of this compound was carefully separated under the optical microscope.

$(Fe,Al)PO_4 \cdot 2H_2O$ (mineral barrandite, or aluminian strengite, or ferrian variscite) crystallizes in the orthorhombic structure (McConnell 1940).

$FeAl_2(PO_4)(OH)_2$ crystallizes in the monoclinic structure with the lattice parameters $a = 716.0 \pm 0.3$, $b = 731.7 \pm 0.3$, $c = 725.9 \pm 0.3$ pm, and $\beta = 120.56 \pm 0.05°$ (Schmid-Beurmann et al. 1997). It was prepared hydrothermally from $FeAlPO_5$, $AlPO_4$, and distilled water at 450°C and 0.2 GPa for 3 days at an oxygen fugacity defined by the Ni/NiO buffer.

$FeAl_2(PO_4)_2(OH)_2 \cdot 6H_2O$ (mineral vauxite) crystallizes in the triclinic structure with the lattice parameters $a = 914.2 \pm 0.3$, $b = 1{,}159.9 \pm 0.3$, $c = 615.8 \pm 0.2$ pm and $\alpha = 98.29 \pm 0.02°$, $\beta = 91.93 \pm 0.03°$, $\gamma = 108.27 \pm 0.03°$ and a calculated density of 2.401 g cm^{-3} (Blanchard and Abernathy 1980) [$a = 913$, $b = 1{,}159$, $c = 614$ pm and $\alpha = 98.3°$, $\beta = 92.0°$, $\gamma = 108.4°$ and the calculated and experimental densities of 2.41 and 2.40 g cm^{-3}, respectively (Baur and Rama Rao 1968)].

$FeAl_2(PO_4)_2(OH)_2 \cdot 8H_2O$ (mineral metavauxite) crystallizes in the monoclinic structure with the lattice parameters $a = 1{,}024.8 \pm 0.2$, $b = 958.6 \pm 0.2$, $c = 696.2 \pm 0.2$ pm, and $\beta = 97.87 \pm 0.02°$ and a calculated density of 2.342 g cm^{-3} (Blanchard and Abernathy 1980) [$a = 1{,}022$, $b = 956$, $c = 694$ pm, and $\beta = 97.9°$ (Baur and Rama Rao 1967)].

$FeAl_2(PO_4)_2(OH)_2·8H_2O$ (mineral paravauxite) crystallizes in the triclinic structure with the lattice parameters $a = 524.2 \pm 0.1$, $b = 1{,}056.9 \pm 0.2$, $c = 697.0 \pm 0.2$ pm and $\alpha = 106.78 \pm 0.03°$, $\beta = 110.81 \pm 0.02°$, $\gamma = 72.29 \pm 0.02°$ (Gatta et al. 2014) [$a = 523.9$, $b = 1{,}056$, $c = 697.2$ pm and $\alpha = 106.80°$, $\beta = 110.78°$, $\gamma = 72.13°$ (Blanchard 1977)].

$FeAl_2(PO_4)_2(OH)_3·5H_2O$ (mineral ferrivauxite) also crystallizes in the triclinic structure with the lattice parameters $a = 919.8 \pm 0.2$, $b = 1{,}160.7 \pm 0.3$, $c = 611.2 \pm 0.2$ pm and $\alpha = 98.237 \pm 0.009°$, $\beta = 91.900 \pm 0.013°$, $\gamma = 108.658 \pm 0.009°$ and a calculated density of 2.39 g cm^{-3} (Raade et al. 2014, 2016).

$FeAl_2(PO_4)_2(OH)_3·7H_2O$ (mineral sigloite) also crystallizes in the triclinic structure with the lattice parameters $a = 519.0 \pm 0.2$, $b = 1{,}041.9 \pm 0.4$, $c = 703.3 \pm 0.3$ pm and $\alpha = 105.00 \pm 0.03°$, $\beta = 111.31 \pm 0.03°$, $\gamma = 70.87 \pm 0.03°$ (Hawthorne 1988) [$a = 526$, $b = 1{,}052$, $c = 706$ pm and $\alpha = 106°58'$, $\beta = 111°30'$, $\gamma = 69°30'$ and the calculated and experimental densities of 2.36 and 2.35 g cm^{-3}, respectively (Hurlbut, and Honea 1962)].

$Fe^{II}Fe^{III}Al_3(PO_4)_4(OH)_5(H_2O)_4·2H_2O$ (mineral tvrdýite) crystallizes in the monoclinic structure with the lattice parameters $a = 2{,}054.3 \pm 0.2$, $b = 510.1 \pm 0.1$, $c = 1{,}887.7 \pm 0.4$ pm, and $\beta = 93.64 \pm 0.01°$ at room temperature and a calculated density of 2.834 g cm^{-3} and $a = 2{,}056.4 \pm 0.4$, $b = 510.10 \pm 0.10$, $c = 1{,}888.3 \pm 0.4$ pm, and $\beta = 93.68 \pm 0.03°$ at 100 K (Sejkora et al. 2015, 2016; Cámara et al. 2017).

$Fe_3Al_4(PO_4)_4(OH)_6·2H_2O$ (mineral gormanite) crystallizes in the triclinic structure with the lattice parameters $a = 1{,}177 \pm 1$, $b = 511 \pm 1$, $c = 1{,}357 \pm 1$ pm and $\alpha = 90°45' \pm 5'$, $\beta = 99°15' \pm 5'$, $\gamma = 90°05' \pm 5'$ and the calculated and experimental densities of 3.13 and 3.12 g cm^{-3}, respectively (Sturman et al. 1981; Fleischer et al. 1982b).

$FeAl_7(PO_4)_4(PO_3OH)_2(OH)_8(H_2O)_8·8H_2O$ (mineral matulaite) crystallizes in the monoclinic structure with the lattice parameters $a = 1{,}060.4 \pm 0.2$, $b = 1{,}660.8 \pm 0.4$, $c = 2{,}064.7 \pm 0.5$ pm, and $\beta = 98.848 \pm 0.007°$ and a calculated density of 2.279 g cm^{-3} (Gatta et al. 2012; Kampf et al. 2012b) [$a = 2{,}040$, $b = 1{,}670$, $c = 1{,}060$ pm, and $\beta = 98.2°$ and a calculated density of 2.330 g cm^{-3} (Fleischer et al. 1980b)].

5.100 ALUMINUM–LITHIUM–STRONTIUM–OXYGEN–PHOSPHORUS

The **$Li_2Sr_2Al(PO_4)_3$** quinary compound, which crystallizes in the monoclinic structure with the lattice parameters $a = 494.5 \pm 0.1$, $b = 2{,}208.8 \pm 0.3$, $c = 863.2 \pm 0.2$ pm, and $\beta = 91.47 \pm 0.01°$ and a calculated density of 3.531 g cm^{-3}, is formed in the Al–Li–Sr–O–P system (Kim et al. 2016b). Single crystals of the title compound were prepared using the molten hydroxide flux method. $SrCO_3$ (2 mM), Al_2O_3 (0.5 mM), $NH_4H_2PO_4$ (3 mM), and $LiOH·H_2O$ (100 mM) were loaded into a 12.5-mm diameter silver tube that was sealed at one end and crimped at the other end. The silver tube was placed in a box furnace, heated to 830°C, held there for 24 h, and then cooled to room temperature by turning off the furnace. The colorless, transparent, block-shaped single crystals were extracted from the flux matrix by sonication in water. The polycrystalline sample was prepared by a solid-state reaction. Stoichiometric amounts of Al_2O_3, $Sr(OH)_2·8H_2O$, $NH_4H_2PO_4$, and a 10% excess of Li_2CO_3 were mixed and preheated at 450°C for 6 h under an Ar atmosphere. Then, the product was reground and heated again at 830°C for 24 h.

5.101 ALUMINUM–SODIUM–MAGNESIUM–CALCIUM–OXYGEN–MANGANESE–IRON–PHOSPHORUS

Two multinary phases are formed in the Al–Na–Mg–Ca–O–Mn–Fe–P system. **(Na,Ca,Mn^{II})(Mn^{II},Fe^{II})(Fe^{III},Fe^{II},Mg)Al(PO$_4$)$_3$** (mineral rosemaryite) crystallizes in the monoclinic structure with the lattice parameters a = 1,200.1 ± 0.2, b = 1,239.6 ± 0.1, c = 632.9 ± 0.1 pm, and β = 114.48 ± 0.01° (Hatert et al. 2006) [a = 1,197.7 ± 0.2, b = 1,238.8 ± 0.2, c = 632.0 ± 0.1 pm, and β = 114°27′ (Fransolet 1995)].

(Na,Ca,Mn^{II})(Mn^{II},Fe^{II})(Fe^{II},Fe^{III},Mg)Al(PO$_4$)$_3$ (mineral wyllieite) also crystallizes in the monoclinic structure with the lattice parameters a = 1,196.7 ± 0.2, b = 1,246.2 ± 0.3, c = 640.9 ± 0.1 pm, and β = 114°38′ ± 1′ (Fransolet 1995) [a = 1,186.8 ± 1.5, b = 1,238.2 ± 1.2, c = 635.4 ± 0.9 pm, and β = 114.52 ± 0.08° and the calculated and experimental densities of 3.601 ± 0.003 and 3.60 g cm^{-3}, respectively (Moore and Ito 1973; Moore and Molin-Case 1974; Fleischer 1974a)].

5.102 ALUMINUM–SODIUM–MAGNESIUM–OXYGEN–MANGANESE–IRON–PHOSPHORUS

The **Na$_2$(Mn^{II},Mg,Fe^{II})$_2$(Al,Fe^{III})(PO$_4$)$_3$** multinary compound (mineral qingheiite), which crystallizes in the monoclinic structure with the lattice parameters a = 1,185.6 ± 0.3, b = 1,241.1 ± 0.3, c = 642.1 ± 0.1 pm, and β = 114.45 ± 0.02° and the calculated and experimental densities of 3.61 and 3.718 g cm^{-3}, respectively, is formed in the Al–Na–Mg–O–Mn–Fe–P system (Zhesheng et al. 1983; Dunn et al. 1984a).

5.103 ALUMINUM–SODIUM–MAGNESIUM–OXYGEN–IRON–PHOSPHORUS

The **Na$_2$Fe^{II}MgAl(PO$_4$)$_3$** multinary compound [mineral qingheiite-(Fe^{II})], which crystallizes in the monoclinic structure with the lattice parameters a = 1,191.0 ± 0.2, b = 1,238.3 ± 0.3, c = 637.2 ± 0.1 pm, and β = 114.43 ± 0.03° and the calculated and experimental densities of 3.54 and 3.6 ± 0.2 g cm^{-3}, respectively, is formed in the Al–Na–Mg–O–Fe–P system (Hatert et al. 2010; Poirier et al. 2011).

5.104 ALUMINUM–SODIUM–STRONTIUM–NIOBIUM–PHOSPHORUS

The **Na$_3$(Sr$_{2.88}$Na$_{0.12}$)[(Al$_{0.94}$Nb$_{0.06}$)P$_4$]** quinary phase, which crystallized in the hexagonal structure with the lattice parameters a = 937.4 ± 0.1 and c = 738.5 ± 0.1 pm, is formed in the Al–Na–Sr–Nb–P system (Somer et al. 1998). It was obtained by the reaction of a mixture of Na$_3$P, SrP, Al, and P (molar ratio 1:3:1:2) with the Nb crucible used for the synthesis at 1,100°C. The compound is obviously Nb stabilized, since the reaction under the same conditions in alumina crucibles leads to the formation of new phases.

5.105 ALUMINUM–SODIUM–STRONTIUM–OXYGEN–FLUORINE–PHOSPHORUS

The **Na$_2$Sr$_2$Al$_2$(PO$_4$)F$_9$** multinary compound (mineral bøggildite), which crystallizes in the monoclinic structure with the lattice parameters a = 525.1 ± 0.3, b = 1,046.4 ± 0.5, c = 1,857.7 ± 0.9 pm, and β = 107.53 ± 0.03° and a calculated density of 3.692 g cm^{-3} (Hawthorne

1982) [$a = 524$, $b = 1{,}048$, $c = 1{,}852$ pm, and $\beta = 107.35°$ (Fleischer 1956)], is formed in the Al–Na–Sr–O–F–P system.

5.106 ALUMINUM–SODIUM–OXYGEN–FLUORINE–PHOSPHORUS

The **NaAl(PO$_4$)F** quinary compound (mineral lacroixite), which crystallizes in the monoclinic structure with the lattice parameters $a = 641.4$, $b = 820.7$, $c = 688.5$ pm, and $\beta = 115.47°$ and an experimental density of 3.29 g cm^{-3} (Lahti and Pajunen 1985) [$a = 641.5 \pm 0.1, 642.0 \pm 0.2, 642.2 \pm 0.1$, $b = 819.0 \pm 0.2, 818.1 \pm 0.6, 813.9 \pm 0.2$, $c = 687.0 \pm 0.2, 688.3 \pm 0.5, 687.5 \pm 0.1$ pm, and $\beta = 115°29' \pm 1', 115°31' \pm 3', 115°34' \pm 1'$ for three various samples, respectively (Fransolet 1989)], is formed in the Al–Na–O–F–P system.

5.107 ALUMINUM–SODIUM–OXYGEN–MANGANESE–PHOSPHORUS

Two quinary compounds are formed in the Al–Na–O–Mn–P system. **Na$_{1.50}$Mn$_{2.48}$Al$_{0.85}$(PO$_4$)$_3$** crystallizes in the monoclinic structure with the lattice parameters $a = 1{,}198.16 \pm 0.10$, $b = 1{,}253.87 \pm 0.13$, $c = 644.07 \pm 0.10$ pm, and $\beta = 114.621 \pm 0.008°$ and a calculated density of 3.614 g cm^{-3} (Hatert 2006). This compound was synthesized under hydrothermal conditions. The starting material was prepared by mixing $NaH_2PO_4 \cdot H_2O$, MnO, and Al_2O_3 in appropriate proportions. H_3PO_4 solution was added to achieve stoichiometry, and the mixture was homogenized in a mortar after evaporation. About 25 mg of the resulting residue was sealed in a gold tube (outer diameter was 2 mm and length was 25 mm), containing 2 mL of distilled water. The gold capsule was then inserted into a Tuttle-type pressure vessel and maintained at a temperature of 800°C and a pressure of 0.1 GPa. After 7 days, the sample, still in the gold tube in the autoclave, was quenched to room temperature in a stream of cold air. The synthesized product consisted of colorless needles of the title compound, together with both irregular colorless crystals of $Na_3Al_2(PO_4)_3$ and an amorphous matrix.

(Na$_{16}$□)Mn$_{25}$Al$_8$(PO$_4$)$_{30}$ (mineral manitobaite) also crystallizes in the monoclinic structure with the lattice parameters $a = 1{,}345.17 \pm 0.07$, $b = 1{,}252.66 \pm 0.07$, $c = 2{,}667.65 \pm 0.13$ pm, and $\beta = 101.582 \pm 0.001°$ and the calculated and experimental densities of 3.628 and 3.621 ± 0.006 g cm^{-3}, respectively (Ercit et al. 2010; Tait et al. 2011; Gatta et al. 2012).

5.108 ALUMINUM–SODIUM–OXYGEN–MANGANESE–IRON–PHOSPHORUS

The **Na$_2$Mn$_5$FeAl(PO$_4$)$_6$** multinary compound (mineral bobfergusonite), which crystallizes in the monoclinic structure with the lattice parameters $a = 1{,}279.6 \pm 0.3$, $b = 1{,}246.5 \pm 0.2$, $c = 1{,}100.1 \pm 0.2$ pm, and $\beta = 97.39 \pm 0.03°$ and a calculated density of 3.66 g cm^{-3} (Tait et al. 2004) [$a = 1{,}277.3 \pm 0.2$, $b = 1{,}248.6 \pm 0.2$, $c = 1{,}103.8 \pm 0.2$ pm, and $\beta = 97.15 \pm 0.01°$ and the calculated and experimental densities of 3.57 and 3.54 ± 0.01 g cm^{-3}, respectively (Ercit et al. 1986a; Hawthorne et al. 1988); $a = 1{,}277.6 \pm 0.2$, $b = 1{,}248.8 \pm 0.2$, $c = 1{,}103.5 \pm 0.2$ pm, and $\beta = 97.21 \pm 0.01°$ (Ercit et al. 1986b)], is formed in the Al–Na–O–Mn–Fe–P system.

5.109 ALUMINUM–SODIUM–OXYGEN–IRON–PHOSPHORUS

Two quinary compounds are formed in the Al–Na–O–Fe–P system. $\mathbf{NaFe^{II}Fe^{III}Al(PO_4)_3}$ (mineral ferrorosemaryite) crystallizes in the monoclinic structure with the lattice parameters a = 1,183.8 ± 0.1, b = 1,234.7 ± 0.1, c = 629.73 ± 0.06 pm, and β = 114.353 ± 0.006° and a calculated density of 3.62 g cm^{-3} (Hatert et al. 2005; Ercit et al. 2006).

$\mathbf{Na_2Fe^{II}{}_5Fe^{III}Al(PO_4)_6}$ (mineral ferrobobfergusonite) also crystallizes in the monoclinic structure with the lattice parameters a = 1,271.56 ± 0.03, b = 1,238.08 ± 0.03, c = 1,093.47 ± 0.03 pm, and β = 97.3329 ± 0.0010° (Yong et al. 2017).

5.110 ALUMINUM–POTASSIUM–CALCIUM–SILICON–OXYGEN–PHOSPHORUS

The $\mathbf{KCa_3Al_2(SiO_4)(Si_2O_7)(PO_4)}$ multinary compound (mineral levantite), which crystallizes in the monoclinic structure with the lattice parameters a = 1,210.06 ± 0.09, b = 511.03 ± 0.04, c = 1,082.52 ± 0.09 pm, and β = 107.237 ± 0.008°, is formed in the Al–K–Ca–Si–O–P system (Galuskin et al. 2017).

5.111 ALUMINUM–POTASSIUM–OXYGEN–COBALT–PHOSPHORUS

The $\mathbf{K(Co^{II},Al)_2(PO_4)_2}$ multinary phase, which crystallizes in the monoclinic structure with the lattice parameters a = 1,331.79 ± 0.15, b = 1,315.23 ± 0.13, c = 868.27 ± 0.12 pm, and β = 100.19 ± 0.01° and a calculated density of 2.795 g cm^{-3}, is formed in the Al–K–O–Co–P system (Chen et al. 1997). Single crystals of the title compound were extracted from an experiment attempting to prepare $K_2CoP_2O_7$. First, an appropriate stoichiometric mixture of K_2CO_3, $CoCl_2{\cdot}6H_2O$ and $(NH_4)_2HPO_4$ was successively preheated at 300°C and 500°C in air for 50 h with intermediate grinding. The resulting product (composition $K_2CoP_2O_7$) was placed in an alumina crucible, which was sealed in a silica tube under argon. The sample was heated gradually to 900°C, where it was kept for 2 days, then cooled at a rate of 5°C h^{-1} to 600°C, and finally quenched at room temperature. Several violet prisms were isolated mechanically from the sample. Microcrystalline $K(Co^{II},Al)_2(PO_4)_2$ was obtained by heating pellets of stoichiometric mixtures of $K_2CoP_2O_7$ and powder Al_2O_3 in an alumina crucible at 850°C for 2 days.

5.112 ALUMINUM–CESIUM–OXYGEN–MOLYBDENUM–PHOSPHORUS

The $\mathbf{CsAl_3Mo_9P_{11}O_{59}}$ quinary compound, which crystallizes in the hexagonal structure with the lattice parameters a = 1,698.9 ± 0.3, c = 1,186.6 ± 0.3 and the calculated and experimental densities of 3.84 and 3.88 g cm^{-3}, respectively, is formed in the Al–Cs–O–Mo–P system (Guesdon et al. 1995). It was prepared in powder form in two steps. In the first step, a mixture of $(NH_4)_2HPO_4$, $Al(NO_3)_3{\cdot}9H_2O$, $CsNO_3$, and MoO_3 was heated to 400°C in air. In the second step, the resulting finely ground powder was mixed with the appropriate amount of Mo (0.66 M). This sample, placed in alumina tube, was sealed in an evacuated silica ampoule, heated to 730°C for 2 days, and quenched at room temperature. A brown powder of this compound was obtained. The crystal growth was performed in two steps in a manner similar to that described earlier, but the sample was heated to 780°C for 1 day, then slowly cooled at 2°C h^{-1} to 650°C, and finally quenched at room temperature. Several orange single crystals were extracted from the latter sample.

5.113 ALUMINUM–CESIUM–OXYGEN–COBALT–PHOSPHORUS

The $\mathbf{Cs_2AlCo_2P_3O_{12}}$ quinary compound, which crystallizes in the cubic structure with the lattice parameter a = 1,382.0 pm, is formed in the Al–Cs–O–Co–P system (Feng et al. 1997).

5.114 ALUMINUM–MAGNESIUM–CALCIUM–ZINC–SILICON–OXYGEN–MANGANESE–IRON–PHOSPHORUS

The $\mathbf{Ca_{14}(Fe^{III},Al)(Al,Zn,Fe^{III},Si,P,Mn,Mg)_{15}O_{36}}$ multinary compound (mineral tululite), which crystallizes in the cubic structure with the lattice parameter a = 1,493.46 ± 0.04 pm, is formed in the Al–Mg–Ca–Zn–Si–O–Mn–Fe–P system (Khoury et al. 2015).

5.115 ALUMINUM–CALCIUM–STRONTIUM–MAGNESIUM–OXYGEN–FLUORINE–IRON–PHOSPHORUS

The $\mathbf{(Ca,Sr)(Mg,Fe^{II})_2Al(P[O,F]_4)_3}$ multinary compound (mineral lasnierite), which crystallizes in the orthorhombic structure with the lattice parameters a = 627.71 ± 0.03, b = 1,768.4 ± 0.3, and c = 816.31 ± 0.04 pm, is formed in the Al–Ca–Sr–Mg–O–F–Fe–P system (Rondeau et al. 2018a,b).

5.116 ALUMINUM–CALCIUM–OXYGEN–FLUORINE–PHOSPHORUS

The $\mathbf{CaAl_2(PO_4)_2F_2}$ quinary compound (mineral abuite), which crystallizes in the orthorhombic structure with the lattice parameters a = 1,181.8 ± 0.2, b = 1,199.3 ± 0.3, and c = 468.72 ± 0.08 pm, is formed in the Al–Ca–O–F–P system (Enju and Uehara 2015).

5.117 ALUMINUM–STRONTIUM–OXYGEN–FLUORINE–PHOSPHORUS

The $\mathbf{SrAl_2(PO_4)_2F_2}$ quinary compound, which crystallizes in the orthorhombic structure with the lattice parameters a = 1,202.6 ± 0.1, b = 1,219.9 ± 0.1, and c = 466.6 ± 0.1 pm and a calculated density of 3.59 g cm^{-3}, is formed in the Al–Ca–O–F–P system (Le Meins and Courbion 1998). It was hydrothermally synthesized in a sealed Pt tube, starting from a mixture composed of SrF_2, γ-Al_2O_3, 85% H_3PO_4, and H_2O (molar ratio 1:0.5:3.7:7.4). Then the Pt tube (degree of filling ~45%) was introduced in a 25-mL autoclave submitted at room temperature to an external pressure of nitrogen (80 MPa). The whole was heated up to 700°C during 24 h. After a slow cooling of 10°C h^{-1} down to 50°C, the crystalline products were filtered, washed with distilled water and alcohol, and finally dried in air at room temperature. Small transparent colorless needles of the title compound were obtained. In softer conditions (Teflon-lined stainless steel autoclave, 180°C, pressure around 3 MPa), this compound can be synthesized in a powder form, but with a slightly modification of reactants and medium: SrF_2 and $AlPO_4$ in water (molar ratio 1:2:55.6).

CHAPTER 6

Systems Based on AlAs

6.1 ALUMINUM–HYDROGEN–SODIUM–MAGNESIUM–CALCIUM–OXYGEN–ARSENIC

The $\mathbf{Na_4MgCa_3Al_4(AsO_3OH)_{12}{\cdot}9H_2O}$ multinary compound (mineral currierite), which crystallizes in the hexagonal structure with the lattice parameters a = 1,220.57 ± 0.09 and c = 920.52 ± 0.07 and the calculated and experimental densities of 3.005 and 3.08 ± 0.02 g cm^{-3}, respectively, is formed in the Al-H-Na-Mg-Ca-O-As system (Kampf et al. 2016, 2017b).

6.2 ALUMINUM–HYDROGEN–SODIUM–CALCIUM–OXYGEN–FLUORINE–ARSENIC

The $\mathbf{NaCa_2Al_2(AsO_4)_2F_4(OH)_2{\cdot}H_2O}$ multinary compound (mineral esperanzaite), which crystallizes in the monoclinic structure with the lattice parameters a = 968.7 ± 0.5, b = 1,073.79 ± 0.06, c = 555.23 ± 0.07 pm, and β = 105.32 ± 0.01° and the calculated and experimental densities of 3.36 ± 0.03 and 3.24 g cm^{-3}, respectively, is formed in the Al–H–Na–Ca–O–F–As system (Foord et al. 1999; Jambor and Roberts 2000).

6.3 ALUMINUM–HYDROGEN–SODIUM–OXYGEN–ARSENIC

Two quinary compounds are formed in the Al-H-Na-O-As system. $\mathbf{NaAl_4[(OH)_4(AsO_4)_3]{\cdot}4H_2O}$ (mineral natropharmacoalumite) crystallizes in the cubic structure with the lattice parameter a = 774.0 ± 0.3 and a calculated density of 2.564 g cm^{-3} (Rumsey et al. 2010; Piilonen et al. 2011).

$\mathbf{Na_3Al_5O_2(OH)_2(AsO_4)}$ crystallizes in the orthorhombic structure with the lattice parameters a = 746.14 ± 0.08, b = 1,746.8 ± 0.3, c = 1,128.2 ± 0.1 pm and a calculated density of 3.73 g cm^{-3} (Yakubovich et al. 2016). To prepare this compound, a gel mixture of high-purity components of Al_2SiO_5 and As_2O_5 (molar ratio 2:1) and an aqueous solution of 25.6 mass% Na_2CO_3 were loaded into a 5-mm Pt capsule. The hydrothermal experiment was done at 600°C and 0.15 GPa for 21 days in an externally heated cold-seal hydrothermal vessel. An outer Cu–Cu_2O oxygen buffer was used to prevent As^V reduction. Colorless transparent prismatic crystals of the title compound were formed. After cooling, the crystals were filtered, washed with water, and dried.

6.4 ALUMINUM–HYDROGEN–SODIUM–OXYGEN–SULFUR–ARSENIC

The $\mathbf{Na_5Al_3[AsO_3(OH)]_4[AsO_2(OH)_2]_2(SO_4)_2{\cdot}4H_2O}$ multinary compound (mineral juansilvaite), which crystallizes in the monoclinic structure with the lattice parameters a = 1,817.75 ± 0.13, b = 862.85 ± 0.05, c = 1,851.38 ± 0.13 pm, and β = 90.389 ± 0.006° and the calculated and experimental densities of 3.005 and 3.01 ± 0.02 g cm^{-3}, respectively, is formed in the Al–H–Na–O–S–As system (Kampf et al. 2015c, 2017c).

6.5 ALUMINUM–HYDROGEN–POTASSIUM–COPPER–BARIUM–ZINC–OXYGEN–IRON–ARSENIC

The $\mathbf{(Ba,K)_{0.5}(Zn,Cu)_{0.5}(Al,Fe)_4(AsO_4)_3{\cdot}5H_2O}$ multinary phase (mineral barium-zinc alumopharmacosiderite), which crystallizes in the tetragonal structure with the lattice parameters a = 1,547.6 ± 0.2 and c = 1,567.5 ± 0.4 and the calculated and experimental densities of 2.82 and 2.8 g cm^{-3}, respectively, is formed in the Al–H–K–Cu–Ba–Zn–O–Fe–As system (Jambor and Roberts 1995a).

6.6 ALUMINUM–HYDROGEN–POTASSIUM–OXYGEN–ARSENIC

The $\mathbf{KAl_4(AsO_4)_3(OH)_4{\cdot}6.5H_2O}$ quinary compound (mineral alumopharmacosiderite), which crystallizes in the cubic structure with the lattice parameter a = 774.5 and a calculated density of 2.676 g cm^{-3}, is formed in the Al–H–K–O–As system (Cabri et al. 1981). When heated to 800°C, the mineral is transformed to $AlAsO_4$.

6.7 ALUMINUM–HYDROGEN–CESIUM–OXYGEN–ARSENIC

The $\mathbf{CsAl(H_2AsO_4)_2(HAsO_4)}$ quinary compound, which crystallizes in the monoclinic structure with the lattice parameters a = 463.4 ± 0.1, b = 1,467.2 ± 0.3, c = 1,515.3 ± 0.3 pm, and β = 93.11 ± 0.03° and a calculated density of 3.756 g cm^{-3}, is formed in the Al–H–Cs–O–As system (Schwendtner and Kolitsch 2007). The title compound was prepared hydrothermally (Teflon-lined stainless steel bomb, 220°C, 9 days, slow furnace cooling) from a mixture of Cs_2CO_3, Al_2O_3 and $H_3AsO_4{\cdot}0.5H_2O$, with a volume ratio of approximately 1:1:3. The Teflon containers were then filled with distilled water to about 80% of their inner volume. The pH value of the starting and final solutions was approximately 1 and 0.5, respectively. This compound formed as colorless prismatic crystals was accompanied by hexagonal tubular crystals of $\mathbf{CsAl(HAsO_4)_2}$ and an uninvestigated amorphous mass.

6.8 ALUMINUM–HYDROGEN–COPPER–CALCIUM–OXYGEN–ARSENIC

The $\mathbf{Ca_2CuAl[AsO_4][AsO_3(OH)]_2(OH)_2{\cdot}5H_2O}$ multinary compound (mineral joteite), which crystallizes in the triclinic structure with the lattice parameters a = 605.30 ± 0.02, b = 1,023.29 ± 0.03, c = 1,291.12 ± 0.04 pm, and α = 87.572 ± 0.002, β = 78.480 ± 0.002°, γ = 78.697 ± 0.002° and a calculated density of 3.056 g cm^{-3}, is formed in the Al–H–Cu–Ca–O–As system (Kampf et al. 2013a, 2013b).

6.9 ALUMINUM–HYDROGEN–COPPER–YTTRIUM–OXYGEN–ARSENIC

The **(Al,Y)Cu$_6$(AsO$_4$)$_3$(OH)$_6$·3H$_2$O** multinary phase (mineral goudeyite), which crystallizes in the hexagonal structure with the lattice parameters a = 1,347.2 ± 0.1 and c = 590.2 ± 0.4 and an experimental density of 3.50 ± 0.03 g cm^{-3} (Wise 1978) [a = 1,350 and c = 590 and an experimental density of 3.5 g cm^{-3} (Sarp et al. 1981)], is formed in the Al–H–Cu–Y–O–As system.

6.10 ALUMINUM–HYDROGEN–COPPER–SILICON–OXYGEN–SULFUR–ARSENIC

The **Cu$_9$Al(HSiO$_4$)$_2$[(SO$_4$)(HAsO$_4$)$_{0.5}$](OH)$_{12}$·8H$_2$O** multinary compound (mineral barrotite), which crystallizes in the trigonal structure with the lattice parameters a = 1,065.0 ± 0.2 and c = 2,195.4 ± 0.7, is formed in the Al–H–Cu–Si–O–S–As system (Sarp and Cerny 2013; Sarp et al. 2014).

6.11 ALUMINUM–HYDROGEN–COPPER–OXYGEN–ARSENIC

Some quinary compounds are formed in the Al–H–Cu–O–As system. **CuAl$_4$(AsO$_4$)$_2$(OH)$_8$(H$_2$O)$_4$** (mineral ceruleite) crystallizes in the monoclinic structure with the lattice parameters a = 720.00 ± 0.14, b = 1,134.5 ± 0.2, c = 985.6 ± 0.2 pm, and β = 105.57 ± 0.03° (Mills et al. 2018) [in the triclinic structure with the lattice parameters a = 1,435.9 ± 0.3, b = 1,468.7 ± 0.3, c = 744.0 ± 0.1 pm, and α = 96.06 ± 0.03°, β = 93.19 ± 0.04°, γ = 91.63 ± 0.04° and the calculated and experimental densities of 2.734 and 2.70 ± 0.02 g cm^{-3}, respectively (Fleischer et al. 1977a)].

Cu$_2$Al(AsO$_4$)(OH)$_4$·4H$_2$O (mineral liroconite) also crystallizes in the monoclinic structure with the lattice parameters a = 1,266.4 ± 0.2, b = 756.3 ± 0.2, c = 991.4 ± 0.3 pm, and β = 91.32 ± 0.02° and a calculated density of 3.04 g cm^{-3} (Burns et al. 1991) [a = 1,270, b = 757, c = 988 pm, and β = 91°23′ and the calculated and experimental densities of 2.95 and 2.96 g cm^{-3}, respectively (Berry and Davis 1947; Berry 1951)].

Cu$_2$Al$_2$(AsO$_4$)(OH,O,H$_2$O)$_6$ (mineral forêtite) crystallizes in the triclinic structure with the lattice paramcters a = 696.9 ± 0.9, b = 767.6 ± 0.9, c = 859.1 ± 1.1 pm, and α = 82.01 ± 0.09, β = 71.68 ± 0.08°, γ = 102.68 ± 0.08° and a calculated density of 3.286 g cm^{-3} (Mills et al. 2012b, 2012c; Cámara et al. 2015).

Cu$_2$Al$_2$(AsO$_4$)$_2$(OH)$_4$·H$_2$O (mineral luetheite) crystallizes in the monoclinic structure with the lattice parameters a = 1,474.3, b = 509.3, c = 559.8 pm, and β = 101°49′ and the calculated and experimental densities of 4.40 and 4.28 ± 0.05 g cm^{-3}, respectively (Fleischer et al. 1977b; Williams 1977).

6.12 ALUMINUM–HYDROGEN–COPPER–OXYGEN–SULFUR–ARSENIC

The **Cu$_{18}$Al$_2$(AsO$_4$)$_4$(SO$_4$)$_3$(OH)$_{24}$·36H$_2$O** multinary compound (mineral chalcophyllite), which crystallizes in the rhombohedral structure with the lattice parameters a = 1,075.6 ± 0.2 and c = 2,867.8 ± 0.4 pm in hexagonal setting and the calculated and experimental densities of 2.684 and 2.69 g cm^{-3}, respectively (Sabelli 1980) [a = 2,053 pm and α = 30°40′ or a = 1,077 and c = 5,752 pm in hexagonal setting and the calculated and experimental densities of 2.60 and 2.67 g cm^{-3}, respectively (Berry and Steacy 1947)], is formed in the Al–H–Cu–O–S–As system.

6.13 ALUMINUM–HYDROGEN–COPPER–OXYGEN–IRON–ARSENIC

The $\mathbf{Cu_2(Fe,Al)_2(AsO_4)_2(OH)_4{\cdot}H_2O}$ multinary phase (mineral chenevixite), which crystallizes in the monoclinic structure with the lattice parameters $a = 570.12 \pm 0.08$, $b = 518.01 \pm 0.07$, $c = 2{,}926.5 \pm 0.2$ pm, and $\beta = 89.99 \pm 0.01°$, is formed in the Al–H–Cu–O–Fe–As system (Burns et al. 2000).

6.14 ALUMINUM–HYDROGEN–MAGNESIUM–CALCIUM–OXYGEN–IRON–COBALT–NICKEL–ARSENIC

The $\mathbf{(Co,Ni,Mg,Ca)_3(Fe,Al)_2(AsO_4)_4{\cdot}11H_2O}$ multinary phase (mineral smolianinovite), which crystallizes in the orthorhombic structure with the lattice parameters $a = 640$, $b = 1{,}172$, $c = 2{,}190$ pm and the calculated and experimental densities of 2.2 and 2.02–2.15 [2.43–2.49 (Yakhontova 1956; Fleischer 1957a)] g cm^{-3}, respectively, is formed in the Al–H–Mg–Ca–O–Fe–Co–Ni–As system (Fleischer 1974c).

6.15 ALUMINUM–HYDROGEN–MAGNESIUM–ZINC–SILICON–OXYGEN–MANGANESE–IRON–ARSENIC

The $\mathbf{Zn_3(Mn,Mg)_{25}(Fe,Al)(As^{III}O_3)_2[(Si,As^{V})O_4]_{10}(OH)_{16}}$ multinary phase (mineral kraisslite), which crystallizes in the orthorhombic structure with the lattice parameters $a = 818.21 \pm 0.01$, $b = 1{,}419.46 \pm 0.03$, $c = 4{,}391.03 \pm 0.08$ pm and a calculated density of 4.083 g cm^{-3} (Cooper and Hawthorne 2012) [in the hexagonal structure with the lattice parameters $a = 822 \pm 1$, $c = 4{,}388 \pm 5$ pm and the calculated and experimental densities of 3.903 and 3.876 g cm^{-3}, respectively (Moore and Ito 1978a)], is formed in the Al–H–Mg–Zn–Si–O–Mn–Fe–As system.

6.16 ALUMINUM–HYDROGEN–MAGNESIUM–ZINC–OXYGEN–MANGANESE–IRON–ARSENIC

The $\mathbf{(Zn,Mn^{II})(Mn^{II},Mg)_{12}(Fe^{III},Al)_2(As^{III}O_3)(As^{V}O_4)_2(OH)_{23}}$ multinary phase (mineral arakiite), which crystallizes in the monoclinic structure with the lattice parameters $a = 1{,}423.6 \pm 0.2$, $b = 820.6 \pm 0.1$, $c = 2{,}422.5 \pm 0.4$ pm, and $\beta = 93.52 \pm 0.01°$, is formed in the Al–H–Mg–Zn–O–Mn–Fe–As system (Cooper and Hawthorne 1999).

6.17 ALUMINUM–HYDROGEN–MAGNESIUM–SILICON–VANADIUM–OXYGEN–MANGANESE–ARSENIC

The $\mathbf{Mn_4(Al,Mg)_6(SiO_4)_2(Si_3O_{10})[(As,V)O_4](OH)_6}$ multinary phase (mineral ardennite), which crystallizes in the orthorhombic structure with the lattice parameters $a = 871.63 \pm 0.02$ and 874.52 ± 0.06, $b = 581.31 \pm 0.01$ and 583.14 ± 0.04, $c = 1{,}851.99 \pm 0.03$ and $1{,}858.91 \pm 0.12$ pm and a calculated density of 3.692 and 3.662 g cm^{-3} for two various samples, respectively (Nagashima and Armbruster 2010) [$a = 580$, $b = 870$, $c = 1{,}849$ pm (Moore 1965); $a = 871.26 \pm 0.08$, $b = 581.08 \pm 0.08$, $c = 1{,}852.14 \pm 0.11$ pm and the calculated and experimental densities of 3.74 and 3.69–3.74 g cm^{-3}, respectively (Donnay and Allmann 1968)], is formed in the Al–H–Mg–Si–V–O–Mn–As system.

6.18 ALUMINUM–HYDROGEN–MAGNESIUM–SILICON–OXYGEN–MANGANESE–IRON–ARSENIC

The **$Mn_3(Mn,Mg,Fe,Al)_{42}[As^{III}O_3]_2(As^VO_4)_4[(Si,As^V)O_4]_6[(As^V,Si)O_4]_2(OH)_{42}$** multinary phase (mineral carlfrancisite), which crystallizes in the trigonal structure with the lattice parameters a = 822.38 ± 0.02 and c = 2,0511.3 ± 0.6 and a calculated density of 3.620 g cm^{-3}, is formed in the Al–H–Mg–Si–O–Mn–Fe–As system (Hawthorne and Pinch 2012; Hawthorne et al. 2013).

6.19 ALUMINUM–HYDROGEN–MAGNESIUM–OXYGEN–ARSENIC

The **$MgAl_2(AsO_4)_2(OH)_2{\cdot}8H_2O$** quinary compound (mineral maghrebite), which crystallizes in the triclinic structure with the lattice parameters a = 543.63 ± 0.17, b = 1,050.0 ± 0.3, c = 707.5 ± 0.2 pm and α = 97.701 ± 0.007, β = 110.295 ± 0.005°, γ = 102.021 ± 0.006° and the calculated and experimental densities of 2.46 and 2.60 ± 0.01 g cm^{-3}, respectively, is formed in the Al–H–Mg–O–As system (Meisser et al. 2012). This mineral dehydrates easily under vacuum.

6.20 ALUMINUM–HYDROGEN–MAGNESIUM–OXYGEN–MANGANESE–ARSENIC

The **$(Mn^{II},Mg,Al)_{15}(As^{III}O_3)(As^VO_4)_2(OH)_{23}$** multinary compound (mineral hematolite), which crystallizes in the rhombohedral structure with the lattice parameters a = 827.5 ± 0.5 and c = 3,660 ± 5 pm in hexagonal setting (Moore and Araki 1978) [a = 1,307 pm and α = 36°53′ or a = 827 and c = 3,651 pm in hexagonal setting and the calculated and experimental densities of 3.48 and 3.49 g cm^{-3}, respectively (Berry and Graham 1948)], is formed in the Al–H–Mg–O–Mn–As system.

6.21 ALUMINUM–HYDROGEN–CALCIUM–OXYGEN–ARSENIC

The **$Ca_5Al_2(AsO_4)_4(OH)_4{\cdot}12H_2O$** quinary compound (mineral tapiaite), which crystallizes in the monoclinic structure with the lattice parameters a = 1,601.6 ± 0.1, b = 577.81 ± 0.03, c = 1,634.1 ± 0.1 pm, and β = 116.704 ± 0.008° and a calculated density of 2.690 g cm^{-3}, is formed in the Al–H–Ca–O–As system (Kampf et al. 2014b, 2015b).

6.22 ALUMINUM–HYDROGEN–CALCIUM–OXYGEN–SULFUR–ARSENIC

The **$(H_3O,Ca)Al_3(SO_4,AsO_4)_2(OH)_6$** multinary phase (mineral schlossmacherite), which crystallizes in the hexagonal structure with the lattice parameters a = 699.8 and c = 1,667, is formed in the Al–H–Ca–O–S–As system (Fleischer et al. 1980b).

6.23 ALUMINUM–HYDROGEN–STRONTIUM–CERIUM–OXYGEN–SULFUR–ARSENIC

The **$(Sr,Ce)Al_3(AsO_4)(SO_4)(OH)_6$** multinary phase (mineral kemmlitzite), which crystallizes in the rhombohedral structure with the lattice parameters a = 683.7 pm and α = 61°51′ or a = 702.7 ± 0.1 and c = 1,651 ± 1 pm in hexagonal setting and the calculated and experimental densities of 3.601 and 3.63 g cm^{-3}, respectively, is formed in the Al–H–Sr–Ce–O–S–As system (Hak et al. 1969; Fleischer 1970).

6.24 ALUMINUM–HYDROGEN–STRONTIUM–OXYGEN–ARSENIC

The $\mathbf{Sr_2Al(AsO_4)_2(OH)}$ quinary compound (mineral grandaite), which crystallizes in the monoclinic structure with the lattice parameters $a = 757.64 \pm 0.05$, $b = 595.07 \pm 0.04$, $c = 880.50 \pm 0.06$ pm, and $\beta = 112.551 \pm 0.002°$ and a calculated density of 4.378 g cm^{-3}, is formed in the Al–H–Sr–O–As system (Cámara et al. 2013, 2014a; Uvarova et al. 2015).

6.25 ALUMINUM–HYDROGEN–BARIUM–OXYGEN–ARSENIC

Two quinary compounds are formed in the Al–H–Ba–O–As system. $\mathbf{BaAl_3(AsO_4)(AsO_3OH)(OH)_6}$ (mineral arsenogorceixite) crystallizes in the hexagonal structure with the lattice parameters $a = 710 \pm 2$ and $c = 1{,}739 \pm 4$ pm and the calculated and experimental densities of 3.71 and 3.65 ± 0.05 g cm^{-3}, respectively (Jambor et al. 1996b).

$\mathbf{Ba_{0.5}Al_4(AsO_4)_3(OH)_4 \cdot 4H_2O}$ (mineral bariopharmacoalumite) crystallizes in the cubic structure.

6.26 ALUMINUM–HYDROGEN–ZINC–OXYGEN–IRON–ARSENIC

The $\mathbf{(Zn,Fe^{II})(Al,Fe^{III})_2(AsO_4)(OH)_5}$ multinary phase (mineral gerdtremmelite), which crystallizes in the triclinic structure with the lattice parameters $a = 516.9 \pm 0.5$, $b = 1{,}303.8 \pm 0.9$, $c = 493.1 \pm 0.4$ pm and $\alpha = 98.78 \pm 0.07$, $\beta = 100.80 \pm 0.06°$, $\gamma = 78.73 \pm 0.06°$ and a calculated density of 3.66 g cm^{-3}, is formed in the Al–H–Zn–O–Fe–As system (Hawthorne et al. 1986).

6.27 ALUMINUM–HYDROGEN–LANTHANUM–CERIUM–NEODYMIUM–SILICON–TITANIUM–OXYGEN–IRON–ARSENIC

The $\mathbf{(Ce,Nd,La)(Fe^{III},Fe^{II},Ti,Al)_3O_2(Si_2O_7)(As^{III}O_3)(OH)}$ multinary phase [mineral cervandonite-(Ce)], which crystallizes in the trigonal subcell with the lattice parameters $a = 650.8 \pm 0.1$ and $c = 1{,}852.0 \pm 0.3$ pm and the monoclinic supercell with the lattice parameters $a = 1{,}128.9 \pm 0.4$, $b = 651.4 \pm 0.2$, $c = 723.8 \pm 0.3$ pm, and $\beta = 121.36 \pm 0.01°$ (Demartin et al. 2008) [$a = 1{,}126.9 \pm 0.2$, $b = 1{,}952.7 \pm 0.3$, $c = 722.6 \pm 0.1$ pm, and $\beta = 121.35 \pm 0.01°$ and a calculated density of 4.9 g cm^{-3} (Armbruster et al. 1988; Jambor and Grew 1990b), is formed in the Al–H–La–Ce–Nd–Si–Ti–O–Fe–As system.

6.28 ALUMINUM–HYDROGEN–LANTHANUM–OXYGEN–ARSENIC

The $\mathbf{LaAl_3(AsO_4)_2(OH)_6}$ quinary compound [mineral arsenoflorencite-(La)], which crystallizes in the trigonal structure with the lattice parameters $a = 703.16 \pm 0.03$ and $c = 1{,}651.51 \pm 0.08$ pm and the calculated and experimental densities of 4.159 and 4.15 ± 0.05 g cm^{-3}, respectively, is formed in the Al–H–La–O–As system (Mills et al. 2010b; Poirier et al. 2011).

6.29 ALUMINUM–HYDROGEN–NEODYMIUM–OXYGEN–ARSENIC

The $\mathbf{NdAl_3(AsO_4)_2(OH)_6}$ quinary compound [mineral arsenoflorencite-(Nd)] is formed in the Al–H–Nd–O–As system (Jambor et al. 1993).

6.30 ALUMINUM–HYDROGEN–URANIUM–OXYGEN–ARSENIC

Some quinary compounds are formed in the Al–H–U–O–As system. $\mathbf{HAl(UO_2)_4(AsO_4)_4 \cdot 20H_2O}$ crystallizes in the tetragonal structure with the lattice parameters $a = 715$

and c = 2,052 pm and a calculated density of 3.20 g cm^{-3} (Walenta 1978; Fleischer et al. 1979c). This compound is more or less stable at room temperature.

$HAl(UO_2)_4(AsO_4)_4 \cdot 32H_2O$ also crystallizes in the tetragonal structure with the lattice parameters a = 715 and c = 2,654 pm and a calculated density of 2.74 g cm^{-3} (Walenta 1978).

$HAl(UO_2)_4(AsO_4)_4 \cdot 40H_2O$ (mineral arsenuranospathite) crystallizes in the orthorhombic structure with the lattice parameters a = 2,992.62 ± 0.07, b = 713.23 ± 0.01, and c = 718.64 ± 0.01 pm and a calculated density of 2.651 g cm^{-3} (Dal Bo et al. 2015) [in the pseudotetragonal structure with the lattice parameters a = 716 and c = 3,037 pm and a calculated density of 2.54 g cm^{-3} (Walenta 1978; Fleischer et al. 1979c)]. Arsenuranospathite is unstable under normal conditions (Walenta 1978; Fleischer et al. 1979c).

6.31 ALUMINUM–HYDROGEN–URANIUM–OXYGEN–FLUORINE–ARSENIC

The **$Al(UO_2)_2(AsO_4)_2(F,OH) \cdot 6.5H_2O$** multinary phase (mineral chistyakovaite), which crystallizes in the monoclinic structure with the lattice parameters a = 1,999 ± 1, b = 979 ± 1, c = 1,962 ± 2 pm, and β = 110.7 ± 0.2° and the calculated and experimental densities of 3.585 and 3.62 ± 0.02 g cm^{-3}, respectively, is formed in the Al–H–U–O–F–As system (Chukanov et al. 2006; Locock et al. 2006a).

6.32 ALUMINUM–HYDROGEN–LEAD–OXYGEN–ARSENIC

The **$PbAl_3(AsO_4)_2(OH)_5 \cdot H_2O$** quinary compound (mineral philipsbornite), which crystallizes in the rhombohedral structure with the lattice parameters a = 701 pm and α = 60.94° or a = 711 and c = 1,705 pm in hexagonal setting and the calculated and experimental densities of 4.33 and >4.1 g cm^{-3}, respectively, is formed in the Al–H–Pb–O–As system (Fleischer et al. 1982b).

6.33 ALUMINUM–HYDROGEN–LEAD–OXYGEN–SULFUR–ARSENIC

The **$PbAl_3(AsO_4)(SO_4)(OH)_6$** multinary compound (mineral hidalgoite), which crystallizes in the rhombohedral structure with the lattice parameters a = 697 ± 2 pm and α = 60.40′ or a = 704 ± 2 and c = 1,699 ± 2 pm in hexagonal setting and the calculated and experimental densities of 4.27 and 3.96 g cm^{-3}, respectively, is formed in the Al–H–Pb–O–S–As system (Smith et al. 1953).

6.34 ALUMINUM–HYDROGEN–BISMUTH–OXYGEN–ARSENIC

The **$BiAl_3(AsO_4)_2(OH)_6$** quinary compound (mineral arsenowaylandite), which crystallizes in the trigonal structure, is formed in the Al–H–Bi–O–As system (Jambor and Roberts 1995a).

6.35 ALUMINUM–HYDROGEN–OXYGEN–IRON–ARSENIC

Two quinary phases are formed in the Al–H–O–Fe–As system. **$(Fe,Al)_2(AsO_4)_2 \cdot 4H_2O$** (mineral aluminian scorodite) has an experimental density of 3.135 g cm^{-3} (Allen et al. 1948).

$[(Al,Fe)_{16}(AsO_4)_9(OH)_{21}(H_2O)_{11}] \cdot 26H_2O$ (mineral liskeardite) crystallizes in the monoclinic structure with the lattice parameters a = 2,472.3 ± 0.1, b = 2,478.9 ± 0.1, c = 779.06 ± 0.02 pm, and β = 89.91 ± 0.01° at 18°C (Grey et al. 2015) [a = 2,457.6 ± 0.5, b = 775.4 ± 0.2,

c = 2,464.1 ± 0.5 pm, and β = 90.19 ± 0.01° at 100 K and the calculated and experimental densities of 1.94 and 1.98 g cm^{-3}, respectively (Grey et al. 2013)]. Above ~100°C, it transformed to a tetragonal phase with the lattice parameters a = 2,060.4 ± 0.2 and c = 773.4 ± 0.1 pm at 118°C (Grey et al. 2015).

6.36 ALUMINUM–SODIUM–POTASSIUM–MAGNESIUM–TITANIUM–OXYGEN–IRON–ARSENIC

The **$(K,Na)_3(Fe^{III},Ti,Al,Mg)_5O_2(AsO_4)_5$** multinary phase (mineral achyrophanite), which crystallizes in the orthorhombic structure with the lattice parameters a = 658.24 ± 0.02, b = 1324.88 ± 0.04, and c = 1,076.13 ± 0.03 pm, is formed in the Al–Na–K–Mg–Ti–O–Fe–As system (Pekov et al. 2018a,b).

6.37 ALUMINUM–SODIUM–POTASSIUM–OXYGEN–ARSENIC

The **$Na_2KAl_3(AsO_4)_4$** quinary compound (mineral ozerovaite), which crystallizes in the orthorhombic structure with the lattice parameters a = 1,061.5 ± 0.2, b = 2,093.7 ± 0.3, and c = 639.3 ± 0.1 pm, is formed in the Al–Na–K–O–As system (Shablinskii et al. 2016).

6.38 ALUMINUM–SODIUM–COPPER–MAGNESIUM–CALCIUM–OXYGEN–IRON–ARSENIC

The **$Na(Cu,Ca)(Mg,Fe,Al)_3(AsO_4)_3$** multinary phase (mineral nickenichite), which crystallizes in the monoclinic structure with the lattice parameters a = 1,188.2 ± 0.4, b = 1,276.0 ± 0.4, c = 664.7 ± 0.2 pm, and β = 112.81 ± 0.02° and a calculated density of 4.06 g cm^{-3} (Auernhammer et al. 1993; Jambor et al. 1994), is formed in the Al–Na–Cu–Mg–Ca–O–Fe–As system.

6.39 ALUMINUM–SODIUM–OXYGEN–FLUORINE–ARSENIC

The **$NaAl(AsO_4)F$** quinary compound (mineral durangite), which melts congruently at 775°C ± 25°C (Foord et al. 1985) and crystallizes in the monoclinic structure with the lattice parameters a = 657.89 ± 0.05, b = 850.71 ± 0.06, c = 702.12 ± 0.05 pm, and β = 115.447 ± 0.004° and a calculated density of 3.915 g cm^{-3} (Downs et al. 2012) [a = 657.4 ± 0.1 and 657.9 ± 0.1, b = 850.5 ± 0.2 and 852.3 ± 0.1, c = 701.9 ± 0.1 and 704.6 ± 0.1 pm, β = 115.34° and 115.47° for two samples and the calculated and experimental densities of 3.92 and 3.90–3.92 g cm^{-3}, respectively (Foord et al. 1985); a = 653 ± 1, b = 846 ± 1, c = 700 ± 2 pm, and β = 115°13′ (Kokkoros 1938); a = 669, b = 866, c = 727 pm, and β = 115°46′ and a calculated density of 3.62 g cm^{-3} (Machatschki 1941)], is formed in the Al–Na–O–F–As system. To obtain this compound, the syrupy arsenic acid and finally powdered Na_3AlF_6 as a viscous mixture were held for 36 h in a steel bomb at about 200°C. Small green crystals of the title compound were obtained.

6.40 ALUMINUM–SODIUM–OXYGEN–COBALT–ARSENIC

The **$Na_4Co_{5.63}Al_{0.91}(AsO_4)_6$** quinary compound, which crystallizes in the monoclinic structure with the lattice parameters a = 1,074.4 ± 0.4, b = 1,484.7 ± 0.3, c = 672.2 ± 0.3 pm, and β = 105.51 ± 0.03°, is formed in the Al–Na–O–Co–As system (Marzouki et al. 2009). To prepare this compound, a mixture of Na_2CO_3, $Co(OAc)_2{\cdot}4H_2O$, and As_2O_5 in a Na/Co/As molar ratio of 4:7:6 was placed in a porcelain boat and first heated at 400°C in air for 24h and then heated gradually to 880°C for 3 days. Some pink parallelepiped crystals were isolated from the sample. Aluminum was diffused from the reaction container. A polycrystalline powder of the title compound was obtained by treating a stoichiometric mixture of the earlier reagents with Al_2O_3 as the aluminum source.

6.41 ALUMINUM–POTASSIUM–COPPER–OXYGEN–ARSENIC

The $\mathbf{K_3Cu_5AlO_2(AsO_4)_4}$ quinary compound (mineral dmisokolovite), which crystallizes in the monoclinic structure with the lattice parameters $a = 1{,}708.48 \pm 0.12$, $b = 571.88 \pm 0.04$, $c = 1{,}653.32 \pm 0.12$ pm, and $\beta = 91.716 \pm 0.006°$ and a calculated density of 4.26 g cm^{-3}, is formed in the Al–K–Cu–O–As system (Pekov et al. 2013, 2015; Belakovskiy et al. 2016b).

6.42 ALUMINUM–POTASSIUM–ZINC–SILICON–OXYGEN–ARSENIC

The $\mathbf{K[(Al,Zn)_2(As,Si)_2O_8]}$ multinary phase (mineral filatovite), which crystallizes in the monoclinic structure with the lattice parameters $a = 877.2 \pm 0.1$, $b = 1{,}337.0 \pm 0.2$, $c = 1{,}469.0 \pm 0.2$ pm, and $\beta = 115.944 \pm 0.006°$ and a calculated density of 2.92 g cm^{-3}, is formed in the Al–K–Zn–Si–O–As system (Filatov et al. 2004; Vergasova et al. 2004; Jambor et al. 2005).

6.43 ALUMINUM–THALLIUM–OXYGEN–IRON–ARSENIC

The $\mathbf{TlFe_{0.22}Al_{0.78}As_2O_7}$ quinary compound, which crystallizes in the triclinic structure with the lattice parameters $a = 629.6 \pm 0.2$, $b = 639.7 \pm 0.2$, $c = 824.2 \pm 0.2$ pm, and $\alpha = 96.74 \pm 0.02$, $\beta = 103.78 \pm 0.02°$, $\gamma = 102.99 \pm 0.03°$ and a calculated density of 5.369 g cm^{-3}, is formed in the Al–Tl–O–Fe–As system (Ouerfelli et al. 2007). Single crystals of the title compound were obtained from a mixture of Tl_2CO_3, $Fe(NO_3)_3{\cdot}9H_2O$, $Al(NO_3)_3{\cdot}9H_2O$, and H_3AsO_4 with a Tl/Fe/Al/As molar ratio of 1:0.1:0.9:2. The synthesis technique consists in completely dissolving the reagents in an aqueous solution. The mixture was slowly evaporated to total dryness and then placed in an alumina crucible. The calcinations of the dry residue were performed through three steps: first at 400°C for 3h, second heated to 700°C for 3h, the molten mixture is then cooled to 650°C for 12h, and finally slowly cooled to room temperature. The obtained green crystals were separated from excess flux by washing the product in boiling water.

CHAPTER 7

Systems Based on AlSb

7.1 ALUMINUM–HYDROGEN–COPPER–ZINC–OXYGEN–ANTIMONY

The $\mathbf{(Zn,Cu)_2Al(OH)_6[Sb^V(OH)_6]}$ multinary phase (mineral zincalstibite-9R), which crystallizes in the trigonal structure with the lattice parameters $a = 534.0 \pm 0.2$ and $c = 8{,}801 \pm 2$ pm, is formed in the Al–H–Cu–Zn–O–Sb system (Mills et al. 2012a).

7.2 ALUMINUM–HYDROGEN–COPPER–CARBON–OXYGEN–SULFUR–ANTIMONY

The $\mathbf{Cu_4Al_2(HSbO_4,SO_4)(CO_3)(OH)_{10}{\cdot}2H_2O}$ multinary phase (mineral camerolaite), which crystallizes in the monoclinic structure with the lattice parameters $a = 1{,}076.5 \pm 0.6$, $b = 290.3 \pm 0.2$, $c = 1{,}252.7 \pm 0.8$ pm, and $\beta = 95.61 \pm 0.04°$ and the calculated and experimental densities of 3.09 and 3.1 ± 0.1 g cm^{-3}, respectively (Jambor and Puziewicz 1992) [$a = 1{,}073 \pm 5$, $b = 289 \pm 1$, $c = 1{,}252 \pm 5$ pm, and $\beta = 95.5 \pm 0.5°$ (Cuchet 1995); in the triclinic structure with the lattice parameters $a = 633.10 \pm 0.13$, $b = 291.30 \pm 0.06$, $c = 1{,}072.7 \pm 0.2$ pm, and $\alpha = 93.77 \pm 0.03°$, $\beta = 96.34 \pm 0.03°$, $\gamma = 79.03 \pm 0.03°$ at 100 ± 2 K for the $\mathbf{Cu_6Al_3(OH)_{18}(H_2O)_2[Sb(OH)_6](SO_4)}$ composition (Mills et al. 2014)], is formed in the Al–H–Cu–C–O–S–Sb system.

7.3 ALUMINUM–HYDROGEN–COPPER–OXYGEN–ANTIMONY

The $\mathbf{Cu_2AlSb(OH)_{12}}$ quinary compound (mineral cualstibite), which crystallizes in the trigonal structure with the lattice parameters $a = 915.0 \pm 0.2$ and $c = 974.5 \pm 0.2$ pm (Bonaccorsi et al. 2007) [$a = 920$ and $c = 973$ and the calculated and experimental densities of 3.25 and 3.18 ± 0.05 g cm^{-3}, respectively (Walenta 1984; Dunn et al. 1985b)], is formed in the Al–H–Cu–O–Sb system.

7.4 ALUMINUM–HYDROGEN–ZINC–OXYGEN–ANTIMONY

The $\mathbf{Zn_2AlSb(OH)_{12}}$ quinary compound (mineral zincalstibite), which crystallizes in the trigonal structure with the lattice parameters $a = 532.1 \pm 0.1$ and $c = 978.6 \pm 0.2$ pm, is formed in the Al–H–Zn–O–Sb system (Bonaccorsi et al. 2007).

7.5 ALUMINUM–LITHIUM–SILVER–OXYGEN–ANTIMONY

The $\mathbf{Li_2Ag_6Al_2Sb_2O_{12}}$ quinary compound is formed in the Al–Li–Ag–O–Sb system (Kumar et al. 2012). It was synthesized by molten ion exchange of $Li_8Al_2Sb_2O_{12}$ with $AgNO_3$.

7.6 ALUMINUM–SODIUM–TELLURIUM–CHLORINE–ANTIMONY

The $\mathbf{NaAl_6Sb_7Te_8Cl_{24}}$ quinary compound, which undergoes a phase transition at 177 K and crystallizes in the triclinic structure with the lattice parameters a = 1,851.72 ± 0.01 and 1,838.67 ± 0.01, b = 1,282.38 ± 0.01 and 1,277.87 ± 0.01, c = 1,378.45 ± 0.01 and 2,742.31 ± 0.01 pm and α = 127.88 ± 0.01° and 127.84 ± 0.01°, β = 133.18 ± 0.01° and 133.18 ± 0.01°, γ = 87.85 ± 0.01 and 87.89 ± 0.01 and a calculated density of 3.43 and 3.50 g cm^{-3} at 293 and 123 K, respectively, is formed in the Al–Na–Te–Cl–Sb system (Eich et al. 2013). To synthesize this compound, in the glove box, weighed amounts of starting materials were filled in glass ampoules, which were evacuated and flame-sealed under vacuum. Tube ovens used for syntheses were aligned at an angle of about 30° to the horizontal to keep the melt compacted in the hot zone. After some days, the ampoules were inspected visually. If the formation of crystals was observed, the hot melt was decanted from the solids. After cooling to ambient temperature, the ampoules were opened in the glove box. Since this compound is very sensitive towards air, samples were stored under Ar atmosphere. Te (0.72 mM), Sb (0.48 mM), $SbCl_3$ (0.4 mM), $AlCl_3$ (1.98 mM), and NaCl (0.18 mM) were filled in a glass ampoule under an Ar atmosphere. The ampoule was evacuated, closed, and heated for 15 days at 125°C. Orange cube-shaped crystals appeared within several days after lowering the temperature to 105°C.

7.7 ALUMINUM–BISMUTH–OXYGEN–IRON–ANTIMONY

The $\mathbf{Al_{0.36}Bi_{1.8}Fe_{0.84}SbO_7}$ solid solution of pyrochlore structure is formed in the Al–Bi–O–Fe–Sb system (Ellert et al. 2016). It crystallizes in the cubic structure with the lattice parameter a = 1,042.35 ± 0.01 pm. It was synthesized by a conventional solid-state method. High-purity reagents Bi_2O_3, Fe_2O_3, Al_2O_3, and Sb_2O_3 were used as starting materials. The stoichiometric mixture of oxides (4 g) was ground in acetone using an agate mortar and pestle for approximately 30 min. The synthesis was carried out in Pt crucibles inside an electrical muffle furnace. The samples were heated in two steps, first at 650°C for 24 h and subsequently at 800°C for 24 h. Each sample was reground after calcination. Subsequently, the samples were annealed at 1,100°C for 5–15 days and regrounded again.

7.8 ALUMINUM–SULFUR–CHLORINE–BROMINE–ANTIMONY

The $\mathbf{[Sb_7S_8Br_2](AlCl_4)_3}$ quinary compound, which crystallizes in the orthorhombic structure with the lattice parameters a = 1,198.9 ± 0.2, b = 1,689.6 ± 0.3, c = 1,737.8 ± 0.4 pm and a calculated density of 3.349 g cm^{-3} at 100 K, is formed in the Al–S–Cl–Br–Sb system (Zhang et al. 2009). Red crystals of this compound were synthesized by reacting Sb with sulfur in 1-ethyl-3-methylimidazolium bromide/$AlCl_3$ ionic liquid (molar ratio 1:11) at 165°C for 10 days.

7.9 ALUMINUM–SELENIUM–CHLORINE–BROMINE–ANTIMONY

Two quinary phases are formed in the Al–Se–Cl–Br–Sb system. **[$Sb_7Se_8Br_2$][$Al(Cl_xBr_{1-x})_4$]$_3$** crystallizes in the orthorhombic structure with the lattice parameters a = 1,251.32 ± 0.05 and 1,237.57 ± 0.03, b = 1,773.94 ± 0.06 and 1,741.16 ± 0.05, c = 1,830.13 ± 0.06 and 904.20 ± 0.02 pm and a calculated density of 4.242 and 4.04 g cm^{-3} for x = 0.6 and 2.32, respectively (Ahmed et al. 2014). This phase was obtained by the following procedure. Sb and Se in amounts that correspond to the composition of the target compound (total mass 190 mg) were added to the bromine-based Lewis acidic ionic liquid 1-*n*-butyl-3-methylimidazolium bromide/$AlBr_3$ (88 mg/500 mg) in a glass ampoule. A small amount of $NbCl_5$ (50 mg) was added to the ionic liquid at room temperature, and the solution turned olive-green. The glass ampoule was sealed and heated at 160°C for 7 days followed by slow cooling (6°C h^{-1}) to room temperature. Over the course of the reaction, the solution changed to red. The ampoule was opened inside a glove box and red plate-like crystals were manually separated from the ionic liquid.

[$Sb_{13}Se_{16}Br_2$][$AlCl_{3.20}Br_{0.80}$]$_5$ crystallizes in the triclinic structure with the lattice parameters a = 908.42 ± 0.05, b = 1,960.7 ± 0.1, c = 2,151.1 ± 0.1 pm, and α = 64.116 ± 0.006°, β = 79.768 ± 0.007°, γ = 88.499 ± 0.007° and a calculated density of 3.95 g cm^{-3} (Ahmed et al. 2014). To prepare this compound, Sb and Se in amounts that correspond to the composition of the target compound (total mass 171 mg) were added to the chlorine-rich Lewis acidic ionic liquid 1-*n*-butyl-3-methylimidazolium bromide/$AlCl_3$ (80 mg/250 mg) in a glass ampoule. $NbCl_5$ was not added. The solid reaction product contained dark-red, air-sensitive crystals, which were scattered among unreacted starting materials.

CHAPTER 8

Systems Based on GaB^V (B^V = N, P, As, Sb)

8.1 SYSTEMS BASED ON GaN

8.1.1 Gallium–Hydrogen–Rubidium–Phosphorus–Oxygen–Nitrogen

The $\mathbf{(NH_4)_{3-x}Rb_xGa_2(PO_4)_3}$ solid solution ($0 \le x \le 0.23$), which crystallizes in the monoclinic structure with the lattice parameters $a = 1{,}337.82 \pm 0.13$, $b = 1{,}032.60 \pm 0.06$, $c = 902.04 \pm 0.07$ pm and $\beta = 111.366 \pm 0.007°$ and a calculated density of 2.8149 ± 0.0004 g cm^{-3} for $x = 0.23$, is formed in the Ga-H-Rb-P-O-N system (Lesage et al. 2004). The crystals of this solid solution were obtained under hydrothermal conditions using 21-mL Teflon-lined stainless steel autoclaves. Ga_2O_3, $(NH_4)_2HPO_4$, and $RbNO_3$ were ground in an agate mortar in the molar ratio Ga/P = 2:3 and $RbNO_3/(NH_4)_2HPO_4$ = 1:1. Approximately, 0.4 mL of deionized water was then added to 0.8 g of the mixture, leading to a highly viscous solution. The obtained solution was heated in an autoclave at 180°C for 25 h and finally slowly cooled at room temperature for 18 h. The crystals of this solid solution were synthesized in the presence of Ga_2O_3 as secondary phase.

8.1.2 Gallium–Hydrogen–Cesium–Phosphorus–Oxygen–Nitrogen

The $\mathbf{(NH_4)_{3-x}Cs_xGa_2(PO_4)_3}$ solid solution ($0 \le x \le 0.54$), which crystallizes in the monoclinic structure with the lattice parameters $a = 1{,}339.2 \pm 0.2$, $b = 1{,}035.79 \pm 0.09$, $c = 904.36 \pm 0.08$ pm and $\beta = 111.412 \pm 0.009°$ and a calculated density of 3.0731 ± 0.0006 g cm^{-3} for $x = 0.54$, is formed in the Ga-H-Cs-P-O-N system (Lesage et al. 2004). The crystals of this solid solution were obtained by the same way as the crystals of $(NH_4)_{3-x}Rb_xGa_2(PO_4)_3$ were prepared, using $CsNO_3$ instead of $RbNO_3$.

8.1.3 Gallium–Hydrogen–Magnesium–Phosphorus–Oxygen–Nitrogen

The $\mathbf{(NH_4)MgGa_2(PO_4)_3(H_2O)_2}$ multinary compound, which crystallizes in the monoclinic structure, is formed in the Ga-H-Mg-P-O-N system (Overweg et al. 1999). This compound was crystallized from a gel composition containing Ga_2O_3, 85% H_3PO_4,

$(NH_4)_2HPO_4$, $Si(OEt)_4$, H_2O, and $Mg(OAc)_2{\cdot}4H_2O$ (molar ratio 1:10:5:0.3:156:1). Typically, 0.5 g of Ga_2O_3 and a proper amount of $Mg(OAc)_2{\cdot}4H_2O$ were mixed together with 7.5 mL of water and a few drops of $Si(OEt)_4$, which acts as a mineralizing agent. $(NH_4)_2HPO_4$ was then added, and the mixture was stirred for 10 min followed by the addition of phosphoric acid and additional stirring until the gel was of uniform composition. Subsequently, the gel was transferred into a Teflon-lined stainless steel autoclave and heated at 190°C for 6 days. After cooling, the solid material was collected by filtration, washed with demineralized water, and dried at 70°C overnight. A purer phase was obtained in a second synthesis if, instead of $Si(OEt)_4$, a small amount of $(NH_4)MgGa_2(PO_4)_3(H_2O)_2$ was used as the seed.

8.1.4 Gallium–Hydrogen–Zinc–Phosphorus–Oxygen–Nitrogen

Some multinary compounds are formed in the Ga–H–Zn–P–O–N system. $\mathbf{(NH_4)ZnGa(PO_4)_2}$ crystallizes in the monoclinic structure with the lattice parameters a = 940.6 ± 0.1, b = 988.1 ± 0.1, c = 861.2 ± 0.1 pm and β = 90.58 ± 0.01° and a calculated density of 2.847 g cm^{-3} and $\mathbf{(NH_4)_2Zn_2Ga(PO_4)_3}$ crystallizes in the cubic structure with the lattice parameters a = 1,345.6 ± 0.1 and a calculated density of 2.843 g cm^{-3} (Zabukovec Logar et al. 2001). These two compounds were obtained by hydrothermal crystallization using Ga_2O_3, $(NH_4)_2HPO_4$, $Zn(OAc)_2{\cdot}2H_2O$, $H_2C_2O_4{\cdot}2H_2O$, Me_4NBr, and distilled water. In the synthesis procedure, four solutions were prepared: 52% $Zn(OAc)_2{\cdot}2H_2O$, 14% $(NH_4)_2HPO_4$, 14% oxalic acid, and 20% Me_4NBr for the gel composition. Solutions of oxalic acid and $(NH_4)_2HPO_4$ were successively added to the dispersion of Ga_2O_3, which was added to the solution of zinc acetate. Finally, Me_4NBr was added dropwise. The system was thoroughly blended with a magnetic mixer to a homogeneous mixture each time before the addition of the next component. At the end of the synthesis, when all the components were added, the gel was also mixed for 10 min. The resulting gel with a molar composition of 0.5 Ga_2O_3 + 1.0 $Zn(OAc)_2$ + 1.0 $(NH_4)_2HPO_4$ + 1.0 $H_2C_2O_4{\cdot}2H_2O$ + 2.0 Me_4NBr + 200.0 H_2O was transferred to a 50-mL stainless steel, Teflon-lined autoclaves, and heated under static conditions at 150°C–180°C in an oven for 1–4 days. The crystallization products were washed with distilled water and dried at 105°C. In all of the final products, there were $(NH_4)ZnGa(PO_4)_2$ and $(NH_4)_2Zn_2Ga(PO_4)_3$ found in the approximate volume ratio of 1:4. The crystals of $(NH_4)_2Zn_2Ga(PO_4)_3$ appeared colorless and cubic in shape, while the $(NH_4)ZnGa(PO_4)_2$ ones were bigger, colorless flat square prisms.

$\mathbf{(NH_4)ZnGa_2(PO_4)_3(H_2O)_2}$ multinary compound crystallizes in the monoclinic structure (Overweg et al. 1999). This compound was obtained by the same way as $(NH_4)MgGa_2(PO_4)_3(H_2O)_2$ was synthesized using $Zn(OAc)_2{\cdot}2H_2O$ instead of $Mg(OAc)_2{\cdot}4H_2O$.

8.1.5 Gallium–Hydrogen–Indium–Phosphorus–Oxygen–Nickel–Nitrogen

The $\mathbf{(NH_4)NiGa_{1.84}In_{0.16}(PO_4)_3(H_2O)_2}$ multinary phase, which crystallizes in the monoclinic structure with the lattice parameters a = 1,324.83 ± 0.08, b = 1,024.68 ± 0.08, c = 890.33 ± 0.07 pm and β = 108.637 ± 0.001° and a calculated density of 3.158 g cm^{-3}, is formed in the Ga–H–In–P–O–Ni–N system (Capitelli et al. 2010). This phase was prepared hydrothermally from a gel of molar composition: 1 Ga_2O_3 + 1 In_2O_3 + 1 $NiCl_2{\cdot}6H_2O$ + 10 H_3PO_4 + 156 H_2O + 0.1 H_3BO_3 + 5 $(NH_4)H_2PO_4$, which was heated at 160°C for 7 days in a

Teflon-lined stainless steel autoclave. The solid product was collected by filtration, washed with deionized water, and dried in air. The product was consisted of light-green prisms of the title compound.

8.1.6 Gallium–Hydrogen–Phosphorus–Oxygen–Nitrogen

Some quinary compounds are formed in the Ga–H–P–O–N system. **$(NH_4)GaPO_4(OH)$** crystallizes in the monoclinic structure with the lattice parameters a = 856.4 ± 0.1, b = 603.87 ± 0.08, c = 448.83 ± 0.06 pm and β = 98.05 ± 0.01° and the calculated and experimental densities of 2.89 and 2.84 ± 0.02 g cm^{-3}, respectively (Bonhomme et al. 2002). This compound was prepared under solvothermal conditions using 23-mL Teflon-lined stainless steel autoclave. The chemicals used in the reactions were as follows: Ga_2O_3 (0.188 g), 85% H_3PO_4 (0,460 g), 29.5% $(NH_4)OH$ (0.4 mL), ethylene glycol (5 mL), and deionized water (5 mL). The gallium oxide was first dispersed in ethylene glycol and deionized water, and then phosphoric acid and ammonium hydroxide were successively added under agitation. The resulting mixture was stirred for 1 h and then placed in the reactor at 172°C for 6 days.

$(NH_4)_3Ga_2(PO_4)_3$ also crystallizes in the monoclinic structure with the lattice parameters a = 1,339.48 ± 0.16, b = 1,031.38 ± 0.09, c = 903.61 ± 0.09 pm and β = 111.323 ± 0.007° and a calculated density of 2.7320 ± 0.0005 g cm^{-3} (Lesage et al. 2004) [a = 1,346.2 ± 0.2, b = 1,030.1 ± 0.1, c = 899.2 ± 0.1 pm and β = 111.28 ± 0.01° and a calculated density of 2.73 g cm^{-3} (Bonhomme et al. 2002)]. The title compound was also obtained under solvothermal conditions using 23-mL Teflon-lined stainless steel autoclave (Bonhomme et al. 2002). The chemicals used in the reactions were as follows: Ga_2O_3 (0.188 g), 85% H_3PO_4 (0,690 g), guanidinium carbonate [$(H_2NCNHNH_2)_2H_2CO_3$] (0.540 g), and 15 mL of ethylene glycol as a solvent. This reaction mixture was heated in the reactor at 170°C for 4 days. The autoclave was then cooled overnight to room temperature. The resulting whitish powders were separated from the solution by vacuum filtration, washed with deionized water and acetone, and dried in air overnight at 50°C. Further synthesis work showed that $(NH_4)GaPO_4(OH)$ and $(NH_4)_3Ga_2(PO_4)_3$ could be obtained using either guanidinium carbonate or ammonium hydroxide as the ammonium source. It rapidly appeared that the main factor governing the prevalence of one phase over the other was the ratio of ethylene glycol to water.

Crystals of $(NH_4)_3Ga_2(PO_4)_3$ could also be synthesized under hydrothermal conditions using 21-mL Teflon-lined stainless steel autoclaves (Lesage et al. 2004). Ga_2O_3 and $(NH_4)_2HPO_4$ were ground in an agate mortar in the molar ratio Ga/P = 2:3. Approximately, 0.4 mL of deionized water was then added to 0.8 g of the phosphate-oxide mixture, leading to a highly viscous solution. Its pH was calculated to be slightly basic (pH = 8.1). The so-obtained solution was heated in the autoclave at 180°C for 25 h and finally slowly cooled to room temperature for 18 h. Colorless crystals were thus obtained mixed with whitish powder. They were separated from the solution by vacuum filtration, washed with deionized water, and dried in air. The final solution turned basic during the reaction (pH ~ 11). Three phases were present under these conditions: Ga_2O_3, $(NH_4)_3Ga_2(PO_4)_3$, and $(NH_4)GaPO_4(OH)$.

According to the data of Chudinova et al. (1987b), **$(NH_4)GaP_2O_7$, $(NH_4)GaHP_3O_{10}$** (two modifications), and **$(NH_4)_2Ga_2P_8O_{24}$** also exist in the Ga–H–P–O–N system. $(NH_4)_2Ga_2P_8O_{24}$ is thermally stable up to ~400°C. All these compounds were synthesized from Ga_2O_3, $(NH_4)H_2PO_4$, $(NH_4)_2HPO_4$, and 85% H_3PO_4 heating the mixtures with the various molar ratio in glassy carbon or Pt/Au crucibles at 150°C–350°C.

8.1.7 Gallium–Hydrogen–Phosphorus–Oxygen–Fluorine–Nitrogen

Two compounds, **$(NH_4)Ga(PO_4)F$** and **$(NH_4)_2Ga_2(PO_4)(HPO_4)F_3$**, which crystallizes in the orthorhombic structure with the lattice parameters a = 1,292.07 ± 0.02, b = 644.0 ± 0.1, c = 1,041.47 ± 0.02 pm and a calculated density of 3.057 g cm^{-3} for the first compound and a = 1,249.67 ± 0.02, b = 770.15 ± 0.01, c = 984.63 ± 0.01 pm and a calculated density of 2.929 g cm^{-3} for the second one, are formed in the Ga–H–P–O–F–N system (Loiseau et al. 2000). The syntheses of these compounds were carried out hydrothermally in a 23-mL Teflon-lined stainless steel bomb under autogeneous pressure. They were obtained using guanidinium carbonate, which is decomposed into ammonium cations during the hydrothermal reaction. Second compound can be synthesized with $(NH_4)F$ instead of using guanidine.

For the synthesis of $(NH_4)Ga(PO_4)F$, GaO(OH) (0.713 g), 85% H_3PO_4 (0.5 mL), 48% HF (0.25 mL), guanidinium carbonate (0.315 g), and H_2O (4.25 mL) (molar ratio 1:1:1:0.5:40). GaO(OH) was added to water, and then, H_3PO_4, HF, and guanidinium carbonate were successively added. The resulting pH of the solution was 3, and the mixture was placed in the autoclave at 180°C for 24 h. The crystalline product was filtered off, washed with distilled water, and dried at room temperature. A small amount of $(NH_4)Ga(PO_4)F$ was formed in the presence of a powdered unknown phase.

$(NH_4)_2Ga_2(PO_4)(HPO_4)F_3$ was synthesized in the same hydrothermal conditions and from the same reactant molar ratio as $(NH_4)Ga(PO_4)F$ was prepared, but the mixture was heated in the autoclave at 180°C for 4 days. This compound was obtained as a pure phase. With $(NH_4)F$ replacing the guanidine, the same solid was obtained by using the next mixture: GaO(OH) (0.713 g), 85% H_3PO_4 (0.5 mL), 48% HF (0.25 mL), $(NH_4)F$ (0.257 g), and H_2O (4,25 mL) (molar ratio 1:1:1:1:40).

8.1.8 Gallium–Hydrogen–Phosphorus–Oxygen–Manganese–Nitrogen

The **$(NH_4)Ga_2Mn(PO_4)_3(H_2O)_2$** multinary compound, which crystallizes in the monoclinic structure with the lattice parameters a = 1,354.3 ± 0.4, b = 1,023.02 ± 0.15, c = 889.4 ± 0.3 pm and β = 108.54 ± 0.03° and a calculated density of 3.01 g cm^{-3}, is formed in the Ga–H–P–O–Mn–N system (Chippindale et al. 1998). This compound was prepared solvothermally from a gel of molar composition 1 Ga_2O_3 + 1 $MnCl_2·4H_2O$ + 10 H_3PO_4 + 50 $HOCH_2CH_2OH$ + 0.3 $Si(OEt)_4$ + 5 $(H_2N)_3CCl$, which was heated at 260°C for 7 days in a Teflon-lined stainless steel autoclave. The solid product was collected by filtration, washed with deionized water, and dried in air. The product consisted of large colorless blocks of the title compound with a small amount of light-brown polycrystalline α-$GaPO_4$. Similar solvothermal treatment of a gel of molar composition: 1 Ga_2O_3 + 1 $MnCl_2·4H_2O$ + 10 H_3PO_4 + 100 H_2O + 0.3 $Si(OEt)_4$ + 5 $(NH_4)_2HPO_4$ produced a white polycrystalline $(NH_4)Ga_2Mn(PO_4)_3(H_2O)_2$.

This compound was also obtained by the same way as $(NH_4)MgGa_2(PO_4)_3(H_2O)_2$ was synthesized using $Mn(OAc)_2·2H_2O$ instead of $Mg(OAc)_2·4H_2O$ (Overweg et al. 1999).

8.1.9 Gallium–Hydrogen–Phosphorus–Oxygen–Iron–Nitrogen

The **$(NH_4)Ga_{1.95}Fe_{1.05}(PO_4)_3(H_2O)_2$** multinary compound, which crystallizes in the monoclinic structure with the lattice parameters a = 1,344.77 ± 0.04, b = 1,021.33 ± 0.03, c = 889.73 ± 0.03 pm and β = 108.473 ± 0.001°, is formed in the Ga–H–P–O–Fe–N system (Bieniok et al. 2008). Solvothermal synthesis method was used for the preparation of this metalophosphate. The synthesis was carried out in a Teflon-lined steel autoclave at 190°C and autogenous pressure for 72 h. As appropriate solvent, ethylene glycol ($C_2H_6O_2$) was used. Imidazole ($C_3H_4N_2$) was added as cationic organic template and NH_4^+-supplier. Phosphate gel precursor with initial pH values of 5–7 was prepared with the reactants $FeCl_2$, Ga_2O_3, 85% H_3PO_4, $C_3H_4N_2$, and $C_2H_6O_2$ (molar ratio of FeO, Ga_2O_3, P_2O_5, $C_3H_4N_2$, and $C_2H_6O_2$ was 1:1:5:10:50). After homogenous stirring, the mixture was heated for 3–5 days. After the heating period, the product was cooled down very quickly and washed several times with distilled water and finally acetone. Brown-violet parallelepipeds of the title compound were obtained.

This compound could also be obtained by the same way as $(NH_4)MgGa_2(PO_4)_3(H_2O)_2$ was synthesized using $Fe(OAc)_2$ instead of $Mg(OAc)_2·4H_2O$ (Overweg et al. 1999).

8.1.10 Gallium–Hydrogen–Phosphorus–Oxygen–Cobalt–Nitrogen

The **$(NH_4)Ga_2Co(PO_4)_3(H_2O)_2$** multinary compound, which crystallizes in the monoclinic structure, with the lattice parameters a = 1,332.3 ± 0.3, b = 1,024.5 ± 0.1, c = 888.6 ± 0.2 pm and β = 108.43 ± 0.02°, is formed in the Ga–H–P–O–Co–N system (Chippindale et al. 1996). The title compound was synthesized as single crystals under hydrothermal conditions from two gels (in molar composition): ***A*** – 1 Ga_2O_3 + 1.3 CoO + 10 H_3PO_4 + 156 H_2O + 0.3 $Si(OEt)_4$ + 10 imidazole; ***B*** – 1 Ga_2O_3 + 1 CoO + 10 H_3PO_4 + 156 H_2O + 0.3 $Si(OEt)_4$ + 5 $(NH_4)_2HPO_4$. Ga_2O_3 (0.5 g) and an appropriate quantity of CoO were dispersed in either n-butanol or water by stirring and a small amount of $Si(OEt)_4$ was added. The amine was also added, and the mixture was stirred for 10 min. Finally, 85% H_3PO_4 was added with further stirring until a gel of uniform consistency was obtained. Each gel was sealed in a Teflon-lined stainless steel autoclave and heated at 180°C for 7 days. After cooling, the solid products were collected by filtration, washed with distilled water, and dried in air. The organic amine could be replaced by $(NH_4)_2HPO_4$.

This compound could also be obtained by the same way as $(NH_4)MgGa_2(PO_4)_3(H_2O)_2$ was synthesized using CoO instead of $Mg(OAc)_2·4H_2O$ (Overweg et al. 1999).

8.1.11 Gallium–Hydrogen–Phosphorus–Oxygen–Nickel–Nitrogen

The **$(NH_4)Ga_2Ni(PO_4)_3(H_2O)_2$** multinary compound, which crystallizes in the monoclinic structure with the lattice parameters a = 1,321.58 ± 0.03, b = 1,023.72 ± 0.02, c = 885.94 ± 0.02 pm and β = 108.242 ± 0.001°, is formed in the Ga–H–P–O–Ni–N system (Bieniok et al. 2008). Solvothermal synthesis method was used for the preparation of this metalophosphate. The synthesis was carried out in Teflon-lined steel autoclave at 180°C and autogenous pressure for 120 h. As appropriate solvent, ethylene glycol ($C_2H_6O_2$) was used. Imidazole ($C_3H_4N_2$) was added as cationic organic template and NH_4^+-supplier. Phosphate gel precursor with initial pH values of 5–7 was prepared with the reactants $NiCl_2 \cdot 6H_2O$, Ga_2O_3, 85% H_3PO_4, $C_3H_4N_2$, and $C_2H_6O_2$ (molar ratio of NiO, Ga_2O_3, P_2O_5, $C_3H_4N_2$, and $C_2H_6O_2$ was 1:1:5:10:50). After homogenous stirring, the mixture was heated for 3–5 days. After the heating period, the product was cooled down quickly and washed several times with distilled water and finally acetone. Light yellow lozenges of the title compound were obtained.

8.1.12 Gallium–Hydrogen–Antimony–Oxygen–Sulfur–Nitrogen

Two multinary compounds, $\mathbf{(NH_4)_4Ga_4SbS_9(OH){\cdot}H_2O}$ and $\mathbf{(NH_4)_3Ga_4SbS_9(OH_2){\cdot}2H_2O}$, are formed in the Ga–H–Sb–O–S–N system (Mertz et al. 2009). First of them is characterized by an energy gap of 2.8 eV. Second compound crystallizes in the cubic structure with the lattice parameters $a = 1{,}290.98 \pm 0.04$, a calculated density of 2.400 g cm^{-3} and an energy gap of 2.6 eV.

These compounds were synthesized hydrothermally using 8-mm Pyrex tubes, to which was added Ga_2S_3 (0.4 mM) Sb_2S_3 (0.9 mM), S (0.9 mM), and 0.3 mL of 30% NH_4OH (Mertz et al. 2009). These tubes were then sealed under vacuum and kept in an oven at 160°C for 7–10 days. Colorless and yellow crystals were collected manually from the mixtures of unreacted starting materials.

8.1.13 Gallium–Hydrogen–Antimony–Sulfur–Nitrogen

The $\mathbf{(NH_4)_5Ga_4SbS_{10}}$ quinary compound, which crystallizes in the cubic structure with the lattice parameters $a = 1{,}290.00 \pm 0.05$ and a calculated density of 2.511 g cm^{-3}, is formed in the Ga–H–Sb–S–N system (Mertz et al. 2009). The title compound was synthesized hydrothermally in a 23-mL Teflon-lined autoclave using Ga (1 mM), Sb (1.0 mM), and S (10 mM) and a 6 mL 50% mixture of NH_4OH (30% in water) and triethylamine (30% in water) at 160°C for a week. There are at least three phases within the system that can be identified: Sb_2S_3, $NH_4Sb_5S_8$, and $(NH_4)_5Ga_4SbS_{10}$. Dark red crystals of this compound were removed manually.

8.1.14 Gallium–Lithium–Barium–Fluorine–Nitrogen

The $\mathbf{LiBa_5GaN_3F_5}$ quinary compound, which crystallizes in the orthorhombic structure with the lattice parameters $a = 1{,}546.3 \pm 0.3$, $b = 570.7 \pm 0.1$, $c = 1{,}225.9 \pm 0.3$ pm, a calculated density of 5.531 g cm^{-3} and an energy gap of 1.9 eV, is formed in the Ga–Li–Ba–F–N system (Hintze and Schnick 2010). The synthesis of this compound was carried out in Ta crucibles (30 mm length, 9.5 mm diameter, 0.5 mm wall thickness). Under Ar atmosphere in a glove box, NaN_3 (0.35 mM), Ga (0.138 mM), Ba (0.549 mM), and EuF_3 (0.027 mM) were mixed and filled into the Ta crucible. For the flux, Na (2.174 mM) and Li (0.145 mM) were added. The Ta crucible was sealed under Ar by arc welding. To protect the Ta crucible from oxidation, it was placed into a silica tube under Ar atmosphere. In a tube furnace, the crucible was heated to 760°C at a rate of 50°C h^{-1}. The temperature was maintained for 48 h and then lowered with 3.7°C h^{-1} to 200°C. Once the temperature reached 200°C, the furnace was turned off and cooled down to room temperature. The Ta crucible was opened and Na was separated from the reaction products by evaporation at 320°C under vacuum (0.1 Pa) for 18 h. From the inhomogeneous gray product, red needle-shaped single crystals of the title compound were isolated.

8.1.15 Gallium–Sodium–Phosphorus–Oxygen–Nitrogen

The $\mathbf{Na_3GaP_3O_9N}$ quinary compound, which crystallizes in the cubic structure with the lattice parameter $a = 935 \pm 1$ pm, is formed in the Ga–Na–P–O–N system (Conanec et al. 1996). This compound could be obtained in the temperature range of 600°C–800°C as a result of the next reactions: $6NaPO_3 + Ga_2O_3 + 2NH_3 = 2Na_3GaP_3O_9N + 3H_2O$ (Conanec et al. 1996) or $3Na_2O + 2P_2O_5 + Ga_2O_3 + 2PON = 2Na_3GaP_3O_9N$ (Feldmann 1987b).

8.1.16 Gallium–Potassium–Phosphorus–Oxygen–Nitrogen

The $\mathbf{K_3GaP_3O_9N}$ quinary compound, which crystallizes in the cubic structure with the lattice parameter $a = 976 \pm 1$ pm, is formed in the Ga–K–P–O–N system (Conanec et al. 1996). This compound could be obtained in the temperature range of 600°C–800°C as a result of the next reactions: $6KPO_3 + Ga_2O_3 + 2NH_3 = 2K_3GaP_3O_9N + 3H_2O$ (Conanec et al. 1996) or $3K_2O + 2P_2O_5 + Ga_2O_3 + 2PON = 2K_3GaP_3O_9N$ (Feldmann 1987b).

8.1.17 Gallium–Indium–Phosphorus–Arsenic–Nitrogen

The spinodal decomposition regions for $\mathbf{In_xGa_{1-x}As_yP_zN_{1-y-z}}$ were estimated from 0°C to 1,000°C for $x, y, z \geq 0.002$ (Elyukhin 2017). This solid solution may be lattice matched to GaP, GaAs, and InP, and the spinodal decomposition temperatures are shown in Figure 8.1. Curves 2, 4, and 5 demonstrate that the spinodal decomposition temperatures are lower than the conventional growth temperatures for the alloys with the nitrogen content up to 0.012.

8.1.18 Gallium–Indium–Phosphorus–Antimony–Nitrogen

The spinodal decomposition regions for $\mathbf{In_xGa_{1-x}Sb_yP_zN_{1-y-z}}$ were estimated from 0°C to 1,000°C for $x, y, z \geq 0.002$ (Elyukhin 2017). The spinodal decomposition regions of $In_xGa_{1-x}Sb_yP_zN_{1-y-z}$ lattice matched to GaP (curve 1, $y = 0.01$), GaAs (curve 2, $y = 0.02$), and InP (curve 3, $y = 0.05$) as functions of the nitrogen content are shown in Figure 8.2. The alloys lattice matched to GaP and InP have lower spinodal decomposition temperatures. The spinodal decomposition temperature is 1,000°C for the alloys lattice matched to InAs and GaSb if the nitrogen content is smaller than 0.002. Thus, $In_xGa_{1-x}Sb_yP_zN_{1-y-z}$ lattice matched to GaP and InP are preferable from the thermodynamic stability standpoint.

8.1.19 Gallium–Indium–Arsenic–Antimony–Nitrogen

The spinodal decomposition regions for $\mathbf{In_xGa_{1-x}Sb_yAs_zN_{1-y-z}}$ were estimated from 0°C to 1,000°C for $x, y, z \geq 0.002$ (Elyukhin 2017). The calculations fulfilled for solid solution demonstrate that the enthalpies of the constituents increase the spinodal decomposition region significantly. The spinodal decomposition regions of $In_xGa_{1-x}Sb_yAs_zN_{1-y-z}$ lattice matched to GaAs, InP, and InAs as functions of the nitrogen content are shown in Figure 8.3.

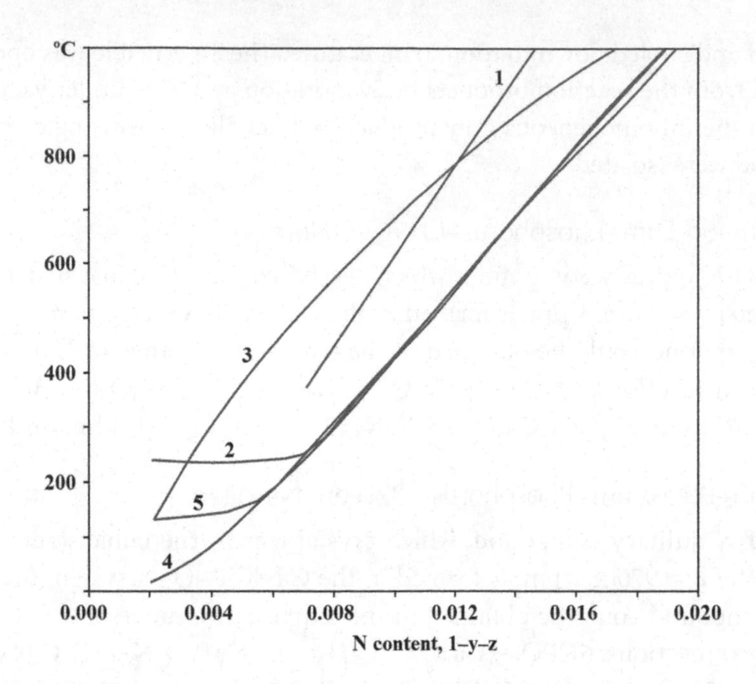

FIGURE 8.1 Spinodal decomposition regions of $In_xGa_{1-x}As_yP_zN_{1-y-z}$ lattice matched to GaP with $y = 0.03$ (curve 1), lattice matched to GaAs with $y = 0.5$ (curve 2), with $y = 0.8$ (curve 3), with $y = 0.2$ (curve 4), and lattice matched to InP with $y = 0.2$ (curve 5). (With permission from Elyukhin, V.A., *J. Cryst. Growth*, **470**, 42, 2017.)

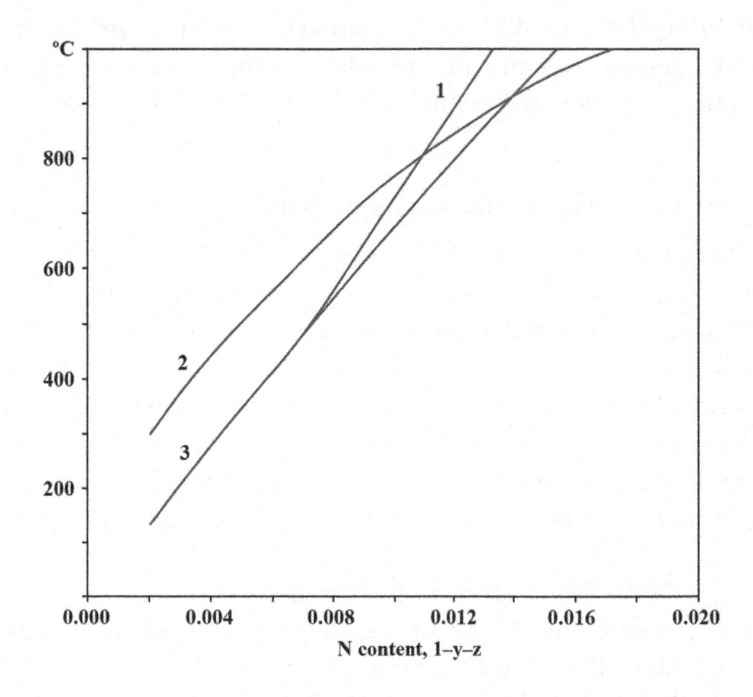

FIGURE 8.2 Spinodal decomposition regions of $In_xGa_{1-x}Sb_yP_zN_{1-y-z}$ lattice matched to GaP with $y = 0.01$ (curve 1), lattice matched to GaAs with $y = 0.02$ (curve 2), and lattice matched to InP with $y = 0.05$ (curve 3). (With permission from Elyukhin, V.A., *J. Cryst. Growth*, **470**, 42, 2017.)

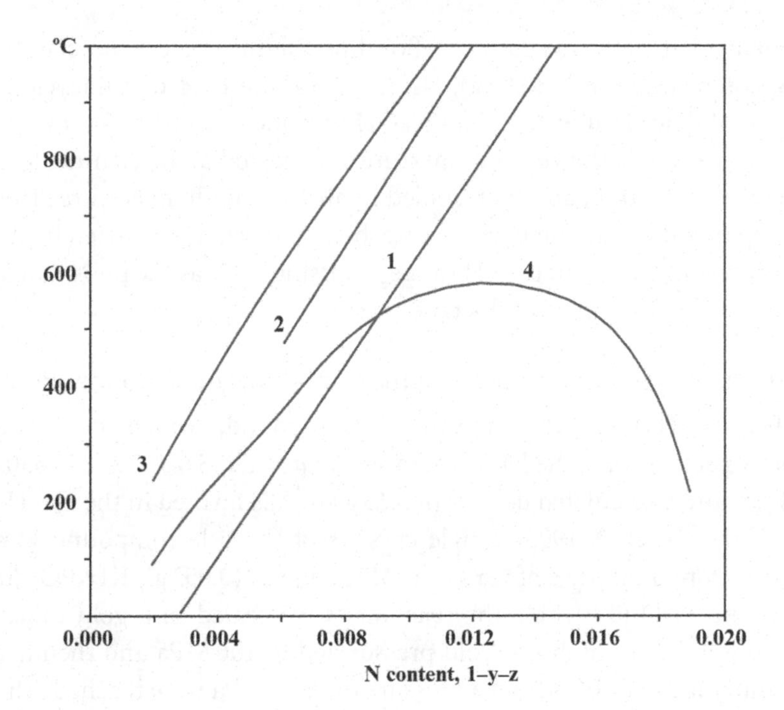

FIGURE 8.3 Spinodal decomposition regions of $In_xGa_{1-x}Sb_yAs_zN_{1-y-z}$ lattice matched to GaAs with $y = 0.002$ and with $y = 0.0157$ (curve 2), lattice matched to InP with $y = 0.05$ (curve 3), and lattice matched to InAs with $y = 0.05$ (curve 4). (With permission from Elyukhin, V.A., *J. Cryst. Growth*, **470**, 42, 2017.)

The spinodal decomposition temperatures of the alloys lattice matched to GaAs increase with increasing Sb content (curves 1: $y = 0.002$ and 2: $y = 0.0157$) at the same nitrogen content. The increase of the Sb content decreases significantly the nitrogen content range of the alloys lattice match to GaAs. The spinodal decomposition regions of $In_xGa_{1-x}Sb_yAs_zN_{1-y-z}$ lattice matched to InP ($y = 0.05$) and InAs ($y = 0.05$) are shown in Figure 8.3, curves 3 and 4, respectively. The spinodal decomposition temperatures of the alloys lattice matched to InP are much higher than those for the alloys lattice matched to InAs for the same Sb and nitrogen contents. Moreover, the nitrogen content range is the most extended one. It is due to the small Ga contents of such alloys. The alloys lattice matched to GaSb have high-temperature spinodal decomposition regions with insignificant nitrogen content.

8.2 SYSTEMS BASED ON GaP

8.2.1 Gallium–Hydrogen–Sodium–Oxygen–Phosphorus

The $\mathbf{NaGaHP_3O_{10}}$ and $\mathbf{Na_3Ga_5(PO_4)_4O_2(OH)_2{\cdot}2H_2O}$ quinary compounds are formed in the Ga–H–Na–O–P system. First compound crystallizes at ~200°C and was synthesized by heating the mixtures of Ga_2O_3, 85% H_3PO_4, and Na_2CO_3 at 150°C–500°C (Chudinova et al. 1979).

$Na_3Ga_5(PO_4)_4O_2(OH)_2{\cdot}2H_2O$ crystallizes in the monoclinic structure with the lattice parameters $a = 791.6 \pm 0.1$, $b = 1{,}348.5 \pm 0.2$, $c = 639.07 \pm 0.08$ pm and $\beta = 99.82 \pm 0.01°$ and a calculated density of 3.62 g cm^{-3} (Attfield et al. 1995). The title compound was synthesized hydrothermally in a 23-mL Teflon-lined stainless steel autoclave from a reaction mixture containing $GaCl_3$ (0.86 g), H_3PO_3 (0.6 g), and 1,4-diazabicyclo[2.2.2]octane (0.9 g) in 15 mL of deionized water. Solid NaOH was added to the reaction mixture until a pH of ~10 was reached. The reaction mixture was heated at 10°C min^{-1} to 190°C, held at this temperature for 100 h, and then cooled in air to room temperature. The crystalline product was separated by suction filtration, washed with water, and dried in air. The material has also been synthesized in the pH range 7–9 using P_2O_5 as the phosphorus source, in the absence of 1,4-diazabicyclo[2.2.2]octane.

8.2.2 Gallium–Hydrogen–Potassium–Germanium–Oxygen–Fluorine–Phosphorus

The $\mathbf{KGa_{0.5}Ge_{0.5}(F,OH)_{0.5}O_{0.5}PO_4}$ multinary compound, which crystallizes in the orthorhombic structure with the lattice parameters a = 1,275.6 ± 0.2, b = 630.6 ± 0.1, c = 1,041.3 ± 0.2 pm and a calculated density of 3.53 g cm^{-3}, is formed in the Ga–H–K–Ge–O–F–P system (Harrison et al. 1995). Single crystals of the title compound were prepared hydrothermally from a mixture of Ga_2O_3 (0.047 g), GeO_2 (0.053 g), KH_2PO_4 (0.681 g), 5 N KF (0.40 mL), and H_2O (0.10 mL). The reactants were sealed in a gold tube, which was installed in a bomb. This bomb was cold pressurized to 100 MPa and then heated until a maximum temperature of 700°C and a pressure of ~250 MPa were reached. The bomb was held at 700°C for 8 h and then cooled to 525°C at a rate of 2°C h^{-1}, to 300°C at 10°C h^{-1}, and then cooled to ambient temperature overnight. After product recovery from the gold tube and workup with water, air stable, transparent, rhomboidal crystals of this compound were recovered.

8.2.3 Gallium–Hydrogen–Potassium–Oxygen–Phosphorus

The $\mathbf{KGaHP_3O_{10}}$ and $\mathbf{KGa(H_2P_2O_7)_2}$ quinary compounds are formed in the Ga–H–K–O–P system (Chudinova et al. 1977, 1978). These compounds were prepared as the crystalline materials by heating the mixtures of Ga_2O_3, 85% H_3PO_4, and K_2CO_3 in glassy carbon crucibles at 100°C–500°C for 5–10 days. $KGaHP_3O_{10}$ crystallizes from the melt in the temperature region from ~140°C up to 300°C and $KGa(H_2P_2O_7)_2$ is stable up to ~330°C, and it decomposes at a higher temperature with the formation of $KGa(PO_3)_4$.

8.2.4 Gallium–Hydrogen–Potassium–Oxygen–Fluorine–Phosphorus

The $\mathbf{KGaF_{1-\delta}(OH)_\delta PO_4}$ ($\delta \approx 0.3$) multinary compound, which crystallizes in the orthorhombic structure with the lattice parameters a = 1,271.7 ± 0.2, b = 630.21 ± 0.08, c = 1,041.3 ± 0.2 pm and a calculated density of 3.54 g cm^{-3}, is formed in the Ga–H–K–O–F–P system (Harrison et al. 1995). Single crystals of the title compound were prepared hydrothermally. A mixture of Ga_2O_3 (0.094 g), 83% H_3PO_4 (0.30 mL), and 5 N KF (0.70 mL) were heat-sealed in a gold tube, which was installed in a bomb. This bomb was cold pressurized to 100 MPa and then heated until a maximum temperature of 700°C and pressure of ~250 MPa were reached. The bomb was held at 700°C for 8 h and then cooled to 525°C at a rate of 2°C h^{-1}, to 300°C at 10°C h^{-1}, and then cooled to ambient temperature overnight. The obtained crystals were filtered, rinsed with H_2O, and air-dried.

8.2.5 Gallium–Hydrogen–Potassium–Oxygen–Nickel–Phosphorus

The $\mathbf{K[NiGa_2(PO_4)_3(H_2O)_2]}$ multinary compound, which crystallizes in the monoclinic structure with the lattice parameters a = 1,320.95 ± 0.13, b = 1,017.33 ± 0.09, c = 881.30 ± 0.09 pm, β = 107.680 ± 0.010° and a calculated density of 3.286 g cm^{-3} at 150 K, is formed in the Ga–H–K–O–Ni–P system (Chippindale et al. 2009). The title compound was prepared hydrothermally from a gel of a composition Ga_2O_3 + $NiCl_2{\cdot}6H_2O$ + 83% H_3PO_4 + H_2O + $Si(OEt)_4$ + KH_2PO_4 (molar ratio 1:1:10:156:0.1:5), which was heated at 160°C for 7 days in a Teflon-lined stainless steel autoclave. The solid product was collected by filtration, washed with deionized water, and dried in air. The product consisted of large yellow faceted blocks of the title compound, which could be easily separated from unidentified white powder.

8.2.6 Gallium–Hydrogen–Rubidium–Oxygen–Phosphorus

Some quinary compounds are formed in the Ga–H–Rb–O–P system. $\mathbf{RbGaHP_3O_{10}}$ and $\mathbf{RbGa(H_2P_2O_7)_2}$ were prepared by heating the mixtures of Ga_2O_3, 85% H_3PO_4, and Rb_2CO_3 at 150°C–500°C (Chudinova et al. 1979).

$\mathbf{Rb_2[Ga_4(HPO_4)(PO_4)_4]{\cdot}0.5H_2O}$ crystallizes in the monoclinic structure with the lattice parameters a = 506.07 ± 0.01, b = 2,164.31 ± 0.06, c = 820.67 ± 0.02 pm, β = 91.768 ± 0.001° and a calculated density of 3.455 g cm^{-3} (Lii 1996a). High-temperature, high-pressure hydrothermal synthesis was performed in gold ampoules contained in an autoclave where pressure was provided by water. Colorless acicular crystals of the title compound with a small amount of colorless crystals of $GaPO_4$ were obtained by heating 3M RbH_2PO_4 (0.375 mL), 3M H_3PO_4 (0.375 mL), and Ga_2O_3 (0.0937 g) (molar ratio Rb/P = 0.5) in a sealed gold ampoule at 600°C and an estimated pressure of 0.24 GPa for 40 h, cooled at 5°C h^{-1} to 275°C, and quenched to room temperature by removing the autoclave from the furnace.

8.2.7 Gallium–Hydrogen–Cesium–Oxygen–Phosphorus

Some quinary compounds are formed in the Ga–H–Rb–O–P system. $\mathbf{CsGaHP_3O_{10}}$ is the most stable phase in this system, with prolonged heating at temperatures above 350°C it gradually turns into $Cs_2Ga_3(PO_4)_{12}$, and has three polymorphic modifications (Chudinova et al. 1987a; Grunze et al. 1987; Anisimova et al. 1995). Two modifications of this compound crystallize in the monoclinic structure with the lattice parameters a = 884.3 ± 0.1, b = 495.23 ± 0.05, c = 1,108.4 ± 0.1 pm, β = 108.793 ± 0.006° (Mi et al. 2003b) and a = 906.1 ± 0.1, b = 871.05 ± 0.09, c = 621.95 ± 0.08 pm, β = 111.993 ± 0.006° (Mi et al. 2003c), respectively. Third modification crystallizes in the orthorhombic structure with the lattice parameters a = 1,176.3 ± 0.1, b = 496.55 ± 0.05, c = 1,572.2 ± 0.1 pm and a calculated density of 3.302 g cm^{-3} (Anisimova et al. 1995).

The title compound was synthesized in aqueous solution by two steps (Mi et al. 2003c). In the first step, the reaction was carried out with a mixture of $GaCl_3$ (1.046 g of Ga dissolved in 5 mL 37% HCl), CsCl (2.525 g), and an excess of HCl (molar ratio Ga/Cs = 1:1). The mixture was heated to the boiling point. While it was cooled down and evaporated in air for several days, transparent colorless crystals were obtained. After filtering from liquid, they were identified to be $CsGaCl_4$. On the second step, the reaction was made with the mixture of $CsGaCl_4$ (3.444 g), $CsOH{\cdot}H_2O$ (1.679 g), and 5 mL of 85% H_3PO_4 (molar ratio 1:1:7). The mixture was heated (open system) to the boiling point on stove and kept heating for 3 days to evaporate the solvent. Three modifications of $CsGaHP_3O_{10}$ crystals were obtained in the reaction product. All of them were colorless and transparent and crystallize in the monoclinic structure. The orthorhombic modification reported by Anisimova et al. (1995) was not found in these products, indicating that it should have different synthetic conditions or a different stability temperature range.

Single crystals of the orthorhombic modification were obtained by fast heating (0.5 h) a mixture of $CsGa(H_2PO_4)_4$, 85% H_3PO_4, and Cs_2CO_3 in the molar ratio $P_2O_5/Cs_2O/Ga_2O_3$ = 15:7.5:1 up to 450°C, followed by carefully controlled crystallization at gradual decrease of the temperature 30°C h^{-1} down to 380°C (Anisimova et al. 1995). According to the data of Chudinova et al. (1987a), this compound is formed in the wide temperature region at a molar ratios $P_2O_5/Cs_2O/Ga_2O_3$ = 15:2.5:(5, 7.5, 10):1.

$CsGa(H_2PO_4)_4$ was synthesized by solving metallic Ga in warm aqua regia, evaporating the excess of HCl and HNO_3 by reaction with 85% H_3PO_4, and neutralization of the phosphoric solution with Cs_2CO_3 up to molar ratio Ga/P/Cs = 1:4:1 (Anisimova et al. 1995, 1997).

$Cs_2GaH_3(P_2O_7)$ is thermally stable up to ~340°C and crystallize in the monoclinic structure with the lattice parameters $a = 507.6 \pm 0.1$, $b = 795.5 \pm 0.1$, $c = 1{,}689.8 \pm 0.3$ pm, $\beta = 96.96 \pm 0.02°$ and a calculated density of 3.37 g cm^{-3} (Grunze et al. 1987, 1988). This compound is formed by heating a mixture of P_2O_5, Cs_2O, and Ga_2O_3 [molar ratios 15:(7.5–10):1] at 150°C–170°C. It is metastable and during heating slowly transforms into $CsGaHP_3O_{10}$ (Chudinova et al. 1987a).

$Cs_2Ga(H_2PO_4)(HPO_4)_2{\cdot}H_3PO_4{\cdot}0.5H_2O$ crystallizes in the tetragonal structure with the lattice parameters $a = 2{,}016.8 \pm 0.5$, $c = 1{,}828.7 \pm 0.5$ pm and a calculated density of 2.613 g cm^{-3} (Anisimova et al. 1997). Colorless needle-shaped crystals of the title compound suitable for X-ray diffraction (XRD) have been obtained by microwave-induced heating a mixture of $CsGa(H_2PO_4)_4$ and Cs_2CO_3 (molar ratio of Cs_2O/Ga_2O_3 = 7.5:1) in an excess of 85% H_3PO_4.

8.2.8 Gallium–Hydrogen–Oxygen–Manganese–Phosphorus

The **$Mn_3Ga_4(H_2O)_6(PO_4)_6$** quinary compound, which crystallize in the monoclinic structure with the lattice parameters $a = 894.68 \pm 0.04$, $b = 1{,}014.81 \pm 0.05$, $c = 1{,}355.40 \pm 0.7$ pm, $\beta = 108.249 \pm 0.001°$ and a calculated density of 3.187 g cm^{-3}, is formed in the Ga–H–O–Mn–P system (Hsu and Wang 2000). To prepare this compound, a mixture of 4,4′-trimethylenedipiperidine (3 mM), $Ga(NO_3)_3{\cdot}xH_2O$ (1.5 mM), $MnCl_2{\cdot}4H_2O$ (1.5 mm), 85% H_3PO_4 (6 mM), $C_2H_6O_2$ (5.8 mL), and H_2O (5.8 mL) was placed in a Teflon-lined digestion bomb with an internal volume of 23 mL and heated at 150°C under autogeneous pressure for 3 days followed by slow cooling to 30°C at 6°C h^{-1}. The product was filtered off, washed with water, rinsed with ethanol, and dried in a desiccator at ambient temperature. The final product contained a small amount of colorless columnar crystals of the title compound and a relatively large amount of colorless needles of **$Mn_5(H_2O)_4(HPO_4)_2(PO_4)_2$**.

8.2.9 Gallium–Indium–Tin–Arsenic–Phosphorus

To estimate the equilibrium compositions of the liquid and solid phases, a computational approach was used, followed by verification of its adequacy for some experimental points (Vasil'yev et al. 1985). The isotherms of the liquidus surface of this quaternary system for the Sn-rich melts were determined.

8.3 SYSTEMS BASED ON GaAs

8.3.1 Gallium–Hydrogen–Cesium–Oxygen–Arsenic

The **$CsGa(H_{1.5}AsO_4)_2(H_2AsO_4)$** quinary compound, which crystallizes in the monoclinic structure with the lattice parameters $a = 471.4 \pm 0.1$, $b = 1{,}467.4 \pm 0.3$, $c = 1{,}516.2 \pm 0.3$ pm, and $\beta = 93.31 \pm 0.03°$ and a calculated density of 3.961 g cm^{-3}, is formed in the Ga–H–Cs–O–As system (Schwendtner and Kolitsch 2005). Large colorless glassy prisms of the title compound were prepared hydrothermally (220°C, 7 days) in a Teflon-lined stainless steel autoclave from a mixture of Cs_2CO_3, Ga_2O_3 (approximate Cs/Ga molar ratio = 1:1), arsenic acid, and distilled water. Enough arsenic acid was added to keep the pH between about 1.5 and 0.5. The Teflon cylinders were filled with distilled water up to approximately 80% of their inner volume. The initial and final pH values were about 1.5 and 1, respectively. The prisms were accompanied by a small amount of very small hexagonal colorless platelets of **$CsGa(HAsO_4)_2$**.

8.3.2 Gallium–Hydrogen–Lead–Oxygen–Sulfur–Arsenic

The **$PbGa_3(AsO_4,SO_4)_2(OH)_6$** multinary phase (mineral gallobeudantite), which crystallizes in the rhombohedral structure with the lattice parameters $a = 722.5 \pm 0.4$ and $c = 1{,}703 \pm 2$ pm in hexagonal setting and a calculated density of 4.61 g cm^{-3}, is formed in the Ga–H–Pb–O–S–As system (Jambor et al. 1996a, 1997b).

8.3.3 Gallium–Copper–Germanium–Selenium–Arsenic

$3GaAs–Cu_2GeSe_3$. According to the data of XRD and metallography, there is a continuous series of solid solutions in this system (Goryunova et al. 1965). The lattice parameter approximately corresponds to the Vegard's law. It was not possible to obtain completely homogeneous and sufficiently large crystalline samples. According to the data of Averkieva et al. (1964, 1968), solid solutions based on GaAs contain up to 40 mol% Cu_2GeSe_3, and the lattice parameter changes linearly versus composition. A narrow region of the solid solution also exists on the base of GaAs.

8.3.4 Gallium–Zinc–Tin–Selenium–Arsenic

GaAs–ZnSe–Sn. Isothermal sections of this quasiternary system at 600°C, 650°C, 700°C, and 800°C are shown in Figure 8.4 and some vertical sections are given in Figure 8.5 (Novikova et al. 1974). The eutectic temperatures in all vertical sections are degenerated from the Sn-rich side and are equal to 218°C–222°C (the eutectic temperature increases with increasing ZnSe content). $(ZnSe)_{1-x}(GaAs)_x$ solid solutions were obtained by the annealing of thoroughly mixed starting binary compounds at 1,050°C for 100 h. These solid solutions were mixed with Sn and annealed at 1,050°C for 50 h.

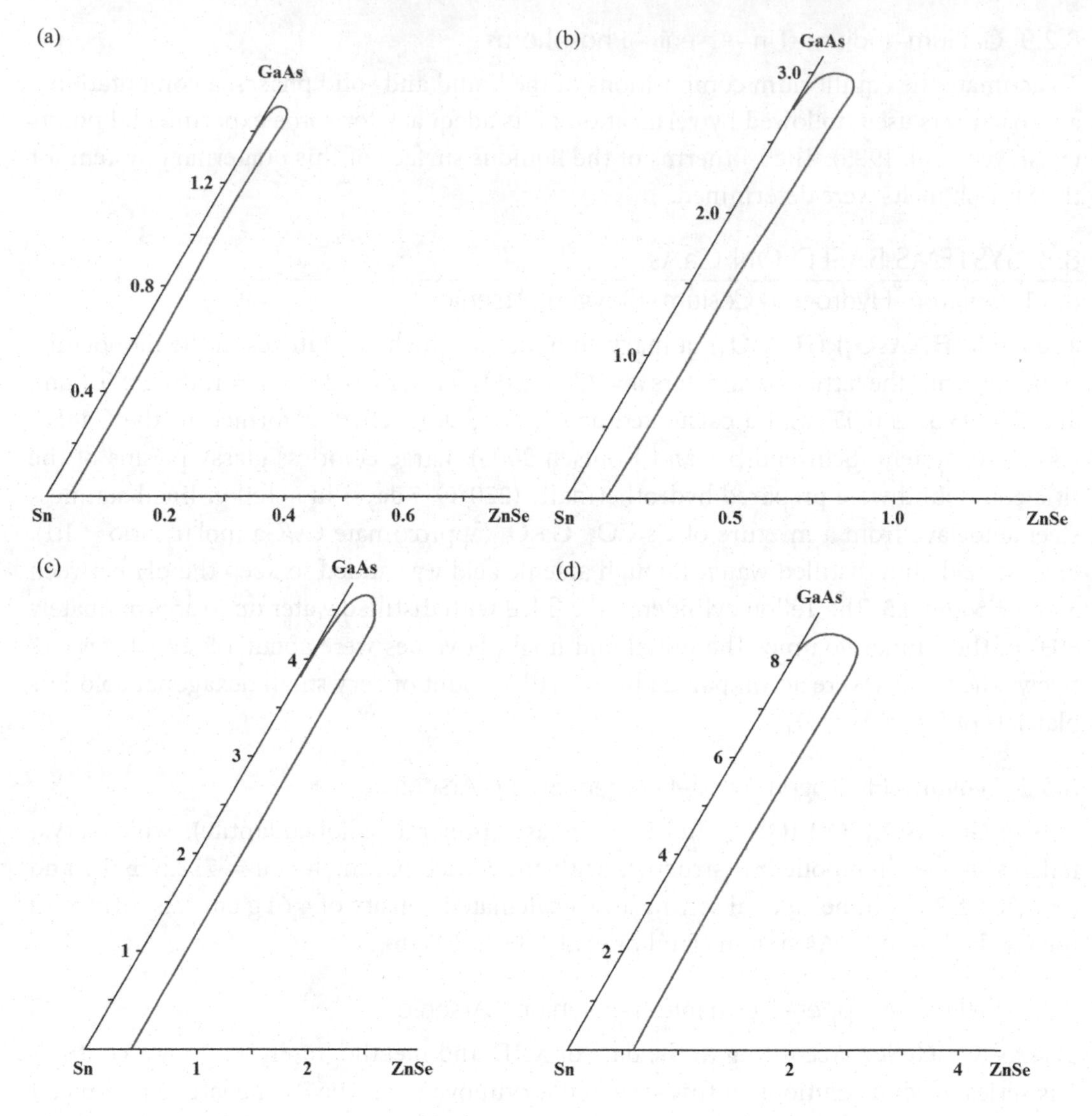

FIGURE 8.4 Isothermal sections of the GaAs–ZnSe–Sn quasiternary system at (a) 600°C, (b) 650°C, (c) 700°C, and (d) 800°C. (With permission from Novikova, E.M. et al., *Deposited in VINITI*, № 3039–74Dep, 1974.)

8.3.5 Gallium–Cadmium–Tin–Antimony–Arsenic

2GaSb–CdSnAs$_2$. According to the data of XRD, solid solution based on $CdSnAs_2$ contain up to 50 mol% 2GaSb (Anshon et al. 1987).

8.3.6 Gallium–Indium–Lead–Antimony–Arsenic

Phase equilibria in the Ga–In–Pb–Sb–As quinary system were investigated experimentally at 500°C and 600°C to prepare the $Ga_{1-x}In_xAs_ySb_{1-y}$ solid solutions isoperiodic with gallium antimonide (Andreev et al. 1999). The compositions of coexisting liquid and solid phases on the GaSb substrate in the heterogeneous system were determined at these temperatures.

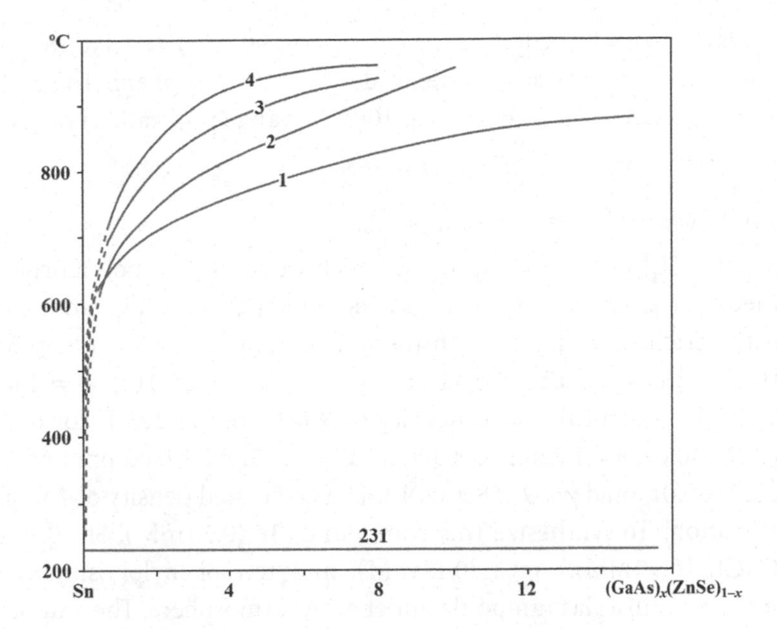

FIGURE 8.5 The $(GaAs)_x(ZnSe)_{1-x}$–Sn vertical sections at (1) $x = 0.8$, (2) 0.6, (3) 0.4, and (4) 0.2. (With permission from Novikova, E.M. et al., *Deposited in VINITI*, № 3039–74Dep, 1974.)

GaAs–GaSb–InAs–InSb–Pb. The melting diagram of this system includes the fields of primary crystallization of lead, homogeneous $Ga_xIn_{1-x}As_ySb_{1-y}$ solid solution, and two $Ga_xIn_{1-x}As_ySb_{1-y}$ solid solutions with different values of x and y that are formed as a result of instability of homogeneous phase due to continuous state changing (Grebenyuk et al. 1999).

8.4 SYSTEMS BASED ON GaSb

8.4.1 Gallium–Lithium–Silver–Oxygen–Antimony

The $\mathbf{Li_2Ag_6Ga_2Sb_2O_{12}}$ quinary compound is formed in the Ga–Li–Ag–O–Sb system (Kumar et al. 2012). It was synthesized by molten ion exchange of $Li_8Ga_2Sb_2O_{12}$ with $AgNO_3$.

8.4.2 Gallium–Sodium–Tellurium–Chlorine–Antimony

The $\mathbf{NaGa_6Sb_7Te_8Cl_{24}}$ quinary compound, which crystallizes in the triclinic structure with the lattice parameters $a = 1{,}218.22 \pm 0.12$, $b = 1{,}276.71 \pm 0.14$, $c = 1{,}830.7 \pm 1.4$ pm and $\alpha = 92.021 \pm 0.004°$, $\beta = 90.129 \pm 0.003°$, $\gamma = 105.793 \pm 0.004$ and a calculated density of 3.84 g cm^{-3} at 100 K, is formed in the Ga–Na–Te–Cl–Sb system (Eich et al. 2013). To synthesize this compound, in the glove box, weighed amounts of starting materials were filled in glass ampoules, which were evacuated and flame-sealed under vacuum. Tube ovens used for syntheses were aligned in an angle of about 30° to the horizontal to keep the melt compacted in the hot zone. After some days, the ampoules were inspected visually. If the formation of crystals was observed, the hot melt was decanted from the solids. After cooling to ambient temperature, the ampoules were opened in the glove box. Since this compound is very sensitive towards air, samples were stored under Ar atmosphere.

Te (0.36 mM), Sb (0.24 mM), $SbCl_3$ (0.12 mM), $GaCl_3$ (0.99 mM), NaCl (0.18 mM), and LiCl (0.18 mM) were added to a glass ampoule under an Ar atmosphere. The ampoules were evacuated and flame-sealed. After 69 days at 100°C, orange prismatic crystals appeared in the black reaction melt.

8.4.3 Gallium–Silver–Tellurium–Chlorine–Antimony

The **$Ag(Sb_7Te_8)[GaCl_4]$** quinary compound, which exists in two polymorphic modifications, is formed in the Ga–Ag–Te–Cl–Sb system (Eich et al. 2014). Both modifications crystallize in the triclinic structure with the lattice parameters a = 1,829.76 ± 0.03, b = 1,278.06 ± 0.02, and c = 2,690.53 ± 0.04 pm and α = 125.37 ± 0.01°, β = 132.06 ± 0.01°, and γ = 91.75 ± 0.01 and a calculated density of 3.92 g cm^{-3} at 123 K for α-modification and a = 1,834.21 ± 0.07, b = 1,278.15 ± 0.04, and c = 1,352.12 ± 0.05 pm and α = 125.38 ± 0.01°, β = 132.23 ± 0.01°, and γ = 91.58 ± 0.01 and a calculated density of 3.90 g cm^{-3} at 163 K for β-modification. To synthesize this compound, Te (0.36 mM), Sb (0.24 mM), $SbCl_3$ (0.12 mM), $GaCl_3$ (0.99 mM), AgCl (0.18 mM), and tetraphenylphosphonium chloride (0.16 mM) were placed in a glass ampoule under an Ar atmosphere. The ampoule was evacuated and flame-sealed. After 19 days at 100°C, black rod-shaped crystals with metallic shiny surfaces appeared in a liquid black melt. The actual orange color of the crystals was revealed as they were ground to a powder. The title compound is air sensitive.

8.4.4 Gallium–Lanthanum–Bismuth–Oxygen–Antimony

The **$LaBi_2GaSb_2O_{11}$** quinary compounds, which crystallizes in the cubic structure with the lattice parameter a = 951 pm, is formed in the Ga–La–Bi–O–Sb system (Sleight and Bouchard 1973). Single crystals of this compound were prepared from a Bi_2O_3 flux. The mixture of Bi_2O_3, Ga_2O_3, La_2O_3, and Sb_2O_3 (molar ratio 12:1:1:1) was heated in air to 1,000°C in an open Pt crucible, cooled at 6°C h^{-1} to 800°C, and the furnace cooled. After washing the product with nitric acid, the crystals of the title compound remained.

8.4.5 Gallium–Gadolinium–Bismuth–Oxygen–Antimony

The **$GdBi_2GaSb_2O_{11}$** quinary compounds, which crystallizes in the cubic structure with the lattice parameter a = 946 pm, is formed in the Ga–Gd–Bi–O–Sb system (Sleight and Bouchard 1973). Single crystals of this compound were prepared from a Bi_2O_3 flux. The mixture of Bi_2O_3, Ga_2O_3, Gd_2O_3, and Sb_2O_3 (molar ratio 12:1:1:1) was heated in air to 1,000°C in an open Pt crucible, cooled at 6°C h^{-1} to 800°C, and the furnace cooled. After washing the product with nitric acid, the crystals of the title compound remained.

8.4.6 Gallium–Tin–Lead–Bismuth–Antimony

The solubility of GaSb in the Sn–Pb–Bi eutectic (25 mass% Sn + 25 mass% Pb + 50 mass% Bi; melting temperature 93.75°C) within the temperature region from 175°C to 650°C was investigated by standing the GaSb ingots for 14–15 h in the melt of the solvent at a given temperature (Golubev et al. 1972). The solubility was determined by GaSb mass loss. The linear dependence of the logarithm of solubility on the reciprocal temperature indicates the absence of chemical reactions between the eutectic and gallium antimonide.

8.4.7 Gallium–Bismuth–Oxygen–Iron–Antimony

The $\mathbf{Ga_{1.2x}Bi_{1.8}Fe_{1.2(1-x)}SbO_7}$ quinary solid solution of pyrochlore structure is formed in the Ga–Bi–O–Fe–Sb system. It crystallizes in the cubic structure with the lattice parameter a = 1,044.348 ± 0.002, 1,044.295 ± 0.003, 1,042.896 ± 0.002, 1,042.363 ± 0.006, 1,041.321 ± 0.004, and 1,039.678 ± 0.003 pm for x = 0, 0.1, 0.3, 0.5, 0.7, and 0.9, respectively (Ellert et al. 2016) [a = 1,044.37 ± 0.01, 1,043.81 ± 0.02, 1,042.56 ± 0.02, 1,041.33 ± 0.01, and 1,039.67 ± 0.02 pm and E_g = 1.90, 1.95, 2.00, 2.11, and 2.25 eV for x = 0.1, 0.3, 0.5, 0.7, and 0.9, respectively (Egorysheva et al. 2017)]. It was synthesized by a conventional solid-state method (Ellert et al. 2016; Egorysheva et al. 2017). High-purity reagents Bi_2O_3, Fe_2O_3, Ga_2O_3, and Sb_2O_3 were used as starting materials. The stoichiometric mixture of oxides (4 g) was ground in acetone using an agate mortar and pestle for approximately 30 min. The synthesis was carried out in Pt crucibles inside an electrical muffle furnace. The samples were heated in two steps, first at 650°C for 24 h and subsequently at 800°C for 24 h. Each sample was reground after calcination. Subsequently, the samples were annealed at 930°C–980°C for 5–15 days and reground again.

CHAPTER 9

Systems Based on InB^V (B^V = N, P, As, Sb)

9.1 SYSTEMS BASED ON InN

9.1.1 Indium–Hydrogen–Oxygen–Phosphorus–Nitrogen

Three quinary compounds are formed in the In–H–O–P–N system. **$(NH_4)In(OH)PO_4$** crystallizes in the tetragonal structure with the lattice parameters a = 941.6 ± 0.2, c = 1,115.9 ± 0.3 pm and a calculated density of 3.288 g cm^{-3} (Mao et al. 2002a). It was synthesized under mild hydrothermal conditions. A mixture of H_3BO_3 (0.155 g), 85% H_3PO_4 (1 mL), 38% $(NH_4)OH$ (1 mL), and In (0.115 g) dissolved in 1 mL of 18% HCl (molar ratio of In/B/P = 1:2.5:8) was added to 1.0 mL of distilled water. The mixture was sealed in a glass tube. The filling of the solution was about 30% of the tube volume. The glass tube was placed in an oven and the temperature was increased slowly to 135°C and reacted for 10 days before cooling to room temperature. The reaction was under autogeneous pressure. The colorless ball-shaped crystals of the title compound were grown.

$(NH_4)In(PO_3(OH))_2$ crystallizes in the monoclinic structure with the lattice parameters a = 960.04 ± 0.04, b = 828.20 ± 0.04, c = 966.93 ± 0.03 pm and β = 116.205 ± 0.004° (Mao et al. 2002c). It was also synthesized under mild hydrothermal conditions. The reactions were carried out with mixtures of $InCl_3$, $LiCl{\cdot}H_2O$, $NaBO_2{\cdot}4H_2O$, and $(NH_4)H_2PO_4$ (molar ratio 1:4:9:6) in aqueous solution. The mixture was sealed in a 20-mL glass tube. The pH value of the solution was about 0.2. The filling of the solution was ~30% of the glass tube volume. The glass tube was placed in an oven with subsequent heating at 140°C for 7 days.

$(NH_4)_2In[(PO_4)(HPO_4)]$ crystallizes in the orthorhombic structure with the lattice parameters a = 1,745.9 ± 0.4, b = 920.0 ± 0.2 and c = 972.0 ± 0.2 pm (Li et al. 2002a). It was also synthesized under mild hydrothermal conditions. The reactions were carried out with mixtures of $InCl_3$, H_3BO_3, and $(NH_4)H_2PO_4$ (molar ratio 1:6:8) in aqueous solution. The mixture was sealed in a 20-mL glass tube. The filling of the solution was ~30% of the glass tube volume. The glass tube was placed in an oven with subsequent heating at 140°C for 7 days.

9.1.2 Indium–Hydrogen–Oxygen–Arsenic–Nitrogen

The **$(NH_4)InAs_2O_7$** quinary compound, which crystallizes in the triclinic structure with the lattice parameters $a = 785.8 \pm 0.2$, $b = 864.9 \pm 0.2$, $c = 1{,}051.5 \pm 0.2$ pm and $\alpha = 88.96 \pm 0.03°$, $\beta = 89.94 \pm 0.03°$, $\gamma = 74.34 \pm 0.03°$ and a calculated density of 3.81 g cm^{-3}, is formed in the In–H–O–As–N system (Schwendtner 2006). It was synthesized under hydrothermal conditions in a Teflon-lined steel autoclave at 220°C under autogenous pressure. Small, colorless, pseudo-disphenoidal crystals commonly intergrowth into clusters, were grown from a mixture of NH_4OH, In_2O_3, H_3AsO_4, and distilled water. The autoclave was filled to about 60%–80% of its inner volume and was then heated to 220°C, kept at this temperature for 7 days, and slowly cooled to room temperature overnight. Initial and final pH values were about 1.5 and 1, respectively. The reaction product was washed thoroughly with distilled water, filtered, and slowly dried at room temperature. The obtained material was phase pure.

9.1.3 Indium–Sodium–Phosphorus–Oxygen–Nitrogen

The **$Na_3InP_3O_9N$** quinary compound, which crystallizes in the cubic structure with the lattice parameter $a = 960 \pm 1$ pm, is formed in the In–Na–P–O–N system (Conanec et al. 1996). This compound could be obtained in the temperature range of 600°C–800°C as a result of the next reaction: $6NaPO_3 + In_2O_3 + 2NH_3 = 2Na_3InP_3O_9N + 3H_2O$.

9.1.4 Indium–Potassium–Phosphorus–Oxygen–Nitrogen

The **$K_3InP_3O_9N$** quinary compound, which crystallizes in the cubic structure with the lattice parameter $a = 999 \pm 1$ pm, is formed in the In–K–P–O–N system (Conanec et al. 1996). This compound could be obtained in the temperature range of 600°C–800°C as a result of the next reaction: $6KPO_3 + In_2O_3 + 2NH_3 = 2K_3InP_3O_9N + 3H_2O$.

9.2 SYSTEMS BASED ON InP

9.2.1 Indium–Hydrogen–Lithium–Oxygen–Phosphorus

Two quinary compounds are formed in the In–H–Li–O–P system. **$LiIn_2P_2O_7{\cdot}4H_2O$** was prepared at the interaction of $Li_4P_2O_7$ and $InCl_3$ in the aqueous solution (Deichman et al. 1967a).

$Li_2In[(PO_4)(HPO_4)]$ crystallizes in the monoclinic structure with the lattice parameters $a = 483.45 \pm 0.07$, $b = 823.6 \pm 0.1$, $c = 772.8 \pm 0.1$ pm and $\beta = 103.466 \pm 0.007°$ (Mi et al. 2002a). Colorless crystals of this compound were obtained under mild hydrothermal conditions. The reactions were carried out with mixtures of In (0.574 g) dissolved in 3 mL of 37% HCl, LiH_2PO_4 (10.393 g), and $Li_2B_4O_7$ (1.691 g) (molar ratio 1:20:2). The mixture was filled in a 20-mL Teflon autoclave with a filling degree of about 50%. The autoclave was placed in an oven with subsequent heating at 170°C for 7 days.

9.2.2 Indium–Hydrogen–Sodium–Oxygen–Phosphorus

Some quinary compounds are formed in the In–H–Na–O–P system. **$NaInHP_3O_{10}$** was synthesized from a mixture of 85% H_3PO_4, In_2O_3, and Na_2CO_3 (molar ratio of $P_2O_5/Na_2O/In_2O_3 = 15{:}5{:}1$) (Avaliani et al. 1979). The mixture was heated in a carbon glass crucible with periodic mixing at 150°C for 12 days and then at 220°C for 2 days. The resulting crystalline mass was washed from the melt with water, acetone, and ether.

$Na_2In_2[PO_3(OH)]_4{\cdot}H_2O$ crystallizes in the triclinic structure with the lattice parameters a = 930.13 ± 0.01, b = 949.76 ± 0.01, c = 926.85 ± 0.07 pm and α = 98.710 ± 0.004°, β = 98.953 ± 0.004°, γ = 60.228 ± 0.006° and a calculated density of 3.217 g cm^{-3} (Mi et al. 2001). The dehydrated compound melts at 1,208°C. It was obtained by a mild hydrothermal method. A mixture of $Na_2B_4O_7{\cdot}10H_2O$ (0.9534 g), $Na_2HPO_4{\cdot}12H_2O$ (2.149 g), and 0.115 g of In dissolved in 2 mL of 18% HCl (molar ratio of Na/In/P/B = 17:1:6:10) was added to 1.0 mL of distilled water. The mixture was sealed in glass tubes about 20 cm in length. The filling of the solution was ~30% of the tube volume. The glass tubes were placed in an oven and the temperature was increased slowly to 80°C and prereacted for a week. The temperature was increased to 130°C and reacted for 2 weeks before cooling to room temperature. The reaction was under autogenous pressure. The subsequent synthesis without the boron source also led to the title compound under the same physical conditions (only with $Na_2HPO_4{\cdot}12H_2O$, In, and HCl).

$Na_4[In_8(HPO_4)_{14}(H_2O)_6]{\cdot}12(H_2O)$ crystallizes in the trigonal structure with the lattice parameters a = 1,385.00 ± 0.02 and c = 1,849.30 ± 0.03 pm and a calculated density of 2.87 g cm^{-3} (Attfield et al. 2000). To prepare this compound, $Na_2HPO_3{\cdot}5H_2O$ (1.57 g) was added to 15 mL of deionized H_2O, followed by the addition of $In(NO_3)_3{\cdot}3H_2O$ (2.60 g) and H_3PO_3 (1.23 g). The mixture was stirred, sealed in a 23-mL Teflon-lined steel autoclave, and heated at 10°C min^{-1} from room temperature to 125°C, where it was held for 50 h before being cooled to room temperature. The crystalline product was separated by suction filtration, washed in deionized water, and dried in air. The product was obtained as colorless crystals with a hexagonal prismatic habit. The crystals are stable to exposure to atmospheric conditions.

$Na_5In_2(P_2O_7)_2{\cdot}7H_2O$ could be prepared from aqueous solutions of $In_2(SO_4)_3{\cdot}5H_2O$ and $NaH_2PO_2{\cdot}H_2O$ (Ensslin et al. 1947).

9.2.3 Indium–Hydrogen–Potassium–Oxygen–Phosphorus

Some quinary compounds are formed in the In–H–K–O–P system. **$KIn(OH)PO_4$** crystallizes in the orthorhombic structure with the lattice parameters a = 927.7 ± 0.2, b = 933.9 ± 0.2, c = 1,124.5 ± 0.2 pm and a calculated density of 3.624 g cm^{-3} (Hriljac et al. 1996). To prepare this compound, In_2O_3 (1.0 g), K_2HPO_4 (1.0 g), and saturated aqueous solution of K_2HPO_4 (12.0 g) were sealed in a gold tube. The tube was heated to 900°C under 0.3 GPa over a period of 3 h, held for 12 h, slowly cooled over 32 h to 500°C, and then quenched to room temperature.

$KIn(H_2P_2O_7)_2$ was synthesized from a mixture of 85% H_3PO_4, In_2O_3 and K_2CO_3 (molar ratio of $P_2O_5/K_2O/In_2O_3$ = 15:5:1) (Avaliani et al. 1979). The mixture was heated in a carbon glass crucible with periodic mixing at 150°C for 12 days and then at 185°C for 2 days. The resulting crystalline mass was washed from the melt with water, acetone, and ether.

$KInP_2O_7{\cdot}4H_2O$ and **$K_5In(P_2O_7)_2{\cdot}xH_2O$** were prepared by the interaction of $InCl_3$ with $K_4P_2O_7$ (Deichman et al. 1967b). First compound melts at the temperature higher than 1,100°C and the second melts at 750°C without decomposition.

9.2.4 Indium–Hydrogen–Rubidium–Oxygen–Phosphorus

The **$RbIn(OH)PO_4$** and **$RbInHP_3O_{10}$** are formed in the In–H–Rb–O–P system. The first of them crystallizes in the tetragonal structure with the lattice parameters $a = 940.0 \pm 0.2$ and $c = 1{,}117.9 \pm 0.4$ pm and a calculated density of 4.200 g cm^{-3} (Lii 1996b). High-temperature, high-pressure hydrothermal syntheses were performed in gold ampoules contained in an autoclave where pressure was provided by water pumped by a compressed air-driven intensifier. $RbIn(OH)PO_4$ was synthesized from aqueous solutions of 2 M $Rb_2(HPO_4)$ (0.18 mL), 3 M $Rb(H_2PO_4)$ (0.15 mL) (molar ratio of Rb/P = 1.5), and In_2O_3 (55.5 mg) which were sealed in a gold ampoule and heated to 550°C at 220 MPa for 8 h. The autoclave was cooled at 5°C h^{-1} to 250°C and quenched to room temperature by removing the autoclave from the furnace. The product was filtered off, washed with water, rinsed with ethanol, and dried in a desiccator at ambient temperature. The product contained pale yellow polycrystalline material and small colorless crystals of $RbIn(OH)PO_4$.

$RbInHP_3O_{10}$ was synthesized from a mixture of 85% H_3PO_4, In_2O_3, and Rb_2CO_3 [molar ratios of $P_2O_5/Rb_2O/In_2O_3$ = 15:(2.5–5):1] (Avaliani et al. 1979). The mixture was heated in a carbon glass crucible with periodic mixing at 150°C for 12 days and then at 300°C for 2 days. The resulting crystalline mass was washed from the melt with water, acetone, and ether.

9.2.5 Indium–Hydrogen–Cesium–Oxygen–Phosphorus

Two quinary compounds are formed in the In–H–Cs–O–P system. **$CsInHP_3O_{10}$** crystallizes in the monoclinic structure with the lattice parameters $a = 1{,}205.4 \pm 0.5$, $b = 893.6 \pm 0.3$, $c = 943.2 \pm 0.4$ pm and $\beta = 111.77 \pm 0.05°$ (Murashova and Chudinova 2001). To prepare this compound, Cs_2CO_3, In_2O_3, and 85% H_3PO_4 (molar ratio of Cs/In/P = 5:1:15) were used as starting chemicals (Avaliani et al. 1979; Murashova and Chudinova 2001). Syntheses were performed at 350°C–400°C in glassy carbon crucibles. Small colorless crystals of the title compound grew in just 5 days. After 3 weeks, they began to convert into well-faced crystals of **$Cs_3In_3P_{12}O_{36}$**.

$Cs[In_2(PO_4)(HPO_4)_2(H_2O)_2]$ also crystallizes in the monoclinic structure with the lattice parameters $a = 658.0 \pm 0.2$, $b = 1{,}809.2 \pm 0.1$, $c = 1{,}018.0 \pm 0.1$ pm and $\beta = 97.92 \pm 0.02°$ and a calculated density of 3.76 g cm^{-3} (Dhingra and Haushalter 1994). It was synthesized by the hydrothermal method from an aqueous mixture of $InCl_3$, 85% H_3PO_4, $CsVO_3$, tetramethyl ammonium hydroxide, and H_2O (molar ratio 2:6:1:1:125), which was heated at 200°C for 2 days in a 23-mL polytetrafluoroethylene-coated acid digestion bomb. Large transparent pale yellow crystals were filtered off, washed thoroughly with H_2O, and dried at room temperature. The product was not in single phase but had 15 mass% of pale blue impurities of vanadium phosphate.

9.2.6 Indium–Hydrogen–Calcium–Oxygen–Phosphorus

Three quinary compounds are formed in the In–H–Ca–O–P system. **$CaIn_2(PO_4)_2(HPO_4)$** crystallizes in the monoclinic structure with the lattice parameters $a = 657.08 \pm 0.06$, $b = 2{,}023.7 \pm 0.2$, $c = 665.72 \pm 0.07$ pm, and $\beta = 91.20 \pm 0.01°$ and a calculated density of 4.170 g cm^{-3} (Tang and Lachgar 1996). The title compound was synthesized as a single-phase product by hydrothermal synthesis technique. Five milliliters of distilled water was added to a mixture of $InCl_3$ (1 mM), CaO (0.5 mM), and 85% H_3PO_4 (~8 mM). The resulting colorless solution (pH = 1) was placed in a 23-mL Teflon-lined stainless steel autoclave and heated at 220°C for 2 days followed by slow cooling to room temperature at a rate of 10°C h^{-1}. The colorless needle-like crystals were vacuum filtered, washed with water and acetone, and dried at room temperature.

$Ca_2In(PO_4)(HPO_4)_2{\cdot}H_2O$ also crystallizes in the monoclinic structure with the lattice parameters a = 757.3 ± 0.1, b = 1,583.8 ± 0.1, c = 931.26 ± 0.07 pm and β = 113.55 ± 0.01° and a calculated density of 3.243 g cm^{-3} (Tang et al. 1999). To prepare this compound, CaO (0.2244 g), 2 mL of 0.5 M $InCl_3$, 2 mL of 1 M $Me_4NOH{\cdot}5H_2O$, and 8 mL of 2 M H_3PO_4 were mixed in a 23-mL Teflon-lined stainless steel autoclave. The resulting gel had a molar ratio of $CaO/InCl_3/Me_4NOH/H_3PO_4$ = 4:1:4:16. The reaction was carried out in a tubular furnace at 190°C for 3 days and then furnace cooled to room temperature. The product was filtered, washed with water and acetone, and dried at room temperature. It consisted of colorless monoclinic column-like crystals of the title compound.

$Ca_2In(PO_4)(P_2O_7)$ also crystallizes in the monoclinic structure with the lattice parameters a = 647.4 ± 0.2, b = 676.1 ± 0.4, c = 1,927 ± 1 pm and β = 99.17 ± 0.04° (Tang et al. 1999). This compound is formed as a result of $Ca_2In(PO_4)(HPO_4)_2{\cdot}H_2O$ dehydration at 900°C.

9.2.7 Indium–Hydrogen–Calcium–Oxygen–Iron–Phosphorus

Two multinary compounds are formed in the In–H–Ca–O–Fe–P system. **$Ca_2(In_{0.5}Fe_{0.5})(PO_4)(HPO_4)_2{\cdot}H_2O$** crystallizes in the monoclinic structure with the lattice parameters a = 754.8 ± 0.2, b = 1,567.0 ± 0.3, c = 924.1 ± 0.2 pm and β = 113.62 ± 0.03° and a calculated density of 3.120 g cm^{-3} (Tang et al. 1999). This compound was prepared using the same method as described for $Ca_2In(PO_4)(HPO_4)_2{\cdot}H_2O$ obtaining. The molar ratio of the starting materials was $CaO/Fe(NO_3)_3/InCl_3/Me_4NOH/H_3PO_4$ = 8:1:1:4:32. The product consisted of colorless block-like crystals.

$Ca_2(In_{0.5}Fe_{0.5})(PO_4)(P_2O_7)$ also crystallizes in the monoclinic structure with the lattice parameters a = 644.80 ± 0.03, b = 672.7 ± 0.4, c = 1,946 ± 2 pm and β = 99.95 ± 0.04° (Tang et al. 1999). This compound is formed as a result of $Ca_2(In_{0.5}Fe_{0.5})(PO_4)(HPO_4)_2{\cdot}H_2O$ dehydration at 900°C.

9.2.8 Indium–Hydrogen–Lead–Oxygen–Phosphorus

The **$InPb[PO_4][PO_3(OH)]$** quinary compound, which crystallizes in the monoclinic structure with the lattice parameters a = 966.6 ± 0.5, b = 666.0 ± 0.4, c = 1,073 ± 1 pm, and β = 109.50 ± 0.01° and a calculated density of 5.22 g cm^{-3} is formed in the In–H–Pb–O–P system (Belokoneva et al. 2001a). It was synthesized under hydrothermal conditions from a mixture of In_2O_3, PbO, P_2O_5, B_2O_3, and H_2O at 250°C and 7 MPa.

9.2.9 Indium–Lithium–Titanium–Oxygen–Phosphorus

The **$Li_{1+x}Ti_{2-x}In_xP_3O_{12}$** ($0.4 < x < 1.1$) quinary phase, which crystallizes in the orthorhombic structure with the lattice parameters a = 864.7 ± 0.2, b = 880.7 ± 0.2, c = 2,432.8 ± 0.3 pm (for $x = 1.08$) and a calculated density of 3.35 g cm^{-3}, is formed in the In–Li–Ti–O–P system (Tran Qui and Hamdoune 1988).

9.2.10 Indium–Lithium–Niobium–Oxygen–Phosphorus

The **$Li_2Nb_{0.5}In_{1.5}(PO_4)_3$** quinary compound, which crystallizes in the monoclinic structure with the lattice parameters a = 864.8 ± 0.2, b = 884.28 ± 0.07, c = 1,231.6 ± 0.2 pm, and β = 90.13 ± 0.02° and the calculated and experimental densities of 3.649 and 3.59 ± 0.06 g cm^{-3}, is formed in the In–Li–Nb–O–P system (Cieren et al. 1996). The synthesis of this compound was carried out in the solid state from the ground mixture of the four reagents Li_2CO_3, Nb_2O_5, In_2O_3, and $(NH_4)_2HPO_4$ in stoichiometric proportions. After 12h of heating at 350°C, the mixture was finely crushed in agate mortar and pelletized. The compound was obtained after an air heat treatment of 12 h to 900°C in a Pt crucible. The crystals of the title compound were obtained by flux growth with $Li_4P_2O_7$ as a flux. For this, a mixture of $Li_4P_2O_7$ and $Li_2Nb_{0.5}In_{1.5}(PO_4)_3$ in the 1: 4 mass ratio was heated to 1,200°C and then cooled at a rate of 6°C h^{-1} up to 900°C.

9.2.11 Indium–Sodium–Cadmium–Oxygen–Phosphorus

The **$NaCdIn_2(PO_4)_3$** quinary compound, which crystallizes in the monoclinic structure with the lattice parameters a = 1,251.7 ± 0.3, b = 1,296.6 ± 0.3, c = 657.1 ± 0.2 pm, and β = 115.36 ± 0.02° (Hatert et al. 2002) [a = 1,251.9 ± 0.2, b = 1,295.9 ± 0.3, c = 657.5 ± 0.1 pm, and β = 115.17 ± 0.01° (Antenucci et al. 1993)], is formed in the In–Na–Cd–O–P system. It was synthesized by the solid-state reactions in air (Antenucci et al. 1993; Hatert et al. 2002). Stoichiometric quantities of $NaHCO_3$, $CdCO_3$, In_2O_3, and $(NH_4)H_2PO_4$ were dissolved in concentrated HNO_3. The dry residue was progressively heated for 15–20 h in a Pt crucible to 950°C. The title compound was obtained by quenching the product in air.

9.2.12 Indium–Sodium–Oxygen–Manganese–Phosphorus

The **$NaMnIn_2(PO_4)_3$** quinary compound, which crystallizes in the monoclinic structure with the lattice parameters a = 1,228.2 ± 0.2, b = 1,294.8 ± 0.2, c = 655.2 ± 0.1 pm, and β = 115.21 ± 0.01°, is formed in the In–Na–O–Mn–P system (Hatert et al. 2003). It was synthesized through the solid-state reactions in air. Stoichiometric quantities of $NaHCO_3$, MnO, In_2O_3, and $(NH_4)H_2PO_4$ were dissolved in concentrated HNO_3 and the resulting solution was evaporated to dryness. The dry residue was progressively heated in a Pt crucible at a heating rate of 500°C h^{-1} to either 900°C or 1,000°C and was then maintained at this temperature for 13–16 h. $NaMnIn_2(PO_4)_3$ was obtained by quenching the product in air.

9.3 SYSTEMS BASED ON InAs

9.3.1 Indium–Hydrogen–Sodium–Oxygen–Arsenic

The **$NaIn(HAsO_4)_2\cdot 6H_2O$** quinary compound is formed in the In–H–Na–O–As (Ezhova et al. 1977). **$NaInAs_2O_7$** is formed as a result of dehydration of this compound at 700°C.

9.3.2 Indium–Hydrogen–Potassium–Oxygen–Arsenic

The **$KIn(HAsO_4)_2 \cdot 3H_2O$** quinary compound is formed in the In–H–K–O–As (Ezhova et al. 1977). **$KInAs_2O_7$** is formed as a result of dehydration of this compound at 700°C.

9.3.3 Indium–Hydrogen–Rubidium–Oxygen–Arsenic

The **$RbIn(HAsO_4)_2 \cdot 3H_2O$** quinary compound is formed in the In–H–Rb–O–As (Ezhova et al. 1977). **$RbInAs_2O_7$** is formed as a result of dehydration of this compound at 700°C.

9.3.4 Indium–Hydrogen–Cesium–Oxygen–Arsenic

The **$CsIn(HAsO_4)_2 \cdot 2H_2O$** quinary compound is formed in the In–H–Cs–O–As (Ezhova et al. 1977). $CsInAs_2O_7$ is formed as a result of dehydration of this compound at 700°C. The title compound was obtained using hydrothermal techniques. A mixture of $InCl_3$, $As_2O_5 \cdot xH_2O$ ($x \approx 3$), $Me_4NOH \cdot 5H_2O$, $CsNO_3$, and water in a molar ratio 4:1:2:4:776 was sealed under vacuum in a thick-wall Pyrex tube (~20% filled) (Tang et al. 2001). The reaction was carried out at 160°C for 12 days followed by slow cooling to room temperature. The products were filtered, washed with water and acetone, and dried at room temperature.

9.3.5 Indium–Hydrogen–Lead–Oxygen–Arsenic

The **$PbIn(AsO_4)(AsO_3OH)$** quinary compound, which crystallizes in the monoclinic structure with the lattice parameters $a = 495.5 \pm 0.1$, $b = 859.1 \pm 0.2$, $c = 1{,}587.4 \pm 0.3$ pm and $\beta = 92.38 \pm 0.03°$ and an experimental density of 5.911 g cm^{-3}, is formed in the In–H–Pb–O–P system (Kolitsch and Schwendtner 2005). The tiny colorless indistinct (grain-like) crystals of the title compound was prepared by a hydrothermal method (Teflon-lined stainless steel bomb, 220°C, 7 days, and slow furnace cooling) from mixtures of distilled water, arsenic acid, In_2O_3, and PbO. The final pH value of the reacted solutions was about 2.

9.3.6 Indium–Copper–Zinc–Germanium–Selenium–Arsenic

$InAs$–$CuInSe_2$–Cu_2GeSe_3–$ZnGeAs_2$. A large volume of this segment of the In–Cu–Zn–Ge–Se–As multinary system extending from InAs was shown to be solid solutions with cubic structure of the sphalerite-type (Pamplin and Hasoon 1985). No new compounds or superlattices were found, and only a nonequilibrium material was prepared in the center of the segment.

9.3.7 Indium–Zinc–Cadmium–Tin–Arsenic

$InAs$–$CdSnAs_2$–$ZnSnAs_2$. The compositions $ZnCdIn_4Sn_2As_8$, $ZnCd_2In_2Sn_3As_8$, and $Zn_2CdIn_2Sn_3As_8$ are all single phase with the sphalerite-type structure (Pamplin and Shah 1965). Equilibrium is reached after only a few days annealing at 595°C. The lattice parameters for these materials are 600.5 ± 0.1, 600.6 ± 0.1, and 594.9 ± 0.1, respectively. The existence of the sphalerite-type structure for the aforementioned materials in this quasiternary system leads to expect a solid solution at all compositions inside the triangle $InAs$–$CdSnAs_2$–$ZnSnAs_2$.

9.4 SYSTEM BASED ON InSb

9.4.1 Indium–Cerium–Iron–Cobalt–Arsenic

The $\mathbf{Ce_{0.33}In_yFe_{1.5}Co_{2.5}Sb_{12}}$ quinary phase is formed in the In–Ce–Fe–Co–As system (Benyahia et al. 2017). Indium is soluble in polycrystalline $Ce_{0.33}In_yFe_{1.5}Co_{2.5}Sb_{12}$ until $y = 0.08 \pm 0.03$. Beyond this concentration, InSb nano-inclusions (~50–200 nm) form at the grain boundaries during the shaping stage. Upon In addition, the lattice parameter a increases from 907.4 pm for $y = 0$ to 907.9 pm for $y = 0.1$. The samples of this phase were prepared via an optimized melting reaction technique starting from a Fe–Co master alloy of $Fe_{1.5}Co_{2.5}$ composition reacted with elemental Ce, In, and Sb in stoichiometric amounts. The reactions were carried out in vitreous carbon crucibles sealed in a quartz tube under a partial Ar pressure. The ampoules were heated at a rate of 100°C h^{-1} up to 1,150°C, maintained at this temperature during 24 h, cooled down to 550°C (100°C h^{-1}), heated to 750°C (100°C h^{-1}), and annealed at this temperature for 96 h.

Appendix A

Ternary Alloys Based on III-V Semiconductors

A.1 SYSTEMS BASED ON BN

A.1.1 Boron–Hydrogen–Nitrogen

BN forms an intercalation compound with N_2H_4 (Croft 1956). The method of N_2H_4 reaction with crystalline BN consisted of sealing the reactants in evacuated Pyrex glass tube and heating the latter at 98°C for 12 h.

A.1.2 Boron–Lithium–Nitrogen

Li–BN intercalation compounds were observed from the milled sample after heat treating (Kim et al. 2016a). The shifted peaks of *h*-BN to the lower angles imply an expanded BN lattice through Li insertion. The exothermic differential thermal analysis (DTA) peaks observed between 200°C and 500°C might be the synthesis temperatures of these compounds. A vibratory ball mill was used for mechanical milling. Three grams of Li and BN mixture (molar ratio of 1:2.2) were set in a stainless vessel under argon atmosphere. After ball milling, the samples were subjected to heat treatment at 700°C for 2 h in argon atmosphere. The lattice parameters of Li–BN intercalation compounds were $a = 446.5 \pm 0.01$, $c = 750.4 \pm 0.01$ pm and $a = 446.5 \pm 0.01$, $c = 375.1 \pm 0.01$ with and without two-layer stacking periodicity structure, respectively. It is difficult to determine the exact phases of such compounds, as both **Li(BN)$_3$** and **Li$_2$(BN)$_3$** phases exhibit similar X-ray diffraction (XRD) peak patterns.

The crystal structures and phase stabilities of various Li–BN intercalation compounds [i.e., **Li(BN)**, $Li_2(BN)_3$, $Li(BN)_3$, **Li(BN)$_4$**, **Li$_2$(BN)$_9$**, and **Li(BN)$_9$**] were studied using density functional theory calculation with/without utilizing disorder calculation method, and compared with experimental results (Kim et al. 2018). It was concluded that $Li(BN)_3$ and $Li(BN)_9$ are the possible structures.

Li–BN intercalation compound was also successfully synthesized by the annealing of powder or bulk *h*-BN and Li in a BN or Ta crucible under Ar atmosphere at 1,250°C for 10 h (Sumiyoshi et al. 2010, 2012) or by the annealing of Li_3N and *h*-BN mixture (molar ratio 0.3:1.1) at 950°C in an Ar-filled glove box (Sumiyoshi et al. 2015). The starting mixture was sealed into a stainless steel tube using arc welding. The welded stainless steel tube was annealed at a temperature ranging from 430°C to 1,230°C for 10 h. The annealed tube was opened in an Ar-filled glove box. The exact composition of this compound is unknown because its single-phase sample could not be produced. The stacking periodicity of the BN atomic layer in this phase changes from the two-layer stacking periodicity of *h*-BN to a one-layer stacking because of Li intercalation (Sumiyoshi et al. 2015). Lithium deintercalation was confirmed by the dispersion of the sample in purified water (Sumiyoshi et al. 2010, 2012).

Two hypothetical ternary intercalation compounds $Li(BN)_3$ and $\mathbf{Li_4(BN)_6}$, which could crystallize in the hexagonal structure with the lattice parameters $a = 436.9$, $c = 788.0$ pm and $a = 444.9$, $c = 743.4$ pm, could exist in the B–Li–N system (Altintas et al. 2011).

According to the data of Freeman and Larkindale (1969a), no reaction was observed between Li and BN at temperatures up to 500°C.

A.1.3 Boron–Sodium–Nitrogen

The reaction between BN and Na was studied by sealing the reactant in evacuated Pyrex glass, or silica, tube and heating to various temperatures (Freeman and Larkindale 1969a). The mixture was heated at 400°C for 24 h. After cooling, the tube was opened, and its contents were immediately immersed in ethanol to remove unreacted sodium. The product of reaction was then recovered by filtration and washed with ethanol until the washing was free of Na. The solid product was given a final wash with distilled water and the washing tested for alkalinity. The resulted product was apparently homogeneous.

A.1.4 Boron–Potassium–Nitrogen

The reaction between BN and K was studied by the same way as in the case of BN and Na interaction, but the mixture was heated at 300°C and propan-2-ol was used instead of ethanol (Freeman and Larkindale 1969a). The $\mathbf{BN_{13}K}$ intercalation compound was obtained.

A.1.5 Boron–Magnesium–Nitrogen

The $\mathbf{Mg_3(BN_2)N}$ ternary compound crystallizes in the hexagonal structure with the lattice parameters $a = 354.008 \pm 0.004$ and $c = 1,603.35 \pm 0.02$ pm and an energy gap of 2.7 eV (Schölch et al. 2015). The synthesis of this compound was carried out by solid-state metathesis reaction shown in the next scheme: $3MgCl_2 + 2Li_3(BN_2) + Mg_3N_2 = 2Mg_3(BN_2)N + 6LiCl$. The starting materials $MgCl_2$, Mg_3N_2, and Li_3BN_2 were mixed in an agate mortar in a molar ratio of 3:1:2 with a total mass of ca. 300 mg. The mixture was fused into a silica ampoule under vacuum. It was heated to 700°C within 6 h and remained at this temperature for 24 h before it was allowed to cool to room temperature over a period of 6 h. The synthesis was also explored with other sources of nitrogen such as Li_3N and NaN_3, according to the reactions $3MgCl_2 + Li_3(BN_2) + 3NaN_3 = Mg_3(BN_2)N + 3LiCl + 3NaCl + 4N_2$ and $3MgCl_2 + Li_3(BN_2) + Li_3N = Mg_3(BN_2)N + 6LiCl$.

A.1.6 Boron–Aluminum–Nitrogen

BN–AlN. The diagram of BN–AlN at 8 GPa that has been proposed in Bartnitskaya et al. (1980) is evidently incorrect as it violates the Gibbs phase rule, which states that, in a binary system, four phases (liquid, AlN, *h*-BN, and *c*-BN) cannot establish invariant equilibrium at a constant pressure (Turkevich et al. 2004). Moreover, the statement that the solubility of aluminum nitride in *c*-BN is higher than that in *h*-BN is difficult to believe. A hypothetical phase diagram of this system, constructed by Turkevich et al. (2004), is presented in Figure A.1. It is seen that *c*-BN solubility in AlN is much lower than that of *h*-BN, the intermediate compounds are absent, and the liquid phase appears at temperatures well above 2,000°C.

The $B_xAl_{1-x}N$ films were deposited by ion-beam assisted deposition onto Si substrates at substrate temperatures of 600°C and 675°C (Edgar et al. 1997). At these conditions, the films containing up to 9 mol% BN were obtained. The *c*-lattice parameter of $B_xAl_{1-x}N$ solid solution decreases linearly with BN content up to approximately 10 mol% BN. The $B_xAl_{1-x}N$ epitaxial growth on the SiC substrates by plasma-assisted molecular beam epitaxy (MBE) was investigated by Nakajima et al. (2005). Boron atoms were supplied by pyrolysis of decaborane ($B_{10}H_{14}$). The maximum boron composition of single-phase $B_xAl_{1-x}N$ layers was 1.8 at%. The $B_xAl_{1-x}N$ films with wurtzite-type structure were fabricated using flow-rate modulation epitaxy (Akasaka and Makimoto 2006). These films were directly grown on SiC substrates at 1,000°C. The source gases were triethylboron, trimethylaluminum, and NH_3. The films containing ca. 1.5 at% B were obtained. The same gases were used by Li et al. (2017) to grow the $B_xAl_{1-x}N$ films using metal-organic chemical vapor deposition (MOCVD). To enhance boron incorporation, flow-modulated epitaxy was used. Relatively high growth temperatures (up to 1,010°C) were used, and the films containing up to 14.4 ± 1.0 at% B were prepared.

The $B_xAl_{1-x}N$ layers were also grown by metal-organic vapor phase epitaxy (MOVPE) at 13.3 kPa using H_2 as carrier gas (Li et al. 2015). Thin layers were grown at low temperature, which was varied from 650°C to 800°C. Then, the temperature was ramped up to 1,020°C

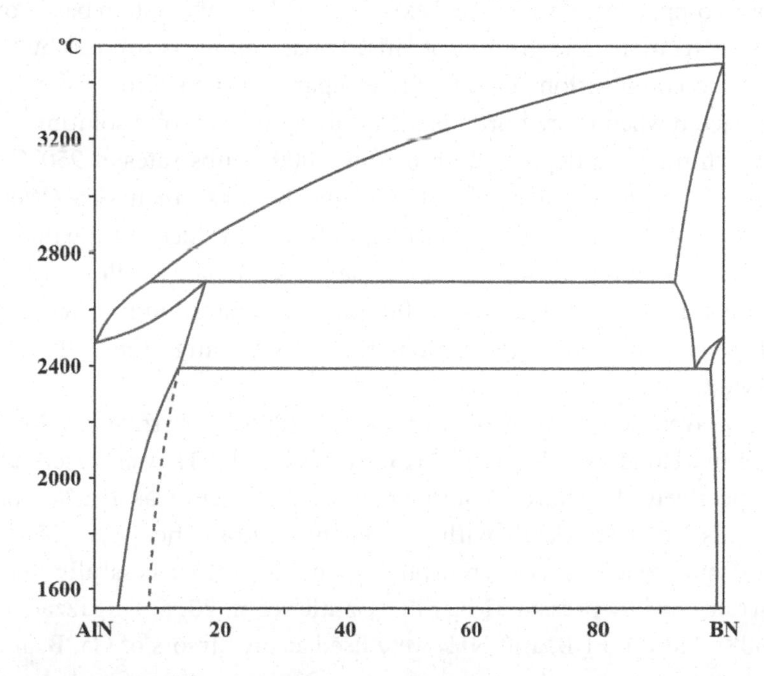

FIGURE A.1 Hypothetical phase diagram of the AlN–BN system at 6 GPa. (With permission from Turkevich, V., et al., *Diamond Relat. Mater.*, **13**(1), 64, 2004.)

and the samples were annealed for 5 min before cooling down. By continuous growth, the layers with up to 5.6 at% B were obtained. Flow-modulated epitaxy was also applied, and $B_xAl_{1-x}N$ layers containing as high as 12 at% B were prepared on both AlN and GaN templates.

The microstructure of $B_xAl_{1-x}N$ films grown by flow-modulated epitaxy at 1,010°C, with B/(B + Al) gas-flow ratios ranging from 0.06 to 0.18, has been investigated by Wang et al. (2017). The boron content obtained from XRD patterns ranges from $x = 0.02$ to 0.09. On the other hand, boron content deduced from the aluminum signal in the Rutherford backscattering spectra ranges from $x = 0.06$ to 0.16, closely following gas-flow ratios. Transmission electron microscopy indicates the sole presence of wurtzite crystal structure in the $B_xAl_{1-x}N$ films and a tendency towards columnar growth for B/(B + Al) gas-flow ratios below 0.12. For higher ratios, the films exhibit a tendency towards twin formation and finer microstructure. Electron energy loss spectroscopy has been used to profile spatial variations in the composition of the films. The Rutherford backscattering spectra suggest that the incorporation of B is highly efficient for proposed growth method, while the XRD data indicate that the epitaxial growth may be limited by a solubility limit in the crystal phase at about 9 mol%, for the range of B/(B + Al) gas-flow ratios that was used.

A.1.7 Boron–Gallium–Nitrogen

BN–GaN. The large structural dissimilarity between BN and GaN results in a high interaction parameter of 296 kJ/mol, indicating that the $B_xGa_{1-x}N$ solid solution deviates greatly from an ideal solid solution (Wei and Edgar 2000). Therefore, the unstable regions in which inhomogeneous compositions are formed exist in the BN–GaN system based on the vapor/solid distribution relation. The growth of single-phase alloys is limited by the existence of a wide unstable composition region. Phase separation or spinodal decomposition of $B_xGa_{1-x}N$ is expected when more boron is introduced into the solid solution.

The $B_xGa_{1-x}N$ films were deposited on 6*H*-SiC (0001) substrates at 950°C–1,000°C by low-pressure MOVPE, using diborane, Me_3Ga, and NH_3 as precursors (Wei et al. 1999, 2000). A single-phase $B_xGa_{1-x}N$ alloy with 1.5 at% B was produced at the diborane/Me_3Ga ratio of 0.005. The phase separation into pure GaN and $B_xGa_{1-x}N$ alloy with $x = 0.30$ was deposited for the diborane/Me_3Ga ratio of 0.1, and the separation into wurtzite $B_xGa_{1-x}N$ and the B-rich phase occurred for this ratio in the 0.01–0.2 range. Only BN was formed for the diborane/Me_3Ga ratio > 0.2.

The $B_xGa_{1-x}N$ layers were also grown on 6*H*-SiC (0001) substrates by MOVPE with a horizontal reactor (Honda et al. 2000). Et_3B, Me_3Ga, and NH_3 were used as B, Ga, and N sources, respectively. The boron composition of the layers was 1 at%, and the bandgap energy increases monotonically with boron composition. The $B_xGa_{1-x}N$ materials with good structural quality and surface morphology have been successfully grown on GaN template substrates by low-pressure MOVPE (Gautier et al. 2006; Ougazzaden et al. 2007; Orsal et al. 2008). Me_3Ga, Et_3B, and NH_3 were used as precursors of Ga, B, and N, respectively. All the growths were performed under pure N_2 process gas. Single-crystal layers of $B_xGa_{1-x}N$ with B content as high as 3.6% have been obtained.

The $B_xGa_{1-x}N$ films with a wurtzite structure and different compositions of boron up to 1.8 at% were grown on the *c*-axis oriented AlN/sapphire template substrates by low-pressure MOVPE at 1,000°C (Ougazzaden et al. 2008). Me_3Ga, Et_3B, and NH_3 were used as precursors. A significant bow ($C=9.2\pm0.5$ eV) of the band gap was observed. A decrease in the optical band gap by 150 meV with respect to that of GaN was measured for the increase in the boron composition from 0 to 1.8 at%.

A.2 SYSTEMS BASED ON AlN

A.2.1 Aluminum–Gallium–Nitrogen

AlN–GaN. The $Al_xGa_{1-x}N$ epitaxial films on sapphire substrates were grown by using laser MBE technique (Tyagi et al. 2018). It was found that the Al molar fraction increases with increase in growth temperature and a maximum Al content of about 23 at% was achieved at a growth temperature of 700°C. The growth of good-quality $Al_xGa_{1-x}N$ layers has been achieved in the temperature range of 600°C–700°C.

Amorphous alloys af AlN and GaN deposited at 100 K at compositions ranging from pure AlN to pure GaN with optical band gaps which vary linearly with composition from 3.27 eV (GaN) to 5.95 eV (AlN) have been synthesized by Chen et al. (2000).

A.2.2 Aluminum–Indium–Nitrogen

AlN–InN. Using *ab initio* methodology, the structural, electronic, mechanical, lattice, and thermodynamic properties of novel compounds within this system under pressure have been presented by Chang et al. (2017). The structural search via evolutionary algorithm unearthed six new metastable phases. Two phases were observed at 2.5 GPa, and four more phases were identified at 5.0 GPa. **$AlIn_4N_5$** crystallizes in the tetragonal structure with the lattice parameters $a=581.9$ and $c=476.6$ pm and the energy gap 0.796 eV at 5 GPa. **$AlIn_5N_6$** crystallizes in the trigonal structure with the lattice parameters $a=587.3$ and $c=641.5$ pm and the energy gap 0.771 eV at 5 GPa. **$Al_2In_2N_4$** crystallizes in the tetragonal structure with the lattice parameters $a=462.2$ and $c=558.8$ pm and the energy gap 2.931 eV at 2.5 GPa and $a=460.9$ and $c=555.9$ pm and the energy gap 2.770 eV at 5 GPa. **$Al_4In_2N_6$** crystallizes in the orthorhombic structure with the lattice parameters $a=509.3$, $b=552.4$, and $c=550.7$ pm and the energy gap 4.262 eV at 5 GPa. **$Al_7In_2N_9$** crystallizes in the monoclinic structure with the lattice parameters $a=546.1$, $b=782.4$, $c=545.7$ pm, and $\beta=103.18°$ and the energy gap 4.995 eV at 5 GPa.

A.2.3 Aluminum–Scandium–Nitrogen

AlN–ScN. Reactive magnetron sputter epitaxy was used to deposit thin solid films of $Al_xSc_{1-x}N$ ($0\le x\le 1$) onto MgO(111) substrates with ScN(111) seed layers (Höglund et al. 2009). Stoichiometric films were deposited from elemental Al and Sc targets at substrate temperatures of 600°C. It was shown that rock-salt structure solid solutions with AlN molar fractions up to ca. 60 mol% can be synthesized. For higher AlN contents, the system phase separates into cubic and wurtzite structure domains.

Thin films of $Al_xSc_{1-x}N$, with *x* varied from 0.28 to 1.00, were also deposited by reactive magnetron sputter epitaxy onto MgO(111) and Al_2O_3(0001) substrates kept at 800°C from elemental Al and Sc targets (Höglund et al. 2010). ScN (≈22 mol%) have been dissolved into AlN, forming a disordered single-crystal solid solution of wurtzite-type. The lattice parameter *a* increases almost linearly from 311 to 321 pm for *x* varied between 1.0 and 0.80, while no significant change in the *c*-parameter was observed. ScN contents between 23 and ≈50 mol% yielded a nanocrystalline mixture of ScN and AlN phases. The solid solution with cubic structure was formed at higher ScN contents. *Ab initio* calculations of mixing enthalpies and lattice parameters of bulk solid solutions with cubic and hexagonal structures predict the transition from cubic to wurtzite structures at $x \approx 0.45$, which deviates from the experimentally reported transition at $x = 0.60$.

The reactive co-sputtering of $Al_{1-x}Sc_xN$ on sapphire(0001) substrates at 850°C yields single-phase epitaxial layers for $x \leq 0.20$ (Deng et al. 2013). The crystalline quality decreases with increasing x. The optical band gap of wurtzite $Al_xSc_{1-x}N$ decreases according to $E_g(x) = 6.15–9.32x$ eV for $x \leq 0.20$. At larger x, the structural instability and phase separation leads to an increasing dominance of absorption associated with a 2.5 eV direct transition. Lattice parameter *a* increases linearly, and lattice parameter *c* remains nearly constant for compositions $0 \leq x \leq 0.20$.

Bulk single crystals of $Al_xSc_{1-x}N$ containing up to 0.55 at% Sc were grown by the physical vapor transport technique (Dittmar et al. 2018). The crystals show high structural perfection comparable to AlN crystals.

The $Al_{1-x}Sc_xN$ solid solutions were also studied using density functional theory with special quasi-random structure methodology (Zhang et al. 2013). It was found they are stable in hexagonal phase up to $x \approx 0.56$ above which rock-salt structures are more stable. Epitaxial strain stabilization can prevent spinodal decomposition up to $x \approx 0.4$ (on AlN or GaN). The alloys were found to retain wide band gaps, which stay direct up to $x =$ 0.25. The band gap decreases with increasing x. The thermodynamic calculation shows that unstrained $Al_{1-x}Sc_xN$ solid solutions are only stable with respect to spinodal decomposition at $x = 0.06$ at 830°C.

A.2.4 Aluminum–Titanium–Nitrogen

The integration of *ab initio* thermodynamics, experiments, and CALPHAD modeling for a comprehensive understanding of the metastable and ternary phases in the Al–Ti–N ternary system was accomplished by Zhang et al. (2017). The supersaturated $Al_xTi_{1-x}N$ solid solutions with both rock salt and wurtzite structures were well described by using the subregular solution model. Thereby, the formation, the phase separation, and the metasolubility of $Al_xTi_{1-x}N$ coatings were successfully elaborated by the metastable quasibinary AlN–TiN section and the physical vapor deposition phase diagram. Moreover, a number of isopleth sections near the Ti-rich corner and the selected isothermal sections from 700°C to 1,700°C were calculated using the optimized database and compared with available experiments and previous thermodynamic assessments. Some of the calculated isothermal sections are given in Figure A.2.

AlN–TiN. New experimental measurements of phase equilibria close to the *c*-TiN-rich region of the metastable *c*-AlN/*c*-TiN quasibinary miscibility gap were conducted by Zhou et al. (2017). Based on the new experimental phase equilibrium data as well as

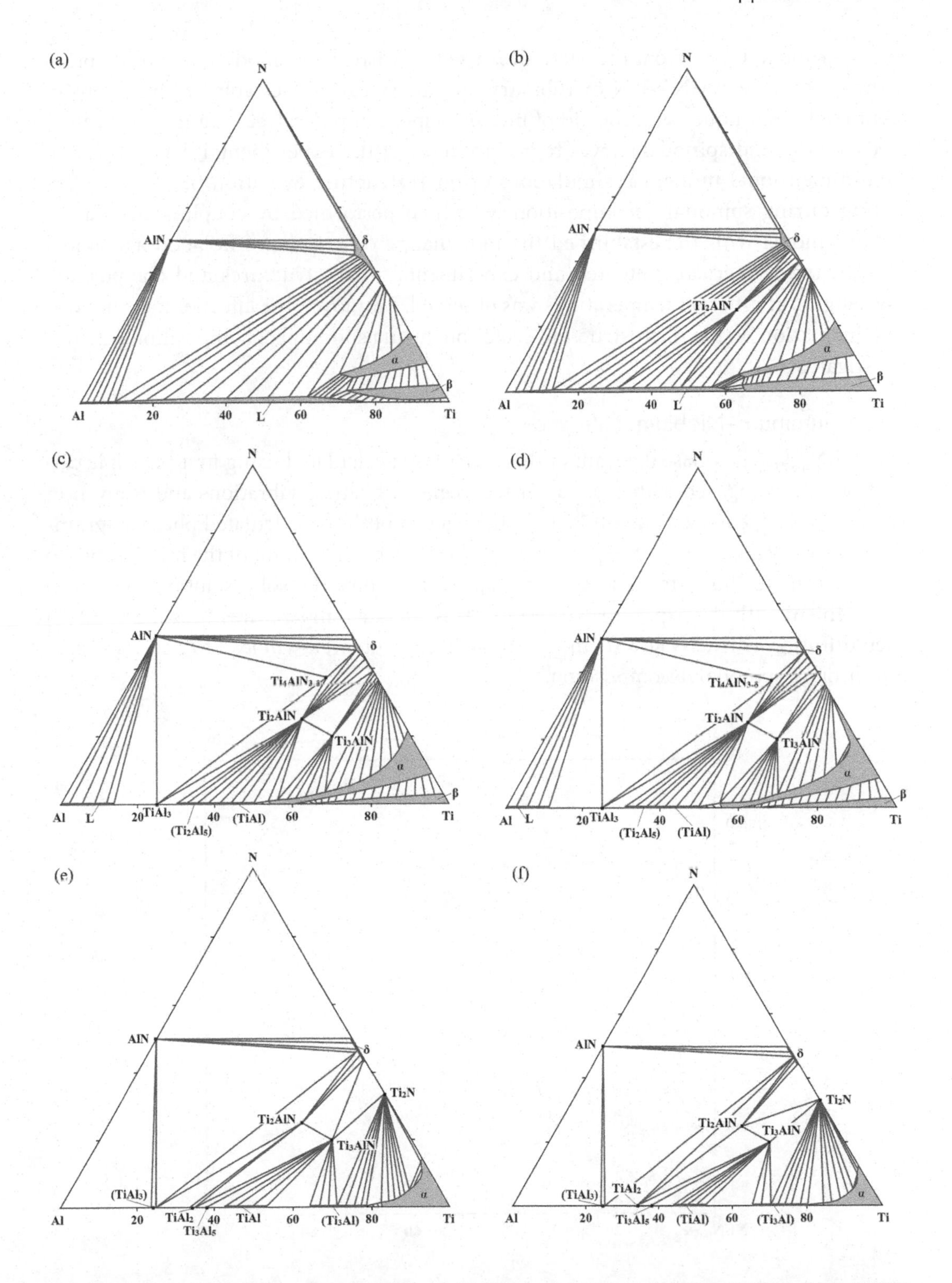

FIGURE A.2 Isothermal sections of the Al–Ti–N ternary system at (a) 1,700°C, (b) 1,600°C, (c) 1,325°C, (d) 1,300°C, (e) 800°C, and (f) 700°C. (With permission from Zhang, Y., et al., *Calphad*, **59**, 142, 2017.)

the experimental data from the literature, a self-consistent thermodynamic description of the metastable *c*-AlN/*c*-TiN quasibinary miscibility gap was established by means of CALPHAD technique, with the aid of first-principle computed free energies. The calculated binodal and spinodal curves reproduce most of the experimental data. Two- and three-dimensional numerical simulations of microstructure evolution in *c*-$Al_{0.47}Ti_{0.53}N$ coating during spinodal decomposition were then performed by coupling the Cahn–Hilliard model with the established thermodynamic description. The good agreement between the numerical simulated and experimental microstructures and composition wavelengths at different temperatures was observed. Moreover, the effect of the composition fluctuation on the microstructure evolution during spinodal decomposition was also investigated.

A.2.5 Aluminum–Niobium–Nitrogen

AlN–$NbN_{0.875}$. The phase diagram of this system was calculated using first-principle calculation and compared with that calculated neglecting lattice vibrations and is given in Figure A.3 (Pogrebnjak et al. 2017). The difference in both the calculated phase diagrams is dramatic. By allowing for the phonon contribution, the maximum of the miscibility gap reduces from 10,700°C to 5,700°C. For small Al fractions, the solid solutions will form in agreement with the experimental results. At moderate temperatures (less than 730°C) when diffusion will activate a further increase in Al content, it will lead to phase segregation through spinodal decomposition.

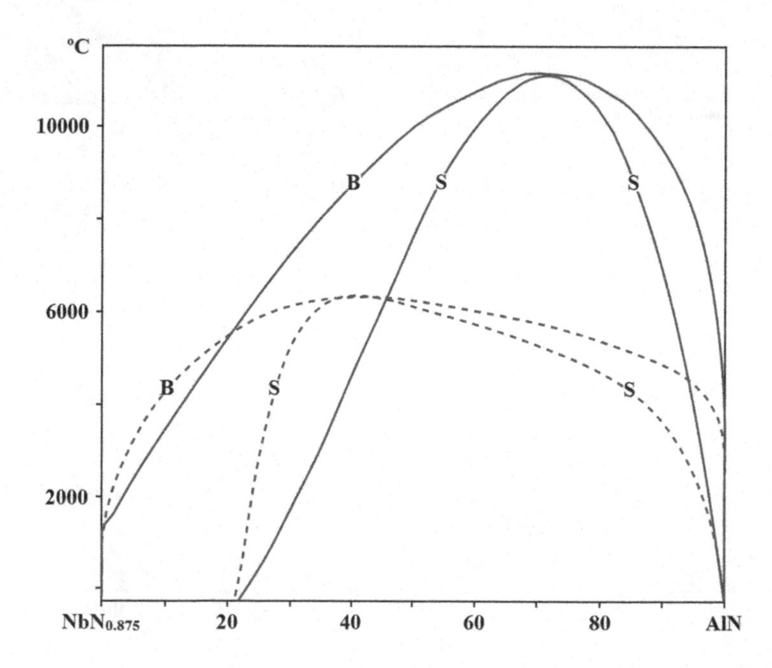

FIGURE A.3 Calculated phase diagram of the AlN–$NbN_{0.875}$ system. The dashed and solid lines correspond to the bimodal (B) and spinodal (S) with and without vibrational contribution, respectively. (With permission from Pogrebnjak, A., et al., *J. Alloys Compd.*, **718**, 260, 2017.)

A.3 SYSTEMS BASED ON AlP

A.3.1 Aluminum–Oxygen–Phosphorus

According to the data of Meer van der (1976), **$Al(PO_3)_3$** crystallizes in the monoclinic structure with the lattice parameters $a = 1{,}042.3 \pm 0.3$, $b = 1{,}868.7 \pm 0.2$, $c = 922.2 \pm 0.1$ pm, and $\beta = 81.63°$ and the calculated and experimental densities of 2.96 and 2.90 g cm^{-3}, respectively. In the [010] direction a subcell structure is present with a period of $b/3$. The crystals suitable for XRD were prepared by melting the mixture of Al_2O_3 and H_3PO_4 in a Pt crucible.

A.4 SYSTEMS BASED ON AlAs

A.4.1 Aluminum–Oxygen–Arsenic

The **$AlAsO_4$** ternary compound was obtained at the heating of **$KAl_4(OH)_4(AsO_4)_3{\cdot}6.5H_2O$** (mineral alumopharmacosiderite), and it crystallizes in the orthorhombic structure with the lattice parameters $a = 716$, $b = 715$, and $c = 711$ pm (Cabri et al. 1981).

A.5 SYSTEMS BASED ON GaN

A.5.1 Gallium–Scandium–Nitrogen

GaN–ScN. The films of $Ga_{1-x}Sc_xN$ solid solutions were grown by reactive sputtering between 300°C and 402°C on quartz substrates (Little and Kordesch 2001). These films were amorphous or microcrystalline. They have optical band gap between 2.0 and 3.5 eV that varies linearly with alloy composition (the band gap decreases with increasing x).

Using radio-frequency MBE over the Sc fraction range x = 0–1, $Ga_{1-x}Sc_xN$ wurtzite-like solid solutions were obtained for $x \leq 0.17$ and cubic rack-salt-like solid solutions were grown for $x \geq 0.54$ (Constantin et al. 2004). The films were grown on sapphire(0001) substrates. The lattice parameters a and c of the solid solutions with wurtzite-like structure increase linearly with increasing x.

These films x with up to ≈0.08 were grown at 800°C on GaN-on-sapphire templates using MBE with NH_3 as a reactive nitrogen source (Moram et al. 2009). It was shown that the addition of Sc resulted in an increase in both the a and the c lattice parameters. The epitaxial wurtzite-structure films were also grown using the same templates and the method by Knoll et al. (2012). It was shown that the Sc content remained below 2 at% for all films. Their growth rates decreases significantly as the Sc flux increased and remained low compared to the growth rates of GaN grown under similar conditions.

Using first principles total energy calculations within the full-potential linearized augmented plane wave method, the structural and electronic properties of $Ga_{1-x}Sc_xN$, with Sc concentrations varying from 0 to 100 at% have been investigated by Moreno-Armenta et al. (2003). It was found that for Sc concentrations less than ≈65 at%, the favored structure is a wurtzite-like one, while for Sc concentrations greater than ≈65 at%, the favored structure is a NaCl-like structure. The fundamental energy gap is large and direct for wurtzite-like crystals. For the rock-salt crystals the fundamental gap is small and indirect, but with an additional larger direct gap.

The $Ga_{1-x}Sc_xN$ solid solutions were also studied using density functional theory with special quasi-random structure methodology (Zhang et al. 2013). It was found they are stable in hexagonal phase up to $x \approx 0.66$ above which rock-salt structures are more stable.

Epitaxial strain stabilization can prevent spinodal decomposition up to $x = 0.27$ (on GaN). The alloys were found to retain wide band gaps, which stay direct up to $x = 0.5$. The band gap decreases with increasing x up to $x = 0.5$. The thermodynamic calculation shows that unstrained $Ga_{1-x}Sc_xN$ solid solutions are only stable with respect to spinodal decomposition at $x = 0.11$ at 830°C.

A.6 SYSTEMS BASED ON GaAs

A.6.1 Gallium–Bismuth–Arsenic

The $\mathbf{GaBi_xAs_{1-x}}$ solid solution was grown on GaAs(001)-oriented substrates by MBE with solid sources of Ga, Bi, and As at a substrate temperature between 350°C and 410°C (Yoshimoto et al. 2003). The relationship between GaBi molar fraction (x) and the lattice constant showed a good linearity. To achieve Bi incorporation into the epilayer, arsenic flux was adjusted in a limited range on the brink of As shortage on the growing surface. The Bi incorporation was saturated at a large Bi flux, probably due to a low miscibility of Bi with GaAs. The value of x increased up to 0.045 with decrease in the substrate temperature to 350°C.

Good quality of $GaBi_xAs_{1-x}$ epilayers on GaAs substrates were obtained by the MOVPE when the growth was performed at 365°C (Oe 2002). The lattice parameters of the alloys increases with the addition of Bi up to $x = 0.026$. The lattice parameters of $GaBi_{0.026}As_{0.974}$ alloy can be set to be 566.7 pm. Pr^i_3Ga, Me_3Bi, and Bu^tAsH_2 have been used as Ga, Bi, and As sources, respectively.

A.7 SYSTEMS BASED ON GaSb

A.7.1 Gallium–Yttrium–Antimony

The $\mathbf{Ga_{1.24}Y_5Sb_{2.77}}$ ternary compound, which crystallizes in the orthorhombic structure with the lattice parameters $a = 796.29 \pm 0.01$, $b = 1,517.5 \pm 0.2$, and $c = 797.95 \pm 0.1$ pm, is formed in the Ga–Y–Sb system (Yan et al. 2017). The starting materials for obtaining it were pieces of yttrium, gallium, and antimony. A mixture of Y, Ga, and Sb (molar ratio 5:1:3) was loaded into a sealed Ta ampoule under Ar atmosphere. The ampoule was heated at 1,300°C for 15 h in a high-frequency induction furnace and then cooled down to room temperature by turning off the furnace. Single crystals were selected from the crushed sample.

A.7.2 Gallium–Bismuth–Antimony

The $\mathbf{GaBi_xSb_{1-x}}$ films with 1.0–13 at% Bi were grown using MBE by varying the growth temperatures, the Bi flux, and the (Bi + Sb)/Ga ratio (Yue et al. 2018). Among the optimization of three growth parameters to enhance the Bi incorporation into GaSb, reducing the (Bi + Sb)/Ga ratio is the most effective method. The Bi content is inversely proportional to the Sb/Ga flux ratio under the near-stoichiometric growth condition. The band gap energy decreases effectively with increase in the Bi content in the range $0.01 \leq x \leq 0.056$, where the linear decreasing rate is 37 meV/at% Bi.

A.7.3 Gallium–Palladium–Antimony

Two ternary compounds, $\mathbf{Ga_{0.47 \pm 0.01}Pd_2Sb_{0.53 \pm 0.01}}$ and $\mathbf{Ga_{0.62 \pm 0.03}Pd_3Sb_{0.38 \pm 0.03}}$, are formed in the Ga–Pd–Sb system. First of them crystallizes in the hexagonal structure with the lattice parameters $a = 703.74 \pm 0.03$ and $c = 331.94 \pm 0.02$ pm (Matselko et al. 2018b), while the second one crystallizes in the tetragonal structure with the lattice parameters $a = 550.36 \pm 0.02$ and $c = 799.90 \pm 0.03$ pm (Matselko et al. 2018a). These compounds were synthesized from Ga, Pd, and Sb by arc melting in a glove box (Ar atmosphere; O_2 and H_2O content below 1 ppm) (Matselko et al. 2018a, 2018b). Samples with nominal compositions $Ga_{15}Pd_{67}Sb_{18}$ for the first compound and $Ga_{14}Pd_{75}Sb_{11}$ for the second one, placed in alumina crucible and evacuated quartz glass tube, were annealed at 850°C (48 h) and then at 500°C (408 h) and subsequently quenched in water.

A.8 SYSTEMS BASED ON InN

A.8.1 Indium–Antimony–Nitrogen

InN–InSb. The $InSb_{1-x}N_x$ films were epitaxially grown on the GaAs, GaSb, and InSb substrates by using MOCVD (Jin et al. 2015, 2016, 2017). The precursors used for the growth were Me_3Sb, Me_3In, and dimethylhydrazine. The input Sb/In and N/(Sb + N) source flux ratios were kept at 4.2 and 0.9, respectively. The films obtained on the GaAs, GaSb, and InSb substrates contained up to 3.4, 1.9, and 0.26 at% N, respectively.

A.9 SYSTEMS BASED ON InAs

A.9.1 Indium–Bismuth–Arsenic

InAs–InBi. The $InBi_xAs_{1-x}$ films were growth by low-pressure MOVPE using Me_3In, Me_3Bi, and Bu^tAsH_2 as precursors of In, Bi, and As, respectively (Okamoto and Oe 1998, 1999). The substrates were the samples of InAs, the growth pressure was 8 kPa, and the substrate temperature was set at 365°C. The films with 3–4 mol% InBi were grown. The Bi-content dependence of energy gap for these films could be expressed by the next equation: $E_g = 0.42 - 0.042x$.

A.10 SYSTEMS BASED ON InSb

A.10.1 Indium–Potassium–Antimony

The $\mathbf{K_{10}In_5Sb_9}$ ternary compound, which crystallizes in the monoclinic structure with the lattice parameters $a = 1{,}674.4 \pm 0.7$, $b = 1{,}255.8 \pm 0.5$, $c = 1{,}769.1 \pm 0.8$ pm, and $\beta = 116.9 \pm 0.1°$, is formed in the In–Ga–Sb system (Blase et al. 1993). It was synthesized from the elements at 600°C in sealed Nb ampoules.

A.10.2 Indium–Thallium–Antimony

The $\mathbf{In_{1-x}Tl_xSb}$ films were grown using low-pressure MOCVD and cyclopentadienylthallium as the source for Tl (Choi et al. 1993). Me_3In and Me_3Sb were used as sources of In and Sb, respectively. The films were obtained on both GaAs and InSb substrates at a growth temperature of 455°C and a pressure of 10 kPa.

A.10.3 Indium–Bismuth–Antimony

InSb–InBi. The thin films of the $InBi_xSb_{1-x}$ solid solutions have been prepared by MBE on the InSb and GaAs substrates (Noreika et al. 1982). Bismuth incorporation into InSb during growth is strongly dependent on Bi surface concentration, and is influenced by substrate temperature and surface nonstoichiometry. Approximately 3 at% Bi could be incorporated substitutionally in Sb sites under In-rich growth conditions. Increased Bi surface accumulation and interstitial incorporation were observed in Sb-rich surface conditions as the relative flux of Bi was increased. Excess surface Bi forms liquid alloys with excess In during growth above 280°C which alters incorporation.

Large Bi incorporation up to 5.8 at% in such films was achieved by the low-pressure MOCVD technique using Me_3In, Me_3Sb, and Me_3Bi as precursors of In, Sb, and Bi, respectively (Lee et al. 1997). The films were grown on InSb and GaAs substrates. The highest Bi concentration was achieved with the growth temperature of 456°C, growth pressure of 10 kPa, and $(Me_3Sb + Me_3Bi)/Me_3In$ ratio of 13.

Appendix B

Quaternary Alloys Based on III-V Semiconductors

B.1 SYSTEMS BASED ON BN

B.1.1 Boron–Copper–Chlorine–Nitrogen

BN forms an intercalation compound with CuCl and $CuCl_2$ (Croft 1956; Freeman and Larkindale 1969b). The method of CuCl reaction with crystalline BN consisted of sealing the reactants in evacuated Pyrex glass tube and heating the latter at 400°C for 12 h (Croft 1956). The intercalation was also attempted by mixing BN and an excess of CuCl or $CuCl_2$ in the glass tubes, which were then sealed under dry air or nitrogen and heated at 450°C for 22 h in the case of CuCl or at 400°C for 20 h in the case of $CuCl_2$ (Freeman and Larkindale 1969b). After the reaction, the tubes were opened, and BN was washed free of excess CuCl or $CuCl_2$ with HCl followed by distilled water.

B.1.2 Boron–Mercury–Chlorine–Nitrogen

BN forms an intercalation compound with Hg_2I_2 (Freeman and Larkindale 1969b). The intercalation was attempted by mixing BN and an excess of Hg_2I_2 in the glass tubes, which was then sealed under a dry air or nitrogen and heated at 280°C for 2 h. After the reaction, the tubes were opened, and BN was washed free of excess Hg_2I_2 with HCl followed by distilled water.

B.1.3 Boron–Aluminum–Gallium–Nitrogen

BN–AlN–GaN. The layers of $B_xAl_yGa_{1-x-y}N$ were grown at 1,270°C by low-pressure metal-organic vapor-phase epitaxy (MOVPE) using Et_3B, Me_3Al, Me_3Ga, and NH_3 as the source materials for B, Al, Ga, and N, respectively (Takano et al. 2002). The B composition could be controlled by adjusting the vapor phase ratio B/(Al + Ga), where this ratio was increased from 1.5 to 8; thus, the boron composition increased from 1 to 13 at%.

B.1.4 Boron–Aluminum–Chlorine–Nitrogen

BN forms intercalation compound with $AlCl_3$ (Croft 1956). The method of $AlCl_3$ reaction with crystalline BN consisted of sealing the reactants in evacuated Pyrex glass tube and heating the latter at 250°C for 12h. According to the data of Freeman and Larkindale (1969b), any indication of intercalation was found at the heating of BN and $AlCl_3$ mixture at 200°C–250°C for up to 5 days.

B.1.5 Boron–Arsenic–Chlorine–Nitrogen

BN forms an intercalation compound with $AsCl_3$ (Croft 1956, Freeman and Larkindale 1969b). The method of $AsCl_3$ reaction with crystalline BN consisted of sealing the reactants in an evacuated Pyrex glass tube and heating the latter at 118°C for 12h (Croft 1956). The intercalation was also attempted by mixing BN and an excess of $AsCl_3$ in the glass tubes, which was then sealed under dry air or nitrogen and heated at 120°C for 24h (Freeman and Larkindale 1969b). After the reaction, the tubes were opened, and BN was washed free of excess of $AsCl_3$ with HCl followed by distilled water.

B.1.6 Boron–Antimony–Chlorine–Nitrogen

BN forms an intercalation compound with $SbCl_3$ (Croft 1956). The method of $SbCl_3$ reaction with crystalline BN consisted of sealing the reactants in evacuated Pyrex glass tube and heating the latter at 250°C for 12h. According to the data of Freeman and Larkindale (1969b), any indication of intercalation was found at the heating of BN and $SbCl_3$ mixture at 250°C for 24h.

B.1.7 Boron–Chlorine–Iron–Nitrogen

BN forms intercalation compound with $FeCl_3$ (Croft 1956; Freeman and Larkindale 1969b). The method of $FeCl_3$ reaction with crystalline BN consisted of sealing the reactants in evacuated Pyrex glass tube and heating the latter at 400°C for 12h (Croft 1956). The intercalation was also attempted by mixing BN and an excess of $FeCl_3$ in the glass tubes, which was then sealed under dry air or nitrogen and heated at 280°C–450°C for up to 5 days (Freeman and Larkindale 1969b). After the reaction, the tubes were opened and BN was washed free of excess of $FeCl_3$ with HCl followed by distilled water.

B.2 SYSTEMS BASED ON AlP

B.2.1 Aluminum–Hydrogen–Oxygen–Phosphorus

$Al(H_2P_3O_{10})\cdot 2H_2O$ quaternary compound as a crystalline phase was obtained by thermal reactions in surplus of H_3PO_4 (Grunze and Grunze 1984). It is formed in the presence of an excess of Li^+, i.e., in an Li/Al ratio of 2.5 at 200°C.

$Al_2PO_4(OH)_3$ quaternary compound (mineral augelite) crystallizes in the monoclinic structure with the lattice parameters $a = 1{,}307.40 \pm 0.10$, $b = 796.90 \pm 0.07$, $c = 509.10 \pm 0.07$ pm and $\beta = 112.282 \pm 0.005°$ and a calculated density of 2.706 g cm^{-3} (Capitelli et al. 2014b).

$Al_3(OH)_4(H_2O)_3(PO_4)(PO_3OH)\cdot H_2O$ quaternary compound (mineral afmite) crystallizes in the triclinic structure with the lattice parameters $a = 738.6 \pm 0.3$, $b = 771.6 \pm 0.3$,

$c = 1{,}134.5 \pm 0.4$ pm and $\alpha = 99.773 \pm 0.005°$, $\beta = 91.141 \pm 0.006°$, $\gamma = 115.58 \pm 0.05°$ and the calculated and experimental densities of 2.394 and 2.39 ± 0.03 g cm^{-3}, respectively (Kampf et al. 2011; Piilonen et al. 2011).

$Al_6(PO_4)_2(PO_3OH)_2(OH)_8{\cdot}4H_2O$ quaternary compound (mineral planerite) also crystallizes in the triclinic structure with the lattice parameters $a = 750.5 \pm 0.2$, $b = 972.3 \pm 0.3$, $c = 781.4 \pm 0.2$ pm and $\alpha = 111.43°$, $\beta = 115.56°$, $\gamma = 68.69°$ and the calculated and experimental densities of 2.71 and 2.68 ± 0.02 g cm^{-3}, respectively (Foord and Taggart, Jr. 1998).

B.2.2 Aluminum–Lithium–Oxygen–Phosphorus

The **$[LiAl(PO_3)_4]_x$** quaternary compound is formed in the Al–Li–O–P system (Grunze and Grunze 1984). This compound was obtained by thermal reactions in surplus of H_3PO_4. It is formed only in the presence of a large excess of Li^+, i.e., in a Li/Al ratio of 10 in a narrow temperature range around 300°C. This compound is thermally stable.

B.2.3 Aluminum–Sodium–Oxygen–Phosphorus

$NaAlP_2O_7$ could be prepared as a result of thermal decomposition of $NaAl(H_2P_2O_7)_2$ (Grunze and Grunze 1984).

$Na_7(AlP_2O_7)_4PO_4$ crystallizes in the tetragonal structure with the lattice parameters $a = 1{,}404.6 \pm 0.3$ and $c = 616.9 \pm 0.2$ pm (Rochère de la et al. 1985). This compound was obtained by crystallization in a flux of alkali phosphates. The starting mixture had the following composition: $2Al(OH)_3 + 6NaH_2PO_4 + Na_4P_2O_7$. The mixture was gradually heated up to 1,000°C and then cooled down at an approximate rate of 80°C h^{-1}. This process led to a glass, which was introduced in a preheated furnace at 800°C and cooled down at a rate of 100°C h^{-1} between 800°C and 700°C, 5°C h^{-1} between 700°C and 520°C, and finally by switching off the furnace. The obtained product was washed in water.

B.2.4 Aluminum–Potassium–Oxygen–Phosphorus

The **$KAlP_2O_7$** and **$K_2Al_2(P_8O_{24})$** quaternary compounds were obtained by thermal reactions in surplus of H_3PO_4 (Grunze and Grunze 1984).

B.2.5 Aluminum–Rubidium–Oxygen–Phosphorus

The **$Rb_2Al_2(P_8O_{24})$** quaternary compound was obtained by thermal reactions in surplus of H_3PO_4 (Grunze and Grunze 1984).

B.2.6 Aluminum–Copper–Selenium–Phosphorus

The **$CuAlP_2Se_6$** quaternary compound, which melts incongruently at 537°C and crystallizes in the trigonal structure with the lattice parameters $a = 627.95 \pm 0.02$ and $c = 1997.13 \pm 0.06$ pm and a calculated density of 4.645 g cm^{-3}, is formed in the Al–Cu–Se–P system (Pfeiff and Kniep 1993). It was obtained by a reaction of CuCl and $AlCl_3$ with $Mg_2P_2Se_6$ followed by cooling from the melt (10 h at 490°C; cooling rate 50°C h^{-1}).

B.2.7 Aluminum–Silver–Selenium–Phosphorus

The $\mathbf{AgAlP_2Se_6}$ quaternary compound, which melts incongruently at 588°C, is formed in the Al–Ag–Se–P system (Pfeiff and Kniep 1993). It was obtained by a reaction of AgCl and $AlCl_3$ with $Mg_2P_2Se_6$ followed by cooling from the melt (10 h at 530°C; cooling rate 50°C h^{-1}).

B.2.8 Aluminum–Gallium–Indium–Phosphorus

The behavior of unstable region for the $\mathbf{Al_yGa_xIn_{1-x-y}P}$ solid solutions was analyzed based on regular approximation for solid solutions (Onabe 1982). The quaternary critical point for these solid solutions exists at 608°C and has a composition $x + y = 0.50$.

B.2.9 Aluminum–Gallium–Arsenic–Phosphorus

The behavior of unstable region for the $\mathbf{Al_xGa_{1-x}As_yP_{1-y}}$ solid solutions was analyzed based on regular approximation for solid solutions (Onabe 1982). The quaternary critical point for these solid solutions exists at –167°C and has a composition $x = 0.50$ and $y = 0.50$.

B.2.10 Aluminum–Gallium–Antimony–Phosphorus

The behavior of unstable region for the $\mathbf{Al_xGa_{1-x}Sb_yP_{1-y}}$ solid solutions was analyzed based on regular approximation for solid solutions (Onabe 1982). The quaternary critical point for these solid solutions exists at 2,275°C and has a composition $x = 0.50$ and $y = 0.50$.

B.2.11 Aluminum–Indium–Arsenic–Phosphorus

The behavior of unstable region for the $\mathbf{Al_xIn_{1-x}As_yP_{1-y}}$ solid solutions was analyzed based on regular approximation for solid solutions (Onabe 1982). The quaternary critical point for these solid solutions exists at 1,121°C and has a composition $x = 0.53$ and $y = 0.45$.

B.2.12 Aluminum–Indium–Antimony–Phosphorus

The behavior of unstable region for the $\mathbf{Al_xIn_{1-x}Sb_yP_{1-y}}$ solid solutions was analyzed based on regular approximation for solid solutions (Onabe 1982). The quaternary critical point for these solid solutions exists at 3,914°C and has a composition $x = 0.59$ and $y = 0.46$.

B.2.13 Aluminum–Arsenic–Antimony–Phosphorus

The behavior of unstable region for the $\mathbf{AlAs_{1-x-y}Sb_xP_y}$ solid solutions was analyzed based on regular approximation for solid solutions (Onabe 1982). The quaternary critical point for these solid solutions exists at 1,992°C and has a composition $x = 0.50$ and $y = 0.50$.

B.3 SYSTEMS BASED ON AlAs

B.3.1 Aluminum–Hydrogen–Oxygen–Arsenic

The $\mathbf{[Al_6(AsO_4)_3(OH)_9(H_2O)_5]\cdot 8H_2O}$ (mineral penberthycroftite) crystallizes in the monoclinic structure with the lattice parameters $a = 778.9 \pm 0.2$, $b = 2{,}477.7 \pm 0.4$, $c = 1{,}573.7 \pm 0.3$ pm and $\beta = 93.960 \pm 0.001°$ at room temperature and $a = 775.3 \pm 0.2$, $b = 2{,}467.9 \pm 0.5$, $c = 1{,}567.9 \pm 0.3$ pm and $\beta = 94.19 \pm 0.03°$ and a calculated density of 2.18 g cm^{-3} at 100 K (Grey et al. 2016a; Belakovskiy et al. 2017).

B.3.2 Aluminum–Gallium–Indium–Arsenic

The behavior of unstable region for the $\mathbf{Al_yGa_xIn_{1-x-y}As}$ solid solutions was analyzed based on regular approximation for solid solutions (Onabe 1982). The quaternary critical point for these solid solutions exists at 482°C and has a composition $x = 0.50$ and $y = 0$.

B.3.3 Aluminum–Gallium–Antimony–Arsenic

The behavior of unstable region for the $\mathbf{Al_xGa_{1-x}Sb_yAs_{1-y}}$ solid solutions was analyzed based on regular approximation for solid solutions (Onabe 1982). The quaternary critical point for these solid solutions exists at 1,231°C and has a composition $x = 0.46$ and $y = 0.49$.

B.3.4 Aluminum–Indium–Antimony–Arsenic

The behavior of unstable region for the $\mathbf{Al_xIn_{1-x}Sb_yAs_{1-y}}$ solid solutions was analyzed based on regular approximation for solid solutions (Onabe 1982). The quaternary critical point for these solid solutions exists at 1,838°C and has a composition $x = 0.59$ and $y = 0.44$.

B.4 SYSTEMS BASED ON GaAs

B.4.1 Gallium–Hydrogen–Oxygen–Arsenic

The $\mathbf{GaAsO_4 \cdot 2H_2O}$ quaternary compound, which crystallizes in the orthorhombic structure with the lattice parameters $a = 1{,}013.3$, $b = 883.0$, and $c = 989.0$ pm, is formed in the Ga–H–O–As system (Pâques-Ledent and Tarte 1969). Its synthesis was carried out by slow precipitation in an acid medium of the Ga salt with the addition of Na_2HAsO_4 solution. The 0.3 M solution of $Ga(NO_3)_3$ (10 mL) was acidified with three or four drops of moderately concentrated H_3PO_4. The pH was then adjusted to 2.4 by the addition of Na_2HAsO_4 dilute solution. The suspension was treated in a sealed tube at 130°C.

B.4.2 Gallium–Zinc–Selenium–Arsenic

$\mathbf{4GaAs\text{–}ZnGa_2Se_4}$. According to the data of XRD, metallography, and measuring of microhardness, a continuous series of solid solutions with sphalerite structure is formed in this system (Boltovets et al. 1969). The lattice parameter of the solid solutions varies with the composition according to a law close to linear. Only near $ZnGa_2Se_4$, there is some deviation from Vegard's law. Near 50 mol% $ZnGa_2Se_4$, there is a rupture on the concentration dependence of energy gap.

B.5 SYSTEMS BASED ON GaSb

B.5.1 Gallium–Lead–Bismuth–Antimony

The solubility of GaSb in the Pb–Bi eutectic (43.5 mass% Pb +56.5 mass% Bi; melting temperature 125°C) within the temperature region from 175°C to 650°C was investigated by standing the GaSb ingots for 14–15 h in the melt of the solvent at a given temperature (Golubev et al. 1972). The solubility was determined by GaSb mass loss. The linear dependence of the logarithm of solubility on the reciprocal temperature indicates the absence of chemical reactions between the eutectic and gallium antimonide.

B.6 SYSTEMS BASED ON InP

B.6.1 Indium–Copper–Selenium–Phosphor

The **$CuInP_2Se_6$** quaternary compound, which melts incongruently at 642°C (Pfeiff and Kniep 1993) [forms due to a sintectic reaction from two liquids at 815°C (Potoriy et al. 2010)] and crystallizes in the monoclinic structure with the lattice parameters $a = 609.56 \pm 0.04$, $b = 1056.45 \pm 0.06$, $c = 1362.30 \pm 0.08$ pm and $\beta = 107.101 \pm 0.003°$ and the calculated and experimental densities of 3.405 and 3.427 g cm^{-3}, respectively (Maisonneuve et al. 1995) [$a = 609.6$, $b = 1056.4$, $c = 1362.3$ pm and $\beta = 107.101°$ and an experimental density of 3.425 g cm^{-3} (Potoriy et al. 2010)], is formed in the In–Cu–Se–P system. It was prepared from a mixture of the elements in stoichiometric proportions, which was heated in evacuated silica tube for two weeks at 600°C (Maisonneuve et al. 1995). Air-stable yellow platelets embedded in a homogeneous powder bulk were obtained. This compound was also prepared by a reaction of CuCl and $InCl_3$ with $Mg_2P_2Se_6$ followed by cooling from the melt (10 h at 490°C; cooling rate 50°C h^{-1}) (Pfeiff and Kniep 1993). Single crystals of the title compound could be grown by the chemical transport reactions (I_2 and CuI were used as transport agents) and by the directional crystallization of the melt (Potoriy et al. 2010).

B.6.2 Indium–Silver–Sulfur–Phosphor

The **$AgInP_2S_6$** quaternary compound, which melts congruently at 780°C and crystallizes in the trigonal structure with the lattice parameters $a = 647.2 \pm 0.8$, $c = 1333 \pm 1$ pm and an experimental density of 3.49 g cm^{-3} and energy gap 2.43 eV, is formed in the Al–Cu–Se–P system (Potoriy et al. 2009).

B.6.3 Indium–Silver–Selenium–Phosphor

The **$AgInP_2Se_6$** quaternary compound, which melts congruently at 673°C, is formed in the In–Ag–Se–P system (Pfeiff and Kniep 1993). It was obtained by a reaction of AgCl and $InCl_3$ with $Mg_2P_2Se_6$ followed by cooling from the melt (10 h at 530°C; cooling rate 50°C h^{-1}).

References

Abeledo de M.E.J., Angelelli V., de Benyacar M.A.R., Gordillo C.E. Sanjuanite, a new hydrated basic sulphate-phosphate of aluminum, *Am. Mineral.*, **53**(1–2), 1–8 (1968).

Adiwidjaja G., Friese K., Klaska K.-H., Schlüter J. The crystal structure of kastningite (Mn,Fe,Mg) $(H_2O)_4[Al_2(OH)_2(H_2O)_2(PO_4)_2]\cdot 2H_2O$ – a new hydroxyl aquated orthophosphate hydrate mineral, *Z. Kristallogr.*, **214**(8), 465–468 (1999).

Aftandilian V.D., Miller H.C., Muetterties E.L. Chemistry of boranes. I. Reactions of boron hydrides with metal and amine salts, *J. Am. Chem. Soc.*, **83**(11), 2471–2474 (1961).

Ahmed E., Breternitz J., Groh M.F., Isaeva A., Ruck M. $[Sb_7Se_8Br_2]^{3+}$ and $[Sb_{13}Se_{16}Br_2]^{5+}$ – double and quadruple spiro cubanes from ionic liquidus, *Eur. J. Inorg. Chem.*, **2014**(19), 3037–3042 (2014).

Akasaka T., Makimoto T. Flow-rate modulation epitaxy of wurtzite AlBN, *Appl. Phys. Lett.*, **88**(4), 041902_1–041902_3 (2006).

Ali T., Bauer E., Hilscher G., Michor H. Anderson lattice in the intermediate valence compound $Ce_3Ni_2B_2N_{3-\delta}$, *Phys. Rev. B*, **83**(11), 115131_1–115131_7 (2011a).

Ali T., Bauer E., Hilscher G., Michor H. Structural, superconducting and magnetic properties of $La_{3-x}R_xNi_2B_2N_{3-\delta}$ with R = Ce, Pr, Nd, *Solid State Phenom.*, **170**, 165–169 (2011b).

Allen V.T., Fahey J.J., Axelrod J.M. Mansfieldite, a new arsenate, the aluminum analogue of scorodite, and the mansfieldite-scorodite series, *Am. Mineralog.*, **33**(3–4), 122–134 (1948).

Altintas B., Parlak C., Bozkurt C., Eryiğit R. Intercalation of graphite and hexagonal boron nitride by lithium, *Eur. Phys. J., B*, **79**(3), 301–312 (2011).

Andreev I.A., Kunitsyna E.V., Solov'ev Y.V., Charykov N.A., Yakovlev Y.P. Using of lead as a neutral solvent for obtaining of the $Ga_{1-x}In_xAs_ySb_{1-y}$ solid solutions [in Russian], *Pis'ma v ZhTF*, **25**(19), 77–81 (1999).

Anisimova N., Bork M., Hoppe R., Meisel M. Crystal structure of a new acentric $CsGaHP_3O_{10}$ form III, *Z. anorg. allg. Chem.*, **621**(6), 1069–1074 (1995).

Anisimova N., Chudinova N., Hoppe R., Serafin M. Preparation and crystal structure of a new acentric caesium gallium hydrogen phosphate containing phosphoric acid, $Cs_2Ga(H_2PO_4)(HPO_4)_2\cdot H_3PO_4\cdot 0.5H_2O$, *Z. anorg. allg. Chem.*, **623**(1–6), 39–44 (1997).

Ankinovich E.A. A new vanadium mineral rusakovite [in Russian], *Zap. Vses. mineralog. obshch.*, **89**(4), 440–447 (1960).

Ankinovich E.A., Bekenova G.K., Shabanova T.A., Zazubina I.S., Sandomirskaya S.M. Mitryaevaite, $Al_{10}[(PO_4)_{8.7}(SO_3OH)_{1.3}]_{\Sigma 10}AlF_3\cdot 30H_2O$, a new mineral species from a Cambrian carbonaceous chert formation, Karatau Range and Zhabagly Mountains, southern Kazakhstan, *Canad. Mineralog.*, **35**(6), 1415–1420 (1997).

Anshon A.V., Drozdov Y.N., Karpovich I.A., Rachkova E.N., Safonov A.A. Solid solutions in the $CdSnAs_2$–GaSb system [in Russian], *Izv. AN SSSR. Neorgan. Mater.*, **23**(9), 1560–1561 (1987).

Antenucci D., Miehe G., Tarte P., Schmahl W.W., Fransolet A.-M. Combined X-ray Rietveld, infrared abd Raman study of a new synthetic variety of alluaudite, $NaCdIn_2(PO_4)_3$, *Eur. J. Mineral.*, **5**(2), 207–213 (1993).

Araki T., Zoltai T. The crystal structure of wavellite, *Z. Kristallogr.*, **127**(1–4), 21–33 (1968).

Aristarain L.F., Hurlbut, Jr. C.S. Teruggite, $4CaO{\cdot}MgO{\cdot}6B_2O_3{\cdot}As_2O_5{\cdot}18H_2O$, a new mineral from Jujuy, Argentina, *Am. Mineralog.*, **53**(11–12), 1815–1827 (1968).

Armbruster T., Bühler C., Graeser S., Stalder H.A., Amthauer G. Cervandonite-(Ce), (Ce,Nd,La) $(Fe^{3+},Fe^{2+},Ti^{4+},Al)_3SiAs(Si,As)O_{13}$, a new Alpine fissure mineral, *Schweiz. Mineral. Petrog. Mitt.*, **68**(2), 125–132 (1988).

Atencio D., Coutinho J.M.V., Mascarenhas Y.P., Ellena J.A. Matioliite, the Mg-analog of burangaite, from Gentil mine, Mendes Pimentel, Minas Gerais, Brazil, and other occurrences, *Am. Mineralog.*, **91**(11–12), 1932–1936 (2006).

Attfield M.P., Cheetham A.K., Natarajan S. The direct synthesis and characterization of the pillared layer indium phosphate $Na_4[In_8(HPO_4)_{14}(H_2O)_6]{\cdot}12(H_2O)$, *Mater. Res. Bull.*, **35**(7), 1007–1015 (2000).

Attfield M.P., Morris R.E., Gutierrez-Puebla E., Monge-Bravo A., Cheetham A.K. Synthesis and structure of a novel microporous gallophosphate, $Na_3Ga_5(PO_4)_4O_2(OH)_2{\cdot}2H_2O$, *J. Chem. Soc., Chem. Commun.*, (8), 843–844 (1995).

Auernhammer M., Effenberger H., Hentschel G., Reinecke T., Tillmanns E. Niekenichite, a new arsenate from the Eifel, Germany, *Mineral. Petrol.*, **48**(1–4), 153–166 (1993).

Avaliani M.A., Chudinova N.N., Tananaev I.V. Synthesis of condensed indium phosphates in polyphosphoric acid melts [in Russian], *Izv. AN SSSR. Neorgan. Mater.*, **15**(9), 1688–1689 (1979).

Averbuch-Pouchot M.T., Durif A., Guitel J.C. Structure cristalline d'un tripolyphosphate acide d'aluminium-ammonium: $AlNH_4HP_3O_{10}$, *Acta Crystallogr.*, **B33**(5), 1436–1438 (1977).

Averkieva G.K., Berdichevskiy G.V., Vaypolin A.A., Goryunova N.A., Prochikhan V.D. On the chemical interaction in the Cu–Ga–Ge–As–Se quinary system [in Russian], *Izv. AN SSSR. Neorgan. Mater.*, **4**(7), 1064–1066 (1968).

Averkieva G.K., Vaypolin A.A., Goryunova N.A. On some ternary compounds of the $A^I{}_2D^{IV}C^{VI}{}_3$ type and based on them solid solutions [in Russian], *Issled. po poluprovodn. Novye poluprovodn. Materialy*, Kishinev, Kartia Moldovenyaske Publish., 44–56 (1964).

Baker W.E. Hinsdalite pseudomorphous after pyromorphite from Dundas, Tasmania, *Pap. Proc. Royal Soc. Tasmania*, **97**, 129–132 (1963).

Balenzano F., Dell'Anna L., Di Pierro M. Francoanellite, $H_6K_3Al_5(PO_4)_8{\cdot}13H_2O$, a new mineral from the caves of Castellana, Puglia, southern Italy, *Neues Jahrb. Mineral., Monatsh.*, 49–57 (1976).

Barnes W.H. The unit cell and space group of childrenite, *Am. Mineralog.*, **34**(1–2), 12–18 (1949).

Bartlett N., Biagioni R.N., McQuillan B.W., Robertson A.S., Thompson A.C. Novel salts of graphite and a boron nitride salt, *J. Chem. Soc., Chem. Commun.*, (5), 200–201 (1978).

Bartnitskaya T.S., Butylenko A.K., Lugovskaya E.S., Timofeeva I.I. Study of the quasibinary cross section of AN–BN at high pressures, *Vysok. Davleniya Svoistva Mater.*, Kiev, Nauk. Dumka Pablish., Kiev, 90–94 (1980).

Baur W.H. Die Kristallstruktur des Edelamblygonits $LiAlPO_4(OH,F)$, *Acta Crystallogr.*, **12**(12), 988–994 (1959).

Baur W.H., Rama Rao B. The crystal structure and the chemical composition of vauxite, *Am. Mineralog.*, **53**(5–6), 1025–1028 (1968).

Baur W.H., Rama Rao B. The crystal structure of metavauxite, *Naturwissenschaften*, **54**(21), 561 (1967).

Baykal A., Kizilyalli M., Kniep R. X-ray powder diffraction and IR study of $NaMg(H_2O)_2[BP_2O_8]{\cdot}H_2O$ and $NH_4Mg(H_2O)_2[BP_2O_8]{\cdot}H_2O$, *J. Mater. Sci.*, **35**(18), 4621–4626 (2000).

Bayliss P. X-ray powder data for nissonite and waylandite, *Powder Diffraction*, **1**(4), 331–333 (1986).

Belakovskiy D.I., Cámara F., Gagne O.C. New mineral names, *Am. Mineralog.*, **102**(3), 694–700 (2017).

Belakovskiy D.I., Camára F., Gagne O.C., Uvarova Y. New mineral names, *Am. Mineralog.*, **101**(6), 1489–1497 (2016a).

Belakovskiy D.I., Gagne O.C., Uvarova Y. New mineral names, *Am. Mineralog.*, **101**(11), 2570–2573 (2016b).

Belakovskiy D., Welch M.D., Cámara F., Gatta G.D., Tait K.T. New mineral names, *Am. Mineralog.*, **97**(8–9), 1523–1530 (2012).

Belokoneva E.L., Gurbanova O.A., Dimitrova O.V., Al'-Ama A.G. Synthesis and crystal structure if the new indium lead phosphate $PbIn[PO_4][PO_3(OH)]$ [in Russian], *Zhurn. neorgan. khimii*, **46**(7), 1121–1126 (2001a).

Belokoneva E.L., Gurbanova O.A., Dimitrova O.V., Stefanovich S.Y. Al'-Ama A.G. Synthesis, crystal structure and properties of new indium borophosphate $In[BP_2O_8]{\cdot}0.8H_2O$ [in Russian], *Zhurn. neorgan. khimii*, **46**(7), 1115–1120 (2001b).

Belova L.N., Gorshkov A.I., Doynikova O.A., Mokhov A.V., Trubkin N.V., Sivtsov A.V. New date on coconinoite [in Russian], *Dokl. AN SSSR*, **329**(6), 772–775 (1993).

Benyahia M., Vaney J.B., Leroy E., Rouleau O., Dauscher A., Lenoir B., Alleno E. Thermoelectric properties in double-filled $Ce_{0.3}In_yFe_{1.5}Co_{2.5}Sb_{12}$ *p*-type skutterudites, *J. Alloys Compd.*, **696**, 1031–1038 (2017).

Berdesinski W. Synthetische Darsfellung von Lüneburgit, *Naturwissenschaften*, **38**(20), 476–477 (1951).

Bernhardt E., Berkei M., Willner H., Schürmann M. Die Reaktionen von $M[BF_4]$ (M = Li, K) und $(C_2H_5)_2O{\cdot}BF_3$ mit $(CH_3)_3SiCN$. Bildung von $M[BF_x(CN)_{4-x}]$ (M = Li, K; x = 1, 2) und $(CH_3)_3SiNCBF_x(CN)_{3-x}$, (x = 0, 1), *Z. anorg. und allg. Chem.*, **629**(4), 677–685 (2003a).

Bernhardt E., Bernhardt-Pitchougina V., Willner H., Ignatiev N. Umpolung at boron by reduction of $[B(CN)_4]^-$ and formation of the dianion $[B(CN)_3]^{2-}$, *Angew. Chem. Int. Ed.*, **50**(50), 12085–12088 (2011a).

Bernhardt E., Bernhardt-Pitchougina V., Willner H., Ignatiev N. Umpolung von Bor im $[B(CN)_4]^-$ Anion durch Reduktion zum Dianion $[B(CN)_3]^{2-}$, *Angew. Chem.*, **123**(50), 12291–12294 (2011b).

Bernhardt E., Finze M., Willner H. Eine effiziente Synthese von Tetracyanoboraten durch Sinterprozesse, *Z. anorg. und allg. Chem.*, **629**(7–8), 1229–1234 (2003b).

Bernhardt E., Willner H. Die Kristallstrukturen der Cyanidofluoridoborate $K[BF_x(CN)_{4-x}]$, x = 0-4, *Z. anorg. und allg. Chem.*, **635**(15), 2511–2514 (2009).

Berry L.G., Davis T. Liroconite, *Am. Mineralog.*, **32**(3–4), 196 (1947).

Berry L.G., Graham A.R. X-ray measurements on brackebuschite and hematolite, *Am. Mineralog.*, **33**(7–8), 489–495 (1948).

Berry L.G. Observations on conichalcite, cornwallite, euchroite, liroconite and olivenite, *Am. Mineralog.*, **36**(5–6), 484–503 (1951).

Berry L.G., Steacy H.R. Euchroite and chalcophyllite, *Am. Mineralog.*, **32**(3–4), 196 (1947).

Berry L.G. Structural crystallography of lazulite, scorzalite and veszelyite, *Am. Mineralog.*, **33**(11–12), 750 (1948).

Beukes G.J., Schoch A.E., Van der Westhuizen W.A., Bok L.D.C., de Bruiyn H. Hotsonite, a new hydrated aluminum-phosphate-sulfate from Pofadder, South Africa, *Am. Mineralog.*, **69**(9–10), 979–983 (1984).

Bieniok A., Brendel U., Lottermoser W., Amthauer G. Metalophosphates with mixed-polyhedral framework structures: Syntheses and characterization, *Z. Kristallogr.*, **223**(3), 186–194 (2008).

Biltz W., Marcus E. Über den Lüneburgit, *Z. anorg. und allg. Chem.*, **77**(1), 124–130 (1912).

Birch W.D., Grey I.E., Mills S.J., Pring A., Bougerol C., Ribaldi-Tunnicliffe A., Wilson N.C., Keck E. Nordgauite, $MnAl_2(PO_4)_2(F,OH)_2{\cdot}5H_2O$, a new mineral from the Hagendorf-Süd pegmatite, Bavaria, Germany: Description and crystal structure, *Mineralog. Mag.*, **75**(2), 269–278 (2011).

Birch W.D., Mills S.J., Schwendtner K., Pring A., Webb J.A., Segnit R., Watts J.A. Parwanite: A new hydrated Na–Mg–Al-phosphate from a lava cave at Parwan, Victoria, Australia, *Aust. J. Mineral.*, **13**(1), 23–30 (2007).

Birch W.D., Pring A., Bevan D.J.M., Kharisun. Wycheproofite: A new hydrated sodium aluminium zirconium phosphate from Wycheproof, Victoria, Australia, and a new occurrence of kosnarite, *Mineralog. Mag.*, **58**(4), 635–639 (1994).

Birch W.D., Pring A., Foord E.E. Selwynite, $NaK(Be,Al)Zr_2(PO_4)_4{\cdot}2H_2O$, a new gainesite-like mineral from Wycheproof, Victoria, Australia, *Canad. Mineralog.*, **33**(1), 55–58 (1995).

Birch W.D., Pring A. Sieleckiite, a new copper aluminium phosphate from Mt Oxide, Queensland, Australia, *Mineralog. Mag.*, **52**(367), 515–518 (1988).

Birsöz B., Baykal A., Toprak M., Köseoglu Y. Synthesis, characterization and magnetic investigation of $(NH_4)_{0.5}Mn_{1.25}(H_2O)_2[BP_2O_8]{\cdot}0.5H_2O$, *Centr. Eur. J. Chem.*, **5**(2), 536–545 (2007).

Blanchard F.N., Abernathy S.A. X-ray powder diffraction data for phosphate minerals: Vauxite, metavauxite, vivianite, Mn-heterosite, scorzalite, and lazulite, *Florida Scientist*, **43**(4), 257–265 (1980).

Blanchard F.N. Calculated and observed X-ray powder diffraction patterns for paravauxite, *Florida Scientist*, **40**(2), 199–202 (1977).

Blanchard F.N. Physical and chemical data for crandallite from Alachua County, Florida, *Am. Mineralog.*, **57**(3–4), 473–484 (1972).

Blanchard F.N. X-ray powder diffraction data for wavelite, *Florida Scientist*, **37**(1), 1–4 (1974).

Blase W., Cordier G., Somer M. Crystal structure of decapotassium nonaantimonidopentaindate, $K_{10}In_5Sb_9$, *Z. Kristallogr.*, **203**(1), 146–147 (1993).

Blount A.M. The crystal structure of crandallite, *Am. Mineralog.*, **59**(1–2), 41–47 (1974).

Boltovets N.S., Drobyazko V.P., Mityurev V.K. Some properties of the solid solutions of the $GaAs{-}ZnGa_2Se_4$ system [in Russian], *Izv. AN SSSR. Neorgan. Mater.*, **5**(4), 803–804 (1969).

Bonaccorsi E., Merlino S., Orlandi P. Zincalstibite, a new mineral, and cualstibite: Crystal chemical and structural relationships, *Am. Mineralog.*, **92**(1), 198–203 (2007).

Bonhomme F., Thoma S.G., Nenoff T.M. Two ammonium templated gallophosphates: Synthesis and structure determination from powder diffraction data of 2D and 3D-GAPON, *Micropor. Mesopor. Mater.*, **53**(1–3), 87–96 (2002).

Bontchev R.P., Do J., Jacobson A.J. The vanadium(V) borophosphate $(NH_4)_5[V_3BP_3O_{19}]{\cdot}H_2O$, *Inorg. Chem.*, **39**(18), 4179–4181 (2000).

Bontchev R.P., Sevov S.C. Hydrothermal synthesis and characterization of two ammonium-templated cobalt aluminophosphates, *J. Mater. Chem.*, **9**(10), 2679–2682 (1999).

Bontchev R.P., Sevov S.C. Synthesis and characterization of a new cobalt aluminophosphate with an open-framework structure, *Chem. Mater.*, **9**(12), 3155–3158 (1997).

Borrego J.M., Conde C.F., Conde A., Stoica M., Roth S., Greneche J.M. Crystallization behavior and magnetic properties of Cu-containing Fe–Cr–Mo–Ga–P–C–B alloys, *J. Appl. Phys.*, **100**(4), 043515_1–043515_6 (2006).

Boy I., Cordier G., Eisenmann B., Kniep R. Oligomere Tetraeder-Anionen in Borophosphaten: Darstellung und Kristall-strukturen von $NaFe[BP_2O_7(OH)_3]$ und $K_2Fe_2[B_2P_4O_{16}(OH)_2]$, *Z. Naturforsch.*, **53B**(2), 165–170 (1998a).

Boy I., Cordier G., Kniep R. Crystal structure of tetrasodium tricopper(II) (diboro-diphosphate-bis(monohydrogenphosphate)) bìs(monohydrogenphosphate), $Na_4Cu_3[B_2P_4O_{15}(OH)_2]{\cdot}2HPO_4$, *Z. Kristallogr., New Cryst. Str.*, **213**(1), 29–30 (1998b).

Boy I., Cordier G., Kniep R. Oligomere Tetraeder-Anionen in Borophosphaten: Sechserringe mit offenen und cyclischen Phosphat-Verzweigungen in der Kristallstruktur von $K_6Cu_2[B_4P_8O_{28}(OH)_6]$, *Z. Naturforsch.*, **53B**(12), 1440–1444 (1998c).

Boy I., Hauf C., Kniep R. $Fe[B_2P_2O_7(OH)_5]$: Ein neues Borophosphat mit unverzweigten Vierer-Einfach Tetraederketten, *Z. Naturforsch.*, **53B**(6), 631–633 (1998d).

Boy I., Kniep R. Crystal structure of lithium copper(II) monoaqua catena-[monoborodiphosphate] dihydrate, $LiCu(H_2O)[BP_2O_8]{\cdot}2H_2O$, *Z. Kristallogr., New. Cryst. Str.*, **216**(1), 7–8 (2001a).

Boy I., Kniep R. Crystal structure of lithium zinc diaqua catena-[monoborodiphosphate]-monohydrate, $LiZn(H_2O)_2[BP_2O_8]{\cdot}H_2O$, *Z. Kristallogr., New. Cryst. Str.*, **216**(1), 9–10 (2001b).

Boy I., Kniep R. K[$B_6PO_{10}(OH)_4$]: Ein Borophosphat mit gestreckten Bändern aus Tetraeder-Vierer-Ringen und offen-zyklischen Verzweigungen über planare $B_2O_3(OH)_2$-Gruppen, *Z. Naturforsch.*, **54B**(7), 895–898 (1999).

Boy I., Schäfer G., Kniep R. $(Ni_{3-x}Mg_x)[B_3P_3O_{12}(OH)_6]\cdot 6H_2O$ ($x \approx 1.5$): A novel borophosphate-hydrate containing isolated six-membered rings of tetrahedra, *Z. anorg. und allg. Chem.*, **627**(2), 139–143 (2001c).

Boy I., Schäfer G., Kniep R. Crystal structure of sodium iron(II) diaqua catena-[monoboro-diphosphate] monohydrate, $NaFe(H_2O)_2[BP_2O_8]\cdot H_2O$, and of potassium iron(II) diaqua catena-[monoboro-diphosphate] hemihydrate, $KFe(H_2O)_2[BP_2O_8]\cdot 0.5H_2O$, *Z. Kristallogr., New. Cryst. Str.*, **216**(1), 13–14 (2001d).

Boy I., Schäfer G., Kniep R. Crystal structure of sodium nickel diaqua catena-[monoboro-diphosphate] monohydrate, $NaNi(H_2O)_2[BP_2O_8]\cdot H_2O$, at 293 K and 198 K, *Z. Kristallogr., New. Cryst. Str.*, **216**(1), 11–12 (2001e).

Boy I., Stowasser F., Schäfer G., Kniep R. $NaZn(H_2O)_2[BP_2O_8]\cdot H_2O$: A novel open-framework borophosphate and its reversible dehydration to microporous sodium zincoborophosphate $Na[ZnBP_2O_8]\cdot H_2O$ with CZP topology, *Chem. Eur. J.*, **7**(4), 834–839 (2001f).

Brennan G.L., Dahl G.H., Schaeffer R. Studies of boron–nitrogen compounds. II. Preparation and reactions of B-trichloroborazole, *J. Am. Chem. Soc.*, **82**(24), 6248–6250 (1960).

Britvin S.N., Pakhomovskiy Y.A., Bogdanova A.N. Krasnovite, $Ba(Al,Mg)(PO_4,CO_3)(OH)_2\cdot 2H_2O$, a new mineral [in Russian], *Zap. Vseros. mineralog. obshch.*, **125**(3), 110–112 (1996).

Bruiyn de H., Beukes G.J., van der Westhuizen W.A., Tordiffe E.A.W. Unit cell dimensions of the hydrated aluminium phosphate-sulphate minerals sanjuanite, kribergite, and hotsonite, *Mineralog. Mag.*, **53**(371), 385–386 (1989).

Brunet F., Chopin C. Bearthite, $Ca_2Al(PO_4)_2OH$: Stability, thermodynamic properties and phase relations, *Contrib. Mineral. Petrol.*, **121**(3), 258–266 (1995).

Brunet F., Morineau D., Schmid-Beurmann P. Heat capacity of lazulite, $MgAl_2(PO_4)_2(OH)_2$, from 35 to 298 K and a (S–V) value for P_2O_5 to estimate phosphate entropy, *Mineralog. Mag.*, **68**(1), 123–134 (2004).

Burns P.C., Eby R.K., Hawthorne F.C. Refinement of the structure of liroconite, a heteropolyhedral framework oxysalt mineral, *Acta Crystallogr.*, **C47**(5), 916–919 (1991).

Burns P.C., Hawthorne F.C. The crystal structure of sinkankasite, a complex heteropolyhedral sheet mineral, *Am. Mineralog.*, **80**(5–6), 620–627 (1995).

Burns P.C., Smith J.V., Steele I.M. Arizona porphyry copper/hydrothermal deposits I. The structure of chenevixite and luetheite, *Mineralog. Mag.*, **64**(1), 25–30 (2000).

Cabri L.J., Fleischer M., Pabst A. New mineral names, *Am. Mineralog.*, **66**(9–10), 1099–1103 (1981).

Cahill C.L., Krivovichev S.V., Burns P.C., Bekenova G.K., Shabanova T.A. The crystal structure of mitryaevaite, $Al_5(PO_4)_2[(P,S)O_3(OH,O)]_2F_2(OH)_2(H_2O)_8\cdot 6.48H_2O$, determined from a micro-crystal using synchrotron radiation, *Canad. Mineralog.*, **39**(1), 179–186 (2001).

Calvo C., Faggiani R., Krishnamachari N. The crystal structure of $Sr_{9.402}Na_{0.209}(PO_4)_6B)_{0.996}O_2$ – a deviant apatite, *Acta Crystallogr.*, **B31**(1), 188–192 (1975).

Campbell F.A. Lazulite from Yukon, Canada, *Am. Mineralog.*, **47**(1–2), 157–160 (1962).

Cámara F., Belakovskiy D.I., Uvarova Y. New mineral names, *Am. Mineralog.*, **102**(4), 916–920 (2017).

Cámara F., Ciriotti M.E., Bittarello E., Nestola F., Bellatreccia F., Massimi F., Radica F., Costa E., Benna P., Piccoli G.C. Grandaite, IMA 2013-059. CNMNC Newsletter No. 18, *Mineralog. Mag.*, **77**(8), 3249–3258 (2013).

Cámara F., Ciriotti M.E., Bittarello E., Nestola F., Massimi F., Radica F., Costa E., Benna P., Piccoli G.C. Arsenic-bearing new mineral species from Valletta mine, Maira Valley, Piedmont, Italy: I. Grandaite, $Sr_2Al(AsO_4)_2(OH)$, description and crystal structure, *Mineralog. Mag.*, **78**(3), 757–774 (2014a).

Cámara F., Gagne O.C., Uvarova Y. New mineral names, *Am. Mineralog.*, **100**(2–3), 658–663 (2015).

Cámara F., Gatta G.D., Uvarova Y., Gagne O.C., Belakovskiy D.I. New mineral names, *Am. Mineralog.*, **99**(10), 2150–2158 (2014b).

Cámara F., Oberti R., Chopin C., Medenbach O. The arrojadite enigma: I. A new formula and a new model for the arrojadite structure, *Am. Mineralog.*, **91**(8–9), 1249–1259 (2006).

Cámara F., Uvarova Y., Gagne O.C., Belakovskiy D.I. New mineral names, *Am. Mineralog.*, **101**(12), 2778–2784 (2016).

Capitelli F., Chita G., Cavallo A., Bellatreccia F., Della Ventura G. Crystal structure of whiteite-(CaFeMg) from Crosscut Creek, Canada, *Z. Kristallogr.*, **226**(9), 731–738 (2011).

Capitelli F., Della Ventura G., Bellatreccia F., Sodo A., Saviano M., Ghiara M.R., Rossi M. Crystal-chemical study of wavellite from Zbirov, Czech Republic, *Mineralog. Mag.*, **78**(4), 1057–1070 (2014a).

Capitelli F., Masi A., El Bali B., Essehli R. Indium-doped gallophosphate $NH_4[NiGa_{1.84}In_{0.16}(PO_4)_3(H_2O)_2]$: Synthesis, X-ray crystal structure and IR spectroscopy, *Z. Kristallogr.*, **225**(9), 359–365 (2010).

Capitelli F., Saviano M., Ghiara M.R., Rossi M. Crystal-chemical investigation of $Al_2(PO_4)(OH)_3$ augelite from Rapid Creek, Yukon, Canada, *Z. Kristallogr.*, **229**(1), 8–16 (2014b).

Cemič L., Schmid-Beurmann P. Lazulite stability relations in the system Al_2O_3–$AlPO_4$–$Mg_3(PO_4)_2$–H_2O, *Eur. J. Mineral.*, **7**(4), 921–929 (1995).

Černá I., Černý P., Ferguson R.B. The fluorine content and some physical properties of the amblygonite-montebrasite minerals, *Am. Mineralog.*, **58**(3–4), 291–301 (1973).

Chang Y.H.R., Yoon T.L., Lim T.L., Tuh M.H. High-pressure phases of $Al_xIn_{1-x}N$ compounds: First principles calculations, *J. Alloys Compd.*, **704**, 160–169 (2017).

Chen H., Chen K., Drabold D.A., Kordesch M.E. Band gap engineering in amorphous $Al_xGa_{1-x}N$: Experiment and *ab initio* calculations, *Appl. Phys. Lett.*, **77**(8), 1117–1119 (2000).

Chen X.-A., Zhao L., Li Y., Guo F., Chen B.-M. Potassium cobalt aluminium mixed phosphate, $K(Co^{II},Al)_2(PO_4)_2$, *Acta Crystallogr.*, **C53**(12), 1754–1756 (1997).

Chen Y., Sun W., Wu L., Jiang Y., Huang Z. Phase relations in SiC–AlN–R_2O_3 (R = Nd, Gd, Yb, Y) systems, *J. Phase Equilibr. Dif.*, **34**(1), 3–8 (2013).

Cheng Y.-B., Thompson D.P. Aluminum-containing nitrogen melilite phases, *J. Am. Ceram. Soc.*, **77**(1), 143–148 (1994).

Chippindale A.M., Cowley A.R., Bond A.D. A new ammonium manganese(II) gallium phosphate hydrate: $NH_4[MnGa_2(PO_4)_3(H_2O)_2]$, *Acta Crystallogr.*, **C54**(11), IUC9800061 (1998).

Chippindale A.M., Cowley A.R., Walton R.I. Solvothermal synthesis and structural characterisation of the first ammonium cobalt gallium phosphate hydrate, $NH_4[CoGa_2P_3O_{12}(H_2O)_2]$, *J. Mater. Chem.*, **6**(4), 611–614 (1996).

Chippindale A.M., Sharma A.V., Hibble S.J. Potassium nickel(II) gallium phosphate hydrate, $K[NiGa_2(PO_4)_3(H_2O)_2]$, *Acta Crystallogr.*, **E65**(5), i38–i39 (2009).

Choi Y.H., Besikci C., Sudharsanan R., Razeghi M. Growth of $In_{1-x}Tl_xSb$, a new infrared material, by low-pressure metalorganic chemical vapor deposition, *Appl. Phys. Lett.*, **63**(3), 361–363 (1993).

Chopin C., Brunet F., Gebert W., Medenbach O., Tillmanns E. Bearthite, $Ca_2Al[PO_4]_2(OH)$, a new mineral from high-pressure terranes of the western Alps, *Schweiz. Mineral. Petrog. Mitt.*, **73**(1), 1–9 (1993).

Chudinova N.N., Avaliani M.A., Guzeeva K.S., Tananaev I.V. Reactions in the K_2O–Ga_2O_3–P_2O_5–H_2O system at 150–500°C [in Russian], *Izv. AN SSSR. Neorgan. Mater.*, **14**(11), 2054–2060 (1978).

Chudinova N.N., Avaliani M.A., Guzeeva K.S., Tananaev I.V. Synthesis of binary condensed phosphates of gallium and alkali metals [in Russian], *Izv. AN SSSR. Neorgan. Mater.*, **15**(12), 2176–2179 (1979).

Chudinova N.N., Grunze I., Guzeeva L.S., Avaliani M.A. Synthesis of binary condensed cesium–gallium phosphates [in Russian], *Izv. AN SSSR. Neorgan. Mater.*, **23**(4), 604–609 (1987a).

Chudinova N.N., Grunze I., Guzeeva L.S. Binary ammonium–gallium phosphates [in Russian], *Izv. AN SSSR. Neorgan. Mater.*, **23**(4), 616–621 (1987b).

Chudinova N.N., Tananaev I.V., Avaliani M.A. Synthesis of binary gallium–potassium polyphosphates in polyphosphoric acid melts [in Russian], *Izv. AN SSSR. Neorgan. Mater.*, **13**(12), 2234–2235 (1977).

Chukanov N.V., Sidorenko G.A., Naumova I.S., Zadov A.E., Kuz'min V.I. Chistyakovaite, a new mineral $Al(UO_2)_2(AsO_4)_2(F,OH)\cdot 6.5H_2O$, *Dokl. Earth Sci.*, **407**(2), 290–293 (2006).

Cid-Dresdner H. Determination and refinement of the crystal structure of turquois, $CuAl_6(PO_4)_4(OH)_8\cdot 4H_2O$, *Z. Kristallogr.*, **121**(2–4), 87–113 (1965).

Cid-Dresdner H. The crystal structure of turquois, *Naturwissenschaften*, **51**(16), 380–381 (1964).

Cid-Dresdner H., Villarroel H.S. Crystallographic study of rashleighite, a member of the turquois group, *Am. Mineralog.*, **57**(11–12), 1681–1691 (1972).

Cieren X., Jaulmes S., Angenault J., Couturier J.C., Quarton M. La phase monoclinique $Li_2Nb_{0.5}In_{1.5}(PO_4)_3$, *Acta Crystallogr.*, **C52**(12), 2967–2969 (1996).

Clark A.M., Couper A.G., Embrey P.G., Fejer E.E. Waylandite: New data, from an occurrence in Cornwall, with a note on 'agnesite', *Mineralog. Mag.*, **50**(358), 731–733 (1986).

Colchester D.M., Leverett P., McKinnon A.R., Sharpe J.L., Williams P.A., Hibbs D.E., Turne P., Hoppe V.H. Cloncurryite, $Cu_{0.56}(VO)_{0.44}Al_2(PO_4)_2(F,OH)_2\cdot 5H_2O$, a new mineral from the Great Australia mine, Cloncurry, Queensland, Australia, an its relationship to nevadaite, *Aust. J. Mineralogy*, **13**(1), 5–13 (2007).

Colombo F., Rius J., Pannunzio-Miner E.V., Pedregosa J.C., Camí G.E., Carbonio R.E. Sanjuanite: Ab initio crystal-structure solution from laboratory powder-diffraction data, complemented by FTIR spectroscopy and DT-TG analyses, *Canad. Mineralog.*, **49**(3), 835–847 (2011).

Conanec R., Feldmann W., Marchand R., Laurent Y. Les phosphates azotés cristallisés de type $Na_3AlP_3O_9N$ et $Na_2Mg_2P_3O_9N$, *J. Solid State Chem.*, **121**(2), 418–422 (1996).

Constantin C., Al-Brithen H., Haider M.B., Ingram D., Smith A.R. ScGaN alloy growth by molecular beam epitaxy: Evidence for a metastable layered hexagonal phase, *Phys. Rev. B*, **70**(19), 193309_1-193309_4 (2004).

Cooper M., Hawthorne F.C. Refinement of the crystal structure of kulanite, *Canad. Mineralog.*, **32**(1), 15–19 (1994a).

Cooper M.A., Hawthorne F.C., Roberts A.C., Foord E.E., Erd R.C., Evans, Jr. H.T., Jensen M.C. Nevadaite, $(Cu^{2+},\square,Al,V^{3+})_6[Al_8(PO_4)_8F_8](OH)_2(H_2O)_{22}$, a new phosphate mineral species from the Gold Quarry mine, Carlin, Eureka County, Nevada: Description and crystal structure, *Canad. Mineralog.*, **42**(3), 741–752 (2004).

Cooper M., Hawthorne F.C. The crystal structure of curetonite, a complex heteropolyhedral sheet mineral, *Am. Mineralog.*, **79**(5–6), 545–549 (1994b).

Cooper M.A., Hawthorne F.C. The crystal structure of goldquarryite, $(Cu^{2+},\square)(Cd,Ca)_2Al_3(PO_4)_4F_2(H_2O)_{10}\{(H_2O),F\}_2$, a secondary phosphate from the Gold Quarry mine, Eureka County, Nevada, U.S.A., *Canad. Mineralog.*, **42**(3), 753–761 (2004).

Cooper M.A., Hawthorne F.C. The crystal structure of kraisslite, $^{[4]}Zn_3(Mn,Mg)_{25}(Fe^{3+},Al)(As^{3+}O_3)_2[(Si,As^{5+})O_4]_{10}(OH)_{16}$, from the Sterling Hill mine, Ogdensburg, Sussex County, New Jersey, USA, *Mineralog. Mag.*, **76**(7), 2819–2836 (2012).

Cooper M.A., Hawthorne F.C. The effect of differences in coordination on ordering of polyvalent cations in close-packed structures: The crystal structure of arakiite and comparison with hematolite, *Canad. Mineralog.*, **37**(6), 1471–1482 (1999).

Croft R.C. New molecular compounds of the layer lattice type. IV. New molecular compounds of boron nitride, *Austral. J. Chem.*, **9**(2), 206–211 (1956).

Cuchet S. Seconde occurrence de camérolaïte, $Cu_4Al_2[(HSbO_4,SO_4)](OH)_{10}(CO_3)\cdot 2H_2O$, Val d'Anniviers, Valais, Suisse. *Schweiz. Mineral. Petrog. Mitt.*, **75**(2), 283–284 (1995).

Cui L., Men H., Makino A., Kubota T., Yubuta K., Qi M., Inoue A. Effect of Cu and P on the crystallization behavior of Fe-rich hetero-amorphous FeSiB alloy, *Mater. Trans.*, **50**(11), 2515–2520 (2009).

Dal Bo F., Hatert F., Baijot M., Philippo S. Crystal structure of arsenuranospathite from Rabejac, Lodève, France, *Eur. J. Mineral.*, **27**(4), 589–597 (2015).

Dal Negro A., Kumbasar I., Ungaretti L. The crystal structure of teruggite, *Am. Mineralog.*, **58**(11-12), 1034-1043 (1973).

Deichman E.N., Tananaev I.V., Ezhova Z.A. On the interaction of lithium diphosphate and indium chloride in the aqueous solution[in Russian], *Izv. AN SSSR. Neorgan. Mater.*, **3**(5), 900–902 (1967a).

Deichman E.N., Tananaev I.V., Ezhova Z.A. On the potassium indium diphosphates [in Russian], *Izv. AN SSSR. Neorgan. Mater.*, **3**(10), 1946–1947 (1967b).

Deliens M., Piret P. La kamitugaïte, $PbAl(UO_2)_5[(P,As)O_4]_2(OH)_9{\cdot}9.5H_2O$, nouveau minéral de Kobokobo, Kivu, Zaïre, *Bull. Minéral.*, **107**(1), 15–19 (1984).

Deliens M., Piret P. Les phosphates d'uranyle et d'aluminum de Kobokobo. VIII. La furongite, *Ann. Soc. Géol. Belg.*, **108**, 365–368 (1985a).

Deliens M., Piret P. Les phosphates d'uranyle et d'aluminium de Kobokobo VII. La moreauite, $Al_3UO_2(PO_4)_3(OH)_2{\cdot}13H_2O$, nouveau mineral, *Bull. Minéral.*, **108**(1), 9–13 (1985b).

Deliens M., Piret P. Les phosphates d'uranyle et d'aluminium de Kobokobo. V. La mundite, nouveau minéral, *Bull. Minéral.*, **104**, 669–671 (1981).

Deliens M., Piret P. Les phosphates d'uranyle et d'aluminum de Kobokobo. II. La phuralumite $Al_2(UO_2)_3(PO_4)_2(OH)_6{\cdot}10H_2O$ et l'upalite $Al(UO_2)_3(PO_4)_2(OH)_3$, nouveaux minéraux, *Bull. Minéral.*, **102**, 333–337 (1979a).

Deliens M., Piret P. Les phosphates d'uranyle et d'aluminum de Kobokobo. IV. La threadgoldite, $Al(UO_2)_2(PO_4)_2(OH){\cdot}8H_2O$, nouveau minéral, *Bull. Minéral.*, **102**, 338–341 (1979b).

Deliens M., Piret P. Les phosphates d'uranyle et d'aluminium de Kobokobo. VI. La triangulite, $Al_3(UO_2)_4(PO_4)_4(OH)_5{\cdot}5H_2O$, nouveau minéral, *Bull. Minéral.*, **105**, 611–614 (1982).

Deliens M., Piret P. Ranunculite, $AlH(UO_2)(PO_4)(OH)_3{\cdot}4H_2O$, a new mineral, *Mineralog. Mag.*, **43**(327), 321–323 (1979c).

Demartin F., Gramaccioli C.M., Graeser S. The crystal structure of cervandonite-(Ce), an interesting example of $As^{3+} \rightarrow$ Si diadochy, *Canad. Mineralog.*, **46**(2), 423–430 (2008).

Demartin F., Gramaccioli C.M., Pilati T., Sciesa E. Sigismundite, $(Ba,K,Pb)Na_3(Ca,Sr)(Fe,Mg,Mn)_{14}Al(OH)_2(PO_4)_{12}$, a new Ba-rich member of the arrojadite group from Spluga Valley, Italy, *Canad. Mineralog.*, **34**(4), 827–834 (1996).

Deng R., Evans S.R., Gall D. Bandgap in $Al_{1-x}Sc_xN$, *Appl. Phys. Lett.*, **102**(11), 112103_1-112103_5 (2013).

Dhingra S.S., Haushalter R.C. Synthesis and crystal structure of the octahedral-tetrahedral framework indium phosphate $Cs[In_2(PO_4)(HPO_4)_2(H_2O)_2]$, *J. Solid State Chem.*, **112**(1), 96–99 (1994).

Dick S., Goßner U., Weiß A., Robl C., Großman G., Ohms G., Zeiske T. Taranakite - the mineral with the longest crystallographic axis, *Inorg. Chim. Acta*, **269**, 47–57 (1998).

Dick S., Zeiske T. Francoanellit $K_3Al_5(HPO_4)_6(PO_4)_2{\cdot}12H_2O$: Struktur und Synthese durch topochemische Entwässerung von Taranakit, *Z. Naturforsch.*, **53B**(7), 711–719 (1998).

Dittmar A., Hartmann C., Wollweber J., Bickermann M., Schmidbauer M., Klimm D. Physical vapor transport growth of bulk $Al_{1-x}Sc_xN$ single crystals, *J. Cryst. Growth*, **500**, 74–79 (2018).

Donnay G., Allmann R. Si_3O_{10} groups in the crystal structure of ardennite, *Acta Crystallogr.*, **B24**(6), 845–855 (1968).

Downs G.W., Yang B.N., Thompson R.M., Wenz M.D., Andrade M.B. Redetermination of durangite, $NaAl(AsO_4)F$, *Acta Crystallogr.*, **E68**(11), i86–i87 (2012).

Driss A., Jouini T. Structure de $Na_2AlBAs_4O_{14}$, un aluminoboroarséniate condensé, *Acta Crystallogr.*, **C44**(5), 791–794 (1988).

Duan C.J., Otten W.M., Delsing A.C.A., Hintzen H.T. Photoluminescence properties of Eu^{2+}-activated sialon S-phase $BaAlSi_5O_2N_7$, *J. Alloys Compd.*, **461**(1–2), 454–458 (2008).

Duan R., Liu W., Cao L., Su G., Xu H., Zhao C. A new copper borophosphate with novel polymeric chains and its structural correlation with raw materials in molten hydrated flux synthesis, *J. Solid State Chem.*, **210**(1), 60–64 (2014).

Dunn P.J., Appleman D.E. Perhamite, a new calcium aluminum silico-phosphate mineral, and a re-examination of viséite, *Mineralog. Mag.*, **41**(320), 437–442 (1977).

Dunn P.J., Cabri L.J., Clark A.M., Fleischer M. New mineral names, *Am. Mineralog.*, **68**(7–8), 849–852 (1983).

Dunn P.J., Chao G.Y., Fleischer M., Ferraiolo J.A., Langley R.H., Pabst A., Zilczer J.A. New mineral names, *Am. Mineralog.*, **70**(1–2), 214–221 (1985a).

Dunn P.J., Chao G.Y., Grice J.D., Ferraiolo J.A., Fleischer M., Pabst A., Zilczer J.A. New mineral names, *Am. Mineralog.*, **69**(5–6), Pt. 1, 565–569 (1984a).

Dunn P.J., Ferraiolo J.A., Fleischer M., Gobel V., Grice J.D., Langley R.H., Shigley J.E., Vanko D.A., Zilczer J.A. New mineral names, *Am. Mineralog.*, **70**(11–12), Pt. 1, 1329–1335 (1985b).

Dunn P.J., Fleischer M., Langley R.H., Shigley J.E., Zilczer J.A. New mineral names, *Am. Mineralog.*, **70**(7–8), 871–881 (1985c).

Dunn P.J., Gobel V., Grice J.D., Puziewicz J., Shigley J.E., Vanko D.A., Zilczer J.A. New mineral names, *Am. Mineralog.*, **70**(3–4), 436–441 (1985d).

Dunn P.J., Grice J.D., Fleischer M., Pabst A. New mineral names, *Am. Mineralog.*, **69**(1–2), 210–215 (1984b).

Dunn P.J., Peacor D.R., White J.S., Ramik R.A. Kingsmountite, a new mineral isostructural with montgomeryite, *Canad. Mineralog.*, **17**(3), 579–582 (1979).

Dunn P.J., Rouse R.C., Campbell T.J., Roberts W.L. Tinsleyite, the aluminum analogue of leucophosphite, from the Tip Top pegmatite in South Dakota, *Am. Mineralog.*, **69**(3–4), 374–376 (1984c).

Dunn P.J., Rouse R.C., Nelen J.A. Englishite: New chemical data and a second occurrence from the Tip Top pegmatite, Custer, South Dakota, *Canad. Mineralog.*, **22**(3), 469–470 (1984d).

Dzikowski T.J., Groat L.A., Jambor J.L. The symmetry and crystal structure of gorceixite, $BaAl_3[PO_3(O,OH)]_2(OH)_6$, a member of the alunite supergroup, *Canad. Mineralog.*, **44**(4), 951–958 (2006).

Edgar J.H., Smith D.T., Eddy C.R., Jr., Carosella C.A., Sartwell B.D. c-Boron–aluminum nitride alloys prepared by ion-beam assisted deposition, *Thin Solid Films*, **298**(1–2), 33–38 (1997).

Egorysheva A.V., Milenov T.I., Rafailov P.M., Gaitko O.M., Avdeev G.V., Dudkina T.D. Optical and vibrational spectra of $Bi_{1.8}Fe_{1.2(1-x)}Ga_{1.2x}SbO_7$ solid solutions with pyrochlore-type structure, *Russ. J. Inorg. Chem.*, **62**(7), 960–963 (2017).

Eich A., Hoffbauer W., Schnakenburg G., Bredow T., Daniels J., Beck J. Double-cube-shaped mixed chalcogen/pentele clusters from $GaCl_3$ melts, *Eur. J. Inorg. Chem.*, **2014**(19), 3043–3052 (2014).

Eich A., Schlüter S., Schnakenburg G., Beck J. $(Sb_7Te_8)^{5+}$ – a double cube shaped polycationic cluster, *Z. anorg. allg. Chem.*, **639**(2), 375–383 (2013).

Ellert O.G., Egorysheva A.V., Maksimov Y.V., Gajtko O.M., Efimov N.N., Svetogorov R.D. Isomorphism in the $Bi_{1.8}Fe_{1.2(1-x)}Ga_{1.2x}SbO_7$ pyrochlores with spin glass transition, *J. Alloys Compd.*, **688**(Pt. A), 1–7 (2016).

Elliott P. Refinement of the crystal structure of sieleckiite and revision of its symmetry, *Mineralog. Mag.*, **81**(4), 917–922 (2017).

Elyukhin V.A. Spinodal decomposition regions of $In_xGa_{1-x}Sb_yAs_zN_{1-y-z}$, $In_xGa_{1-x}Sb_yP_zN_{1-y-z}$ and $In_xGa_{1-x}As_yP_zN_{1-y-z}$ alloys, *J. Cryst. Growth*, **470**, 42–45 (2017).

Embrey P.G. Cahnite from Capo di Bove, Rome, *Mineralog. Mag.*, **32**(251) 666–668 (1960).

Engelhardt H., Borrmann H., Kniep R. Crystal structure of rubidium vanadium(III) catena-[monohydrogen-monoborate-bis(monophosphate)], $RbV[BP_2O_8(OH)]$, *Z. Kristallogr., New Cryst. Str.*, **215**(2), 203–204 (2000a).

Engelhardt H., Borrmann H., Schnelle W., Kniep R. Das erste Vanadium(III)-Borophosphat: Darstellung und Kristallstruktur von $CsV_3(H_2O)_2[B_2P_4O_{16}(OH)_4]$, *Z. anorg. und allg. Chem.*, **626**(7), 1647–1652 (2000b).

Engelhardt H., Kniep R. Crystal structure of caesium iron(III) catena-[monohydrogenmonoborate-bis(monophosphate)], $CsFe[BP_2O_8(OH)]$, *Z. Kristallogr., New Cryst. Str.*, **214**(4), 443–444 (1999).

Engelhardt H., Schnelle W., Kniep R. $Rb_2Co_3(H_2O)_2[B_4P_6O_{24}(OH)_2]$: Ein Borophosphat mit ${}^2_\infty$-Tetraeder-Anionenteilstruktur und Oktaeder-Trimeren $(Co^{II}{}_3O_{12}(H_2O)_2)$, *Z. anorg. und allg. Chem.*, **626**(6), 1380–1386 (2000c).

Enju S., Uehara S. Abuite, IMA 2014-084. CNMNC Newsletter No. 23, *Mineralog. Mag.*, **79**(1), 51–58 (2015).

Ensslin F., Dreyer H., Lessman O. Zur Chemie des Indiums. XI Verbindungen des Indiums mit den Sauerstoffsäuren des Phosphors, *Z. anorg. allg. Chem.*, **254**(5–6), 315–318 (1947).

Ercit T.S., Anderson A.J., Černý P., Hawthorne F.C. Bobfergusonite: A new primary phosphate mineral from Cross Lake, Manitoba, *Canad. Mineralog.*, **24**(4), 599–604 (1986a).

Ercit T.S., Hawthorne F.C., Černý P. The crystal structure of bobfergusonite, *Canad. Mineralog.*, **24**(4), 605–614 (1986b).

Ercit T.S., Piilonen P.C., Locock A.J., Kolitsch U., Rowe R. New mineral names, *Am. Mineralog.*, **91**(7), 1201–1209 (2006).

Ercit T.S., Tait K.T., Cooper M.A., Abdu Y., Ball N.A., Anderson A.J., Černý P., Hawthorne F.C. Manitobaite, $Na_{16}Mn^{2+}{}_{25}Al_8(PO_4)_{30}$, a new phosphate mineral species from Cross Lake, Manitoba, Canada, *Canad. Mineralog.*, **48**(6), 1455–1463 (2010).

Erd R.C., Foster M.D., Proctor P.D. Faustite; a new mineral, the zinc analogue of turquois, *Am. Mineralog.*, **38**(11–12), 964–972 (1953).

Esmaeilzadeh S., Grins J., Shen Z., Edén M., Thiaux M. Study of sialon S-phases $M_2Al_xSi_{12-x}N_{16-x}O_{2+x}$, M = Ba and $Ba_{0.9}Eu_{0.1}$, by X-ray single crystal diffraction, X-ray powder diffraction, and solid-state nuclear magnetic resonance, *Chem. Mater.*, **16**(11), 2113–2120 (2004).

Esmaeilzadeh S., Schnick W. Synthesis and structural investigations of $La_{13}Si_{18}Al_{12}O_{15}N_{39}$, *Solid State Sci.*, **5**(3), 503–508 (2003).

Ewald B., Menezes P., Prots Y., Kniep R. Solvothermal synthesis of alkali-metal borophosphates: Crystal structure of $Rb_3[B_2P_3O_{10}(OH)_3]$, *Z. anorg. und allg. Chem.*, **630**(11), 1721 (2004a).

Ewald B., Prots Y., Kniep R. Crystal structure and properties of $M^{III}(H_2O)_2[BP_2O_8]\cdot H_2O$ (M^{III} = Sc, In), *Z. anorg. und allg. Chem.*, **630**(11), 1720 (2004b).

Ewald B., Prots Y., Kudla C., Grüner D., Cardoso-Gil R., Kniep R. Crystal structure and thermochemical properties of a first scandium borophosphate, $Sc(H_2O)_2[BP_2O_8]\cdot H_2O$, *Chem. Mater.*, **18**(3), 673–679 (2006).

Ewald B., Prots Y., Menezes P., Kniep R. Crystal structure of diaqua-indium catena-monoboro-bisphosphate monohydrate, $In(H_2O)_2[BP_2O_8]\cdot H_2O$, *Z. Kristallogr., New Cryst. Str.*, **219**(4), 351–352 (2004d).

Ewald B., Prots Y., Menezes P., Natarajan S., Zhang H., Kniep K. Chain structures in alkali metal borophosphates: Synthesis and characterization of $K_3[BP_3O_9(OH)_3]$ and $Rb_3[B_2P_3O_{11}(OH)_2]$, *Inorg. Chem.*, **44**(18), 6431–6438 (2005a).

Ewald B., Öztan Y., Prots Y., Kniep R. Crystal structure of diaqua(magnesium, cobalt) bis(hydroxyboro)bisphosphate monohydrate, $Mg_{1-x}Co_x(H_2O)_2[B_2P_2O_8(OH)_2]\cdot H_2O$ ($x \approx 0.25$), *Z. Kristallogr., New Cryst. Str.*, **220**(4), 535–536 (2005b).

Ewald B., Öztan Y., Prots Y., Kniep R. Structural patterns and dimensionality in magnesium borophosphates: The crystal structures of $Mg_2(H_2O)[BP_3O_9(OH)_4]$ and $Mg(H_2O)_2[B_2P_2O_8(OH)_2]\cdot H_2O$, *Z. anorg. und allg. Chem.*, **631**(9), 1615–1621 (2005c).

Ezhova Z.A., Deichman T.N., Tananaev I.V. Investigation of the interaction of indium chloride with arsenates of alkali metals in an aqueous solution at 25°C [in Russian], *Zhurn. neorgan. khimii*, **22**(10), 2696–2703 (1977).

Fanfani L., Nunzi A., Zanazzi P.F. The crystal structure of wardite, *Mineralog. Mag.*, **37**(289), 598–605 (1970).

Fanfani L., Nunzi A., Zanazzi P.F., Zanzari A.R. Additional data on the crystal structure of montgomeryite, *Am. Mineralog.*, **61**(1–2), 12–14 (1976).
Feldmann W. Über Natrium–Aluminium–Nitridophosphat $Na_3AlP_3O_9N$ und Natrium–Magnesium–Nitridophasphat $Na_2Mg_2P_3O_9N$, *Z. Chem.*, **27**(3), 100–101 (1987a).
Feldmann W. Über Nitridophosphate $M^I_3M^{III}P_3O_9N$ (M^I = Na, K; M^{III} = Al, Ga, Cr, Fe, Mn), *Z. Chem.*, **27**(5), 182–183 (1987b).
Feng P., Bu X., Stucky G.D. Hydrothermal syntheses and structural characterization of zeolite analogue compounds based on cobalt phosphate, *Nature*, **388**(6644), 735–740 (1997).
Feng Y., Li M., Shi H., Huang Q., Qiu D. A novel chlorine-containing borophosphate based on (4,3)-connected 3-D borophosphate anion $[B_6P_{11}O_{42}(OH)_2]^{13-}$ with unique B: P ratio and 22-tetrahedral cages, *CrystEngComm*, **15**(11), 2048–2051 (2013).
Fernie J.A., Lewis M.H., Leng-Ward G. Crystallization of Nd–Si–Al–O–N glasses, *Mater. Lett.*, **9**(1), 29–32 (1989).
Filatov S.K., Krivovichev S.V., Burns P.C., Vergasova L.P. Crystal structure of filatovite, $K[(Al,Zn)_2(As,Si)_2O_8]$, the first arsenate of the feldspar group, *Eur. J. Mineral.*, **16**(3), 537–543 (2004).
Fisher D.J. Cacoxenite from Arkansas, *Am. Mineralog.*, **51**(11–12), 1811–1814 (1966).
Fisher D.J. Dickinsonites, fillowite and alluaudites, *Am. Mineralog.*, **50**(10), 1647–1669 (1965).
Fisher D.J., Runner J.J. Morinite from the Black Hills, *Am. Mineralog.*, **43**(5–6), 585–594 (1958).
Fitzpatrick J. Powder X-ray diffraction data of florencite-(Nd), *Powder Diffraction*, **1**(4), 330 (1986).
Fleischer M., Burns R.G., Cabri L.J., Chao G.Y., Hogarth D.D., Pabst A. New mineral names, *Am. Mineralog.*, **64**(11–12), 1329–1334 (1979a).
Fleischer M., Cabri L.J., Chao G.Y., Mandarino J.A., Pabst A. New mineral names, *Am. Mineralog.*, **67**(5–6), Pt. 1, 621–624 (1982a).
Fleischer M., Cabri L.J., Chao G.Y., Pabst A. New mineral names, *Am. Mineralog.*, **63**(3–4), 424–427 (1978a).
Fleischer M., Cabri L.J., Chao G.Y., Pabst A. New mineral names, *Am. Mineralog.*, **65**(7–8), 808–814 (1980a).
Fleischer M., Cabri L.J., Chao G.Y., Pabst A. New mineral names, *Am. Mineralog.*, **65**(9–10), 1065–1070 (1980b).
Fleischer M., Cabri L.J. New mineral names, *Am. Mineralog.*, **66**(11–12), 1274–1280 (1981).
Fleischer M., Cabri L.J., Nickel E.H., Pabst A. New mineral names, *Am. Mineralog.*, **62**(5–6), 593–600 (1977a).
Fleischer M., Cabri L.J., Pabst A. New mineral names, *Am. Mineralog.*, **65**(3–4), 406–408 (1980c).
Fleischer M., Chao G.Y., Mandarino J.A. New mineral names, *Am. Mineralog.*, **64**(5–6), Pt. 1, 652–659 (1979b).
Fleischer M., Chao G.Y., Mandarino J.A. New mineral names, *Am. Mineralog.*, **67**(7–8), 854–860 (1982b).
Fleischer M., Chao G.Y., Pabst A. New mineral names, *Am. Mineralog.*, **64**(3–4), 464–467 (1979c).
Fleischer M., Mandarino J.A. New mineral names, *Am. Mineralog.*, **62**(1–2), 173–176 (1977).
Fleischer M., Mandarino J.A., Pabst A. New mineral names, *Am. Mineralog.*, **65**(1–2), 205–210 (1980d).
Fleischer M. New mineral names, *Am. Mineralog.*, **41**(11–12), 958–960 (1956).
Fleischer M. New mineral names, *Am. Mineralog.*, **42**(3–4), 307–308 (1957a).
Fleischer M. New mineral names, *Am. Mineralog.*, **42**(7–8), 580–586 (1957b).
Fleischer M. New mineral names, *Am. Mineralog.*, **45**(11–12), 1313–1317 (1960).
Fleischer M. New mineral names, *Am. Mineralog.*, **48**(1–2), 209–217 (1963).
Fleischer M. New mineral names, *Am. Mineralog.*, **49**(11–12), 1774–1778 (1964).
Fleischer M. New mineral names, *Am. Mineralog.*, **51**(3–4), Pt. 1, 529–534 (1966).
Fleischer M. New mineral names, *Am. Mineralog.*, **52**(9–10), 1579–1589 (1967).
Fleischer M. New mineral names, *Am. Mineralog.*, **55**(1–2), 317–323 (1970).

Fleischer M. New mineral names, *Am. Mineralog.*, **56**(3–4), Pt. 1, 631–640 (1971).
Fleischer M. New mineral names, *Am. Mineralog.*, **57**(9–10), 1552–1561 (1972).
Fleischer M. New mineral names, *Am. Mineralog.*, **59**(1–2), 208–212 (1974a).
Fleischer M. New mineral names, *Am. Mineralog.*, **59**(7–8), 873–875 (1974b).
Fleischer M. New mineral names, *Am. Mineralog.*, **59**(9–10), 1139–1141 (1974c).
Fleischer M., Pabst A., Cabri L.J. New mineral names, *Am. Mineralog.*, **61**(9–10), 1053–1056 (1976a).
Fleischer M., Pabst A., Mandarino J.A., Chao G.Y., Cabri L.J. New mineral names, *Am. Mineralog.*, **61**(1–2), 174–186 (1976b).
Fleischer M., Pabst A., Mandarino J.A., Chao G.Y. New mineral names, *Am. Mineralog.*, **62**(9–10), 1057–1061 (1977b).
Fleischer M., Pabst A., White J.S. New mineral names, *Am. Mineralog.*, **63**(7–8), 793–796 (1978b).
Foord E.E., Hughes J.M., Cureton F., Maxwell C.H., Falster A.U., Sommer A.J., Hlava P.F. Esperanzaite, $NaCa_2Al_2(As^{5+}O_4)_2F_4(OH)\cdot 2H_2O$, a new mineral species from the La Esperanza mine, Mexico: Descriptive mineralogy and atomic arrangement, *Canad. Mineralog.*, **37**(1), 67–72 (1999).
Foord E.E., Oakman M.R., Maxwell C.H. Durangite from the Black Range, New Mexico, and new data on durangite from Durango and Cornwall, *Canad. Mineralog.*, **23**(2), 241–246 (1985).
Foord E.E., Taggart, Jr. J.E. A reexamination of the turquoise group: The mineral aheylite, planerite (redefined), turquoise and coeruleolactite, *Mineralog. Mag.*, **62**(1), 93–111 (1998).
Förtsch E.B. 'Plumbogummite' from Roughten Gill, Cumberland, *Mineralog. Mag.*, **36**(280), 530–538 (1967).
Frank-Kamenetskiy V.A., Komkov A.I., Nardov V.V. Radiometric data on florencite and koivenite [in Russian], *Zap. Vses. mineralog. obshch.*, **82**(4), 297–301 (1953).
Fransolet A.-M., Abraham K. Une association triplite – montebrasite – griphite dans la pegmatite de Buranga, Rwanda, *Ann. Soc. Géol. Belg.*, **106**, 299–309 (1983).
Fransolet A.-M., Knorring Von O., Fontan F. A new occurrence of samuelsonite in the Buranga pegmatite, Rwanda, *Bull. Geol. Surv. Finland*, **64**(Pt. 1), 13–21 (1992).
Fransolet A.-M., Oustrière P., Fontan F., Pillard F. La mantiennéite, une nouvelle espèce minérale du gisement de vivianite d'Anloua, Cameroun, *Bull. Minéral.*, **107**(6), 737–744 (1984).
Fransolet A.-M. The problem of Na–Li substitution in primary Li–Al phosphates: New data on lacroixite, a relatively widespread mineral, *Canad. Mineralog.*, **27**(2), 211–217 (1989).
Fransolet A.-M. Wyllieite et rosemaryite dans la pegmatite de Buranga, Rwanda, *Eur. J. Mineral.*, **7**(3), 567–575 (1995).
Freeman A.G., Larkindale J.P. Evidence for the formation of boron nitride – alkali metal intercalation compounds, *Inorg. Nucl. Chem. Lett.*, **5**(11), 937–939 (1969a).
Freeman A.G., Larkindale J.P. Preparation, Mössbauer spectra, and structure of intercalation compounds of boron nitride with metal halides, *J. Chem. Soc. A: Inorg., Phys., Theor.*, **0**(0), 1307–1308 (1969b).
Frondel C., Ito J. Composition of palermoite, *Am. Mineralog.*, **50**(5–6), 777–779 (1965).
Frondel C., Lindberg M.L. Second occurrence of brazilianite, *Am. Mineralog.*, **33**(3–4), 135–141 (1948).
Frondel C. Studies of uranium minerals (VIII): Sabugalite, an aluminum-autunite, *Am. Mineralog.*, **36**(9–10), 671–679 (1951).
Fu C., Xu L., Dan Z., Qin F., Makino A., Chang H., Hara N. Annealing effect of amorphous Fe–Si–B–P–Cu precursors on microstructural evolution and redox behavior of nanoporous counterparts, *J. Alloys Compd.*, **726**, 810–819 (2017).
Fukuda Y., Ishida K., Mitsuishi I., Nunoue S. Luminescence properties of Eu^{2+}-doped green-emitting Sr-sialon phosphor and its application to white light-emitting diodes, *Appl. Phys. Exp.*, **2**(1), 012401_1–012401_3 (2009).
Fukuhara M. Phase relations in the Si_3N_4-rich portion of the Si_3N_4–AlN–Al_2O_3–Y_2O_3 system, *J. Am. Ceram. Soc.*, **71**(7), C-359–C-361 (1988).

Funahashi S., Michiue Y., Takeda T., Xie R.-J., Hirosaki N. Substitutional disorder in $Sr_{2-y}Eu_yB_{2-2x}Si_{2+3x}Al_{2-x}N_{8+x}$ ($x \approx 0.12$, $y \approx 0.10$), *Acta Crystallogr.*, **C70**(5), 452–454 (2014).

Galli E., Brigatti M.F., Malferrari D., Sauro F., de Waele J. Rossiantonite, $Al_3(PO_4)(SO_4)_2(OH)_2(H_2O)_{10}{\cdot}4H_2O$, a new hydrated aluminum phosphatesulfate mineral from Chimanta massif, Venezuela: Description and crystal structure, *Am. Mineralog.*, **98**(10), 1906–1913 (2013a).

Galli E., Brigatti M.F., Malferrari D., Sauro F., De Waele J. Rossiantonite, IMA 2012-056. CNMNC Newsletter No. 15, February 2013, page 5, *Mineralog. Mag.*, **77**(1), 1-12 (2013b).

Galliski M.A., Cooper M.A., Hawthorne F.C., Černý P. Bederite, a new pegmatite phosphate mineral from Nevados de Palermo, Argentina: Description and crystal structure, *Am. Mineralog.*, **84**(10), 1674–1679 (1999).

Galuskin E.V., Krüger B., Galuskina I.O., Krüger H., Vapnik Y., Pauluhn A., Olieric V. Levantite, IMA 2017-010. CNMNC Newsletter No. 37, *Mineralog. Mag.*, **81**(3), 737–742 (2017).

Gamyanin G.N., Zayakina N.V., Galenchikova L.T. Arangasite, $Al_2(PO_4)(SO_4)F{\cdot}7.5H_2O$, a new mineral from Alyaskitovy deposit, Eastern Yakutia, Russia, *Geol. Ore Deposits*, **56**(7), 560–566 (2014).

Gamyanin G.N., Zayakina N.V., Galenchikova L.T. Arangasite, IMA 2012-018. CNMNC Newsletter No. 14, *Mineralog. Mag.*, **76**(5), 1281–1288 (2012).

Gatehouse B.M., Miskin B.K. The crystal structure of brazilianite, $NaAl_3(PO_4)_2(OH)_4$, *Acta Crystallogr.*, **B30**(5), 1311–1317 (1974).

Gatta G.D., Cámara F., Tait K.T., Belakovskiy D. New mineral names, *Am. Mineralog.*, **97**(11–12), 2064–2072 (2012).

Gatta G.D., Nénert G., Vignola P. Coexisting hydroxyl groups and H_2O molecules in minerals: A single-crystal neutron diffraction study of eosphorite, $MnAlPO_4(OH)_2{\cdot}H_2O$, *Am. Mineralog.*, **98**(7), 1297–1301 (2013a).

Gatta G.D., Vignola P., Meven M. On the complex H-bonding network in paravauxite, $Fe^{2+}Al_2(PO_4)_2(OH)_2{\cdot}8H_2O$: A single-crystal neutron diffraction study, *Mineralog. Mag.*, **78**(4), 841–850 (2014).

Gatta G.D., Vignola P., Meven M., Rinaldi R. Neutron diffraction in gemology: Single-crystal diffraction study of brazilianite, $NaAl_3(PO_4)_2(OH)_4$, *Am. Mineralog.*, **98**(8–9), 1624–1630 (2013b).

Gautier S., Sartel C., Ould Saad Hamady S., Maloufi N., Martin J., Jomard F., Ougazzaden A. MOVPE growth study of $B_xGa_{(1-x)}N$ on GaN template substrate, *Supelattices Microstruct.*, **40**(4–6), 233–238 (2006).

Ge M.-H., Liu W., Chen H.-H., Li M.-R., Yang X.-X., Zhao J.-T. $NH_4Cd(H_2O)_2(BP_2O_8){\cdot}0.72H_2O$: A new borophosphate with abnormal structure changes caused by hydrogen interactions, *Z. anorg. und allg. Chem.*, **631**(6–7), 1213–1217 (2005).

Ge M.-H., Mi J.-X., Huang Y.-X., Zhao J.-T., Kniep R. Crystal structure of lithium cadmium diaqua catena-[monoborodiphosphate]-monohydrate, $LiCd(H_2O)_2[BP_2O_8]{\cdot}H_2O$, *Z. Kristallogr., New Cryst. Str.*, **218**(3), 273–274 (2003a).

Ge M.-H., Mi J.-X., Huang Y.-X., Zhao J.-T., Kniep R. Crystal structure of sodium cadmium diaqua catena-[monoborodiphosphate]-hydrate, $NaCd(H_2O)_2[BP_2O_8]{\cdot}0.8H_2O$, *Z. Kristallogr., New Cryst. Str.*, **218**(2), 165–166 (2003b).

Gilkes R.J., Palmer B. Synthesis, properties, and dehydroxylation of members of the crandallite-goyazite series, *Mineralog. Mag.*, **47**(343), 221–227 (1983).

Golubev L.V., Khachaturyan O.A., Shmartsev Y.V. Solubility of GaSb in bismuth and based on it eutectic alloys [in Russian], *Elektronnaya tekhnika. Ser. 6. Materialy*, (5), 156–158 (1972).

Gordon S.G. Crystallographic data on wavellite from Llallagua, Bolivia and on cacoxenite from Hellertown, Pennsylvania, *Am. Mineralog.*, **35**(1–2), 132 (1950).

Goryunova N.A., Averkieva G.K., Vaypolin A.A. On the possibility of obtaining single crystals of multicomponent alloys [in Russian], *Fizika*, Leningrad, Publish. House of Lening. Inst. Civil Eng., 52–53 (1965).

Goryunova N.A., Baranov B.V., Valov Y.A., Prochukhan V.D. About solubility of boron phosphide in various melts [in Russian], *Fizika*, Leningrad, Publish. House of Lening. Inst. Civil Eng., 15–19 (1964).

Grand G., Demit J,. Ruste J., Torre J.P. Composition and stability of Y–Si–Al–O–N solid solutions based on α-Si_3N_4 structure, *J. Mater. Sci.*, **14**(7), 1749–1751 (1979).

Grebenyuk A.M., Litvak A.M., Charykov N.A., Puchkov L.V., Yakovlev Y.P., Klepikov V.V., Udovenko A.G., Izotova S.G., Nikitin V.A., Zubkova M.Y. On the possibility of melt–solid phase equilibria calculations in the Pb–InAs–GaAs–InSb–GaSb system [in Russian], *Zhurn. neorgan. khimii*, **44**(1), 113–114 (1999).

Grey I.E., Betterton J., Kampf A.R., Macrae C.M., Shanks F.L., Price J.R. Penberthycroftite, $[Al_6(AsO_4)_3(OH)_9(H_2O)_5]\cdot 8H_2O$, a second new hydrated aluminium arsenate mineral from the Penberthy Croft mine, St. Hilary, Cornwall, UK, *Mineralog. Mag.*, **80**(7), 1149–1160 (2016a).

Grey I.E., Brand H.E.A., Rumsey M.S., Gozukara Y. Ultra-flexible framework breathing in response to dehydrationin liskeardite, $[(Al,Fe)_{16}(AsO_4)_9(OH)_{21}(H_2O)_{11}]\cdot 26H_2O$, a natural open-framework compound, *J. Solid State Chem.*, **228**, 146–152 (2015).

Grey I.E., Keck E., Mumme W.G., Pring A., Macrae C.M., Glenn A.M., Davidson C.J., Shanks F.L., Mills S.J. Kummerite, $Mn^{2+}Fe^{3+}Al(PO_4)_2(OH)_2\cdot 8H_2O$, a new laueite-group mineral from the Hagendorf Süd pegmatite, Bavaria, with ordering of Al and Fe^{3+}, *Mineralog. Mag.*, **80**(7), 1243–1254 (2016b).

Grey I.E., Mumme W.G., Macrae C.M., Caradoc-Davies T., Price J.R., Rumsey M.S., Mills S.J. Chiral edge-shared octahedral chains in liskeardite, $[(Al,Fe)_{32}(AsO_4)_{18}(OH)_{42}(H_2O)_{22}]\cdot 52H_2O$, an open framework mineral with a pharmacoalumite-related structure, *Mineralog. Mag.*, 77(8), 3125–3135 (2013).

Grice J.D., Dunn P.J. Attakolite: New data and crystal-structure determination, *Am. Mineralog.*, 77(11–12), 1285–1291 (1992).

Grice J.D., Dunn P.J., Ramik R.A. Whiteite-(CaMnMg), a new mineral species from the Tip Top pegmatite, Custer, South Dakota, *Canad. Mineralog.*, **27**(4), 699–702 (1989).

Grins J., Esmaeilzadeh S., Shen Z. Structures of filled α-Si_3N_4-type $Ca_{0.27}La_{0.03}Si_{11.38}Al_{0.62}N_{16}$ and $LiSi_9Al_3O_2N_{14}$, *J. Am. Ceram. Soc.*, **86**(4), 727–730 (2003).

Grins J., Shen Z., Nygren M., Ekstrom T. Preparation and crystal structure of $LaAl(Si_{6-z}Al_z)N_{10-z}O_z$, *J. Mater. Chem.*, **5**(11), 2001–2006 (1995).

Groat L.A., Grew E.S., Evans R.J., Pieczka A., Ercit T.S. The crystal chemistry of holtite, *Mineralog. Mag.*, **73**(6), 1033–1050 (2009).

Groat L.A., Raudsepp M., Hawthorne F.C., Ercit T.S., Sherriff B.L., Hartman J.S. The amblygonite-montebrasite series: Characterization by single-crystal structure refinement, infrared spectroscopy, and multinuclear MAS-NMR spectroscopy, *Am. Mineralog.*, **75**(9–10), 992–1008 (1990).

Grunze I., Maksimova S.I., Palkina K.K., Chibiskova N.T., Chudinova N.N. The structure of $Cs_2GaH_3(P_2O_7)_2$ [in Russian], *Izv. AN SSSR. Neorgan. Mater.*, **24**(2), 264–267 (1988).

Grunze I., Grunze H. Darstellung von oligomeren und polymeren Alkali-Aluminium- und Alkali-Eisen-Phosphaten, *Z. anorg. allg. Chem.*, **512**(5), 39–47 (1984).

Grunze I., Palkina K.K., Chudinova N.N., Guzeeva L.S., Avaliani M.A., Maksimova S.I. The structure and thermal transformations of binary cesium–gallium phosphates [in Russian], *Izv. AN SSSR. Neorgan. Mater.*, **23**(4), 610–615 (1987).

Guesdon A., Borel M.M., Leclaire A., Grandin A., Raveau B. An aluminophosphate of molybdenum(V) with a tunnel structure: $Cs_9Mo_9Al_3P_{11}O_{59}$, *J. Solid State Chem.*, **114**(2), 451–458 (1995).

Guesmi A., Driss A. $KCo(H_2O)_2BP_2O_8\cdot 0.48H_2O$ and $K_{0.17}Ca_{0.42}Co(H_2O)_2BP_2O_8\cdot H_2O$: Two cobalt borophosphates with helical ribbons and disordered (K,Ca)/H_2O schemes, *Acta Crystallogr.*, **C68**(8), i55–i59 (2012).

Guesmi A., Driss A. Synthesis, characterization and crystal structure of a new cobalt borophosphate, $NaCoH_2BP_2O_9$, *Adv. Eng. Mater.*, **6**(10), 840–842 (2004).

Guo M., Wang Y.G., Miao X.F. Effect of heating rate on the microstructural and magnetic properties of nanocrystalline $Fe_{81}Si_4B_{12}P_2Cu_1$ alloys, *J. Mater. Sci.*, **46**(6), 1680–1684 (2011).

Gupta P.K.S., Swihart G.H., Dimitrijević R., Hossain M.B. The crystal structure of lüneburgite, $Mg_3(H_2O)_6[B_2(OH)_6(PO_4)_2]$, *Am. Mineralog.*, **76**(7–8), 1400–1407 (1991).

Gurbanova O.A., Belokoneva E.L., Dimitrova O.V. Synthesis and crystal structure of new borophosphate $NaIn[BP_2O_8(OH)]$ [in Russian], *Zhurn. neorgan. khimii*, **47**(1), 10–13 (2002).

Guy B.B., Jeffrey G.A. The crystal structure of fluellite, $Al_2PO_4F_2(OH)·7H_2O$, *Am. Mineral.*, **51**(11–12), 1579–1592 (1966).

Györi B., Emri J., Fehér I. Preparation and properties of novel cyano and isocyano derivatives of borane and the tetrahydroborate anion, *J. Organomet. Chem.*, **255**(1), 17–28 (1983).

Hak J., Johan Z., Kvaček M., Liebscher W. Kemmlitzite, a new mineral of the woodhouseite group, *Neues Jahrb. Mineral., Monatsh.*, 201–212 (1969).

Hampshire S., Park H.K., Thompson D.P., Jack K.H. α'-Sialon ceramics, *Nature*, **274**(5674), 880–882 (1978).

Hanson A.W. The crystal structure eosphorite, *Acta Crystallogr.*, **13**(5), 384–387 (1960).

Harrison W.T.A., Phillips M.L.F., Stucky G.D. Substitution chemistry of gallium for titanium in nonlinear optical KTiOPO4: Syntheses and single-crystal structures of $KGaF_{1-δ}(OH)_δPO_4$ ($δ \approx 0.3$) and $KGa_{0.5}Ge_{0.5}(F,OH)_{0.5}O_{0.5}PO_4$, *Chem. Mater.*, **7**(10), 1849–1856 (1995).

Harrowfield I.R., Segnit E.R., Watts J.A. Aldermanite, a new magnesium aluminium phosphate, *Mineralog. Mag.*, **44**(333), 59–62 (1981).

Hasegawa T., Yamane H. Synthesis and crystal structure analysis of $Li_2NaBP_2O_8$ and $LiNa_2B_5P_2O_{14}$, *J. Solid State Chem.*, **225**, 65–71 (2015).

Hatert F., Antenucci D., Fransolet A.-M., Liégeois-Duyckaerts M. The crystal chemistry of lithium in the alluaudite structure: A study of the $(Na_{1-x}Li_x)CdIn_2(PO_4)_3$ solid solution ($x = 0$ to 1), *J. Solid State Chem.*, **163**(1), 194–201 (2002).

Hatert F., Baijot M., Philippo S., Wouters J. Qingheiite-(Fe^{2+}), $Na_2Fe^{2+}MgAl(PO_4)_3$, a new phosphate mineral from the Sebastião Cristino pegmatite, Minas Gerais, Brazil, *Eur. J. Miner.*, **22**(3), 459–467 (2010).

Hatert F., Hermann R.P., Fransolet A.-M., Long G.J., Grandjean F. A structural, infrared, and Mössbauer spectral study of rosemaryite, $NaMnFe^{3+}Al(PO_4)_3$, *Eur. J. Mineral.*, **18**(6), 775–785 (2006).

Hatert F., Hermann R.P., Long G.J., Fransolet A.-M., Grandjean F. An X-ray Rietveld, infrared, and Mössbauer spectral study of the $NaMn(Fe_{1-x}In_x)_2(PO_4)_3$ alluaudite-type solid solution, *Am. Mineralog.*, **88**(1), 211–222 (2003).

Hatert F., Lefèvre P., Fransolet A.-M., Spirlet M.-R., Rebbouh L., Fontan F., Keller P. Ferrorosemaryite, $NaFe^{2+}Fe^{3+}Al(PO_4)_3$, a new phosphate mineral from the Rubindi pegmatite, Rwanda, *Eur. J. Mineral.*, **17**(5), 749–759 (2005).

Hatert F., Lefèvre P., Fransolet A.-M. The crystal structure of bertossaite, $CaLi_2[Al_4(PO_4)_4(OH,F)_4]$, *Canad. Mineral.*, **49**(4), 1079–1087 (2011).

Hatert F. $Na_{1.50}Mn_{2.48}Al_{0.85}(PO_4)_3$, a new synthetic alluaudite-type compound, *Acta Crystallogr.*, **C62**(1), i1–i2 (2006).

Hauf C., Boy I., Kniep R. Crystal structure of dimagnesium (monohydrogenmonophosphate-dihydrogenmonoborate-monophosphate), $Mg_2[BP_2O_7(OH)_3]$, *Z. Kristallogr., New Cryst. Str.*, **214**(1), 3–4 (1999).

Hauf C., Kniep R. Crystal structure of diammonium *catena*-(monoboro-mono-dihydrogen-diborate monophosphate), $(NH_4)_2[B_3PO_7(OH)_2]$, *Z. Kristallogr.*, **211**(10), 705–706 (1996a).

Hauf C., Kniep R. Crystal structure of lithium *catena*-(monoboro-mono-dihydrogendiborate-monohydrogenphosphate), $Li[B_3PO_6(OH)_3]$, *Z. Kristallogr., New. Cryst. Str.*, **212**(1), 313–314 (1997a).

Hauf C., Kniep R. Crystal structure of tripotassium *catena*-[triboro-monohydrogenphosphate bis(monohydrogenborate)], $K_3[B_5PO_{10}(OH)_3]$, *Z. Kristallogr.*, **211**(10), 707–708 (1996b).

Hauf C., Kniep R. $Rb[B_2P_2O_8(OH)]$ und $Cs[B_2P_2O_8(OH)]$: Die ersten Borophosphate mit dreidimensional vernetzter Anionenteilstruktur, *Z. Naturforsch.*, **52B**(11), 1432–1435 (1997b).

Hausen D.M. Schoderite, a new phosphovanadate mineral from Nevada, *Am. Mineralog.*, **47**(5–6), 637–648 (1962).

Hawthorne F.C., Abdu Y.A., Ball N.A., Pinch W.W. Carlfrancisite: $Mn^{2+}_3(Mn^{2+},Mg,Fe^{3+},Al)_{42}(As^{3+}O_3)_2(As^{5+}O_4)_4[(Si,As^{5+})O_4]_6[(As^{5+},Si)O_4]_2(OH)_{42}$, a new arseno-silicate mineral from the Kombat mine, Otavi Valley, Namibia, *Am. Mineralog.*, **98**(10), 1693–1696 (2013).

Hawthorne F.C., Burke E.A.J., Ercit T.S., Grew E.S., Grice J.D., Jambor J.L., Puziewicz J., Roberts A.C., Vanko D.A. New mineral names, *Am. Mineralog.*, **73**(1–2), 189–199 (1988).

Hawthorne F.C., Dunn P.J., Grice J.D., Puziewicz J., Shigley J.E. New mineral names, *Am. Mineralog.*, **71**(5–6), Pt. 1, 845–847 (1986).

Hawthorne F.C., Jambor J., Bladh K.W., Burke E.A.J., Grice J.D., Phillips D., Roberts A.C., Schedler R.A., Shigley J.E. New mineral names, *Am. Mineralog.*, **72**(9–10), 1023–1028 (1987).

Hawthorne F.C., Pinch W.W. Carlfrancisite, IMA 2012-033. CNMNC Newsletter No. 14, *Mineralog. Mag.*, **76**(5), 1281–1288 (2012).

Hawthorne F.C. Sigloite: The oxidation mechanism in $[M^{3+}_2(PO_4)_2(OH)_2(H_2O)_2]^{2-}$ structures, *Mineral. Petrol.*, **38**(3), 201–211 (1988).

Hawthorne F.C. The crystal structure of bøggildite, *Canad. Mineralog.*, **20**(2), 263–270 (1982).

Hawthorne F.C. The crystal structure of morinite, *Canad. Mineralog.*, **17**(1), 93–102 (1979).

Hawthorne F.C. The crystal structure of tancoite, *Tschermaks Mineral. Petrog. Mitt.*, **31**, 121–135 (1983).

Hecht C., Stadler F., Schmidt P.J., Schmedt auf der Günne J., Baumann V., Schnick W. $SrAlSi_4N_7:Eu^{2+}$ – A nitridoalumosilicate phosphor for warm white light (pc)LEDs with edge-sharing tetrahedral, *Chem. Mater.*, **21**(8), 1595–1601 (2009).

He M., Cui L., Kubota T., Yubuta K., Makino A., Inoue A. Fe-rich soft magnetic FeSiBPCu hetero-amorphous alloys with high saturation magnetization, *Mater. Trans.*, **50**(6), 1330–1333 (2009).

Heyward C., McMillen C.D., Kolis J. Hydrothermal synthesis and structural analysis of new mixed oxyanion borates: $Ba_{11}B_{26}O_{44}(PO_4)_2(OH)_6$, $Li_9BaB_{15}O_{27}(CO_3)$ and $Ba_3Si_2B_6O_{16}$, *J. Solid State Chem.*, **203**, 166–173 (2013).

Hilmas G.E., Tien T.-Y. Effect of AlN and Al_2O_3 additions on the phase relationships and morphology of SiC. Part I. Compositions and properties, *J. Mater. Sci.*, **34**(22), 5605–5612 (1999a).

Hilmas G.E., Tien T.-Y. Effect of AlN and Al_2O_3 additions on the phase relationships and morphology of SiC. Part II. Microstructural observations, *J. Mater. Sci.*, **34**(22), 5613–5621 (1999b).

Hintze F., Schnick W. A novel nitridogallate fluoride $LiBa_5GaN_3F_5$ – Synthesis, crystal structure, and band gap determination, *Solid State Sci.*, **12**(8), 1368–1373 (2010).

Hoffmann S., Jeazet H.B.T., Menezes P.W., Prots Y., Kniep R. $Na_3Pb^{II}[B(O_3POH)_4]$: An alkali-metal lead borophosphate with heterocubane-like units Na_3PbO_4, *Inorg. Chem.*, **47**(22), 10193–10195 (2008).

Höglund C., Bareño J., Birch J., Alling B., Czigány Z., Hultman L. Cubic $Sc_{1-x}Al_xN$ solid solution thin films deposited by reactive magnetron sputter epitaxy onto ScN(111), *J. Appl. Phys.*, **105**(11), 113517_1–113517_7 (2009).

Höglund C., Birch J., Alling B., Bareño J., Czigány Z., Persson P.O.Å., Wingqvist G., Zukauskaite A., Hultman L. Wurtzite structure $Sc_{1-x}Al_xN$ solid solution films grown by reactive magnetron sputter epitaxy: Structural characterization and first-principles calculations, *J. Appl. Phys.*, **107**(12), 123515_1–123515_7 (2010).

Honda T., Shibata M., Kurimoto M., Tsubamoto M., Yamamoto J., Kawanishi H. Band-gap energy and effective mass of BGaN, *Jpn. J. Appl. Phys.*, **39**(4B), Pt. 1, 2389–2393 (2000).

Hoskins B.F., Mumme W.G., Pryce M.W. Holtite, $(Si_{2.25}Sb_{0.75})B[A1_6(Al_{0.43}Ta_{0.27}\Upsilon_{0.30})O_{15}(O,OH)_{2.25}]$: Crystal structure and crystal chemistry, *Mineralog. Mag.*, **53**(372), 457–463 (1989).

Hoyos M.A., Calderon T., Vergara I., Garcia-Solé J. New structural and spectroscopic data for eosphorite, *Mineralog. Mag.*, **57**(387), 329–336 (1993).

Hriljac J.A., Grey C.P., Cheetham A.K., VerNooy P.D., Torardi C.C. Synthesis and structure of $KIn(OH)PO_4$: Chains of hydroxide-bridged $InO_4(OH)_2$ octahedra, *J. Solid State Chem.*, **123**(2), 243–248 (1996).

Hsu K.-F., Wang S.-L. Novel gallium phosphate framework encapsulating trinuclear $Mn_3(H_2O)_6O_8$ cluster: Hydrothermal synthesis and characterization of $Mn_3(H_2O)_6Ga_4(PO_4)_6$, *Inorg. Chem.*, **39**(8), 1773–1778 (2000).

Huang Y.-X., Ewald B., Schnelle W., Prots Y., Kniep R. Chirality and magnetism in a novel series of isotypic borophosphates: $M^{II}[BPO_4(OH)_2]$ (M^{II} = Mn, Fe, Co), *Inorg. Chem.*, **45**(19), 7578–7580 (2006).

Huang Y.-X., Mao S.-Y., Mi J.-X., Wei Z.-B., Zhao J.-T., Kniep R. Crystal structure of sodium gallium (monohydrogenmonophosphate-dihydrogenmonoborate-monophosphate), $NaGa[BP_2O_7(OH)_3]$, *Z. Kristallogr., New Cryst. Str.*, **216**(1), 15–16 (2001a).

Huang Y.-X., Mi J.-X., Mao S.-Y., Wei Z.-B., Zhao J.-T., Kniep R. Crystal structure of sodium indium (monohydrogenmonophosphate-dihydrogenmonoborate-monophosphate), $NaIn[BP_2O_7(OH)_3]$, *Z. Kristallogr., New Cryst. Str.*, **217**(1), 7–8 (2002a).

Huang Y.-X., Prots Y., Kniep R. Crystal structure of cobalt manganese monoaqua catena-[monohydrogenborate-tris(hydrogenphosphate)], $(Co_{0.6}Mn_{0.4})_2(H_2O)[BP_3O_9(OH)_4]$, *Z. Kristallogr., New Cryst. Str.*, **224**(3), 371–372 (2009).

Huang Y.-X., Prots Y., Kniep R. $Zn[BPO_4(OH)_2]$: A zinc borophosphate with the rare moganite-type topology, *Chem. Eur. J.*, **14**(6), 1757–1761 (2008).

Huang Y.-X., Schäfer G., Carrillo-Cabrera W., Cardoso R., Schnelle W., Zhao J.-T., Kniep R. Open-framework borophosphates: $(NH_4)_{0.4}Fe^{II}{}_{0.55}Fe^{III}{}_{0.5}(H_2O)_2[BP_2O_8]\cdot 0.6H_2O$ and $NH_4Fe^{III}[BP_2O_8(OH)]$, *Chem. Mater.*, **13**(11), 4348–4354 (2001b).

Huang Y.-X., Zhao J.-T., Mi J.-X., Borrmann H., Kniep R. Crystal structure of rubidium indium (monophosphate-hydrogenmonoborate-monophosphate), $RbIn[BP_2O_8(OH)]$, *Z. Kristallogr., New Cryst. Str.*, **217**(1), 163–164 (2002b).

Huang Z.-K., Chen I.-W. Rare-earth melilite solid solution and its phase relations with neighboring phases, *J. Am. Ceram. Soc.*, **79**(8), 2091–2097 (1996).

Huang Z.-K., Greil P., Petzow G. Formation of α-Si_3N_4 solid solutions in the system Si_3N_4–AlN–Y_2O_3, *J. Am. Ceram. Soc.*, **66**(6), C-96–C-97 (1983).

Huang Z.-K., Sun W.-Y., Yan D.-S. Phase relations of the Si_3N_4–AlN–CaO system, *J. Mater. Sci. Lett.*, **4**(3), 255–259 (1985).

Huang Z.-K., Tien T.-Y. Solid-liquid reaction in the Si_3N_4–AlN–Y_2O_3 system under 1 MPa of nitrogen, *J. Am. Ceram. Soc.*, **79**(6), 1717–1719 (1996).

Huang Z.-K., Tien T.-Y., Yen T.-S. Subsolidus phase relationships in Si_3N_4–AlN–rare-earth oxide systems, *J. Am. Ceram. Soc.*, **69**(10), C-241–C-242 (1986).

Huang Z.-K., Yan D.-S. Phase relationships in Si_3N_4–AlN–M_xO_y systems and their implications for sialon fabrication, *J. Mater. Sci.*, **27**(20), 5640–5644 (1992).

Huminicki D.M.C., Hawthorne F.C. Hydrogen bonding in the crystal structure of seamanite, *Canad. Mineralog.*, **40**(3), 923–928 (2002).

Hurlbut, Jr. C.S. Childrenite–eosphorite series, *Am. Mineralog.*, **35**(9–10), 793–805 (1950).

Hurlbut, Jr. C.S., Honea R. Sigloite, a new mineral from Llallagua, Bolivia, *Am. Mineralog.*, **47**(1–2), 1–8 (1962).

Hurlbut, Jr. C.S. Wardite from Beryl Mountain, New Hampshire, *Am. Mineralog.*, **37**(9–10), 849–852 (1952).

Hurlbut, Jr. C.S., Weichel E.J. Additional data on brazilianite, *Am. Mineralog.*, **31**(9–10), 507 (1946).

Imani M., Enayati M.H. Investigation of amorphous phase formation in Fe–Co–Si–B–P – Thermodynamic analysis and comparison between mechanical alloying and rapid solidification experiments, *J. Alloys Compd.*, **705**, 462–467 (2017).

Inoue A., Cook J.S. Effect of additional elements (M) on the thermal stability of supercooled liquid in $Fe_{72-x}Al_5Ga_2P_{11}C_6B_4M_x$ glassy alloys, *Mater. Trans., JIM*, **37**(1), 32–38 (1996).

Inoue A., Cook J.S. Fe-based ferromagnetic glassy alloys with wide supercooled liquid region, *Mater. Trans., JIM*, **36**(9), 1180–1183 (1995a).
Inoue A., Cook J.S. Multicomponent Fe-based glassy alloys with wide supercooled liquid region before crystallization, *Mater. Trans., JIM*, **36**(10), 1282–1285 (1995b).
Inoue A., Katsuya A. Multicomponent Co-based amorphous alloys with wide supercooled liquid region, *Mater. Trans., JIM*, **37**(6), 1332–1336 (1996).
Inoue A., Murakami A., Zhang T., Takeuchi A. Thermal stability and magnetic properties of bulk amorphous Fe–Al–Ga–P–C–B–Si alloys, *Mater. Trans., JIM*, **38**(3), 189–196 (1997).
Inoue A., Park R.E. Soft magnetic properties and wide supercooled liquid region of Fe–P–B–Si base amorphous alloys, *Mater. Trans., JIM*, **37**(11), 1715–1721 (1996).
Inoue A., Shinohara Y., Cook J.S. Thermal and magnetic properties of bulk Fe-based glassy alloys prepared by copper mold casting, *Mater. Trans., JIM*, **36**(12), 1427–1433 (1995).
Ishizawa N., Kamoshita M., Fukuda K., Shioi K., Hirosaki N. $Sr_3(Al_{3+x}Si_{13-x})(N_{21-x}O_{2+x})$:$Eu^{2+}$ ($x \sim 0$): A monoclinic modification of Sr-sialon, *Acta Crystallogr.*, **E66**(2), i14 (2010).
Ivanov O.K., Shiryaeva L.L., Khoroshilova L.A., Petrishcheva V.G. Hotsonite: Confirmation of the discovery and the new data (Belyavinski mine, the Urals) [in Russian], *Zap. Vses. mineralog. obshch.*, **119**(1), 121–126 (1990).
Izumi F., Mitomo M., Bando Y. Rietveld refinements for calcium and yttrium containing α-sialons, *J. Mater. Sci.*, **19**(9), 3115–3120 (1984).
Izumi F., Mitomo M., Suzuki J. Structure refinement of yttrium α-sialon from X-ray powder profile data, *J. Mater. Sci. Lett.*, **1**(12), 533–535 (1982).
Jambor J.L., Burke E.A.J., Ercit T.S., Grice J.D. New mineral names, *Am. Mineralog.*, **73**(11–12), 1492–1499 (1988).
Jambor J.L., Burke E.A.J., Grew E.S., Puziewicz J. New mineral names, *Am. Mineralog.*, **78**(5–6), 672–678 (1993).
Jambor J.L., Burke E.A.J. New mineral names, *Am. Mineralog.*, **74**(11–12), 1399–1404 (1989).
Jambor J.L., Burke E.A.J. New mineral names, *Am. Mineralog.*, **76**(11–12), 2020–2026 (1991).
Jambor J.L., Grew E.S. New mineral names, *Am. Mineralog.*, **75**(1–2), 240–246 (1990a).
Jambor J.L., Grew E.S. New mineral names, *Am. Mineralog.*, **75**(7–8), 931–937 (1990b).
Jambor J.L., Grew E.S., Roberts A.C. New mineral names, *Am. Mineralog.*, **90**(2–3), 518–522 (2005).
Jambor J.L., Owens D.R., Grice J.D., Feinglos M.N. Gallobeudantite, $PbGa_3[(AsO_4),(SO_4)]_2(OH)_6$, a new mineral species from Tsumeb, Namibia, and associated new gallium analogues of the alunite–jarosite family, *Canad. Mineralog.*, **34**(6), 1305–1315 (1996a).
Jambor J.L., Pertsev N.N., Roberts A.C. New mineral names, *Am. Mineralog.*, **80**(7–8), 845–850 (1995).
Jambor J.L., Pertsev N.N., Roberts A.C. New mineral names, *Am. Mineralog.*, **81**(1–2), 249–254 (1996b).
Jambor J.L., Pertsev N.N., Roberts A.C. New mineral names, *Am. Mineralog.*, **82**(3–4), 430–433 (1997a).
Jambor J.L., Pertsev N.N., Roberts A.C. New mineral names, *Am. Mineralog.*, **82**(9–10), 1038–1041 (1997b).
Jambor J.L., Puziewicz J. New mineral names, *Am. Mineralog.*, **77**(9–10), 1116–1121 (1992).
Jambor J.L., Puziewicz J., Roberts A.C. New mineral names, *Am. Mineralog.*, **84**(4), 685–688 (1999).
Jambor J.L., Puziewicz J., Roberts A.C. New mineral names, *Am. Mineralog.*, **82**(5–6), 620–624 (1997c).
Jambor J.L., Roberts A.C. New mineral names, *Am. Mineralog.*, **80**(1–2), 184–188 (1995a).
Jambor J.L., Roberts A.C. New mineral names, *Am. Mineralog.*, **80**(9–10), 1073–1077 (1995b).
Jambor J.L., Roberts A.C. New mineral names, *Am. Mineralog.*, **84**(1–2), 193–198 (1999a).
Jambor J.L., Roberts A.C. New mineral names, *Am. Mineralog.*, **84**(9), 1464–1468 (1999b).
Jambor J.L. Roberts A.C. New mineral names, *Am. Mineralog.*, **85**(1), 263–266 (2000).
Jambor J.L., Roberts A.C. New mineral names, *Am. Mineralog.*, **88**(11–12), Pt. 1, 1836–1840 (2003).

Jambor J.L., Roberts A.C. New mineral names, *Am. Mineralog.*, **89**(11–12), 1826–1834 (2004).

Jambor J.L., Roberts A.C., Puziewicz J. New mineral names, *Am. Mineralog.*, **79**(5–6), 570–574 (1994).

Jambor J.L., Vanko D.A. New mineral names, *Am. Mineralog.*, **75**(5–6), 706–713 (1990).

Jambor J.L., Vanko D.A. New mineral names, *Am. Mineralog.*, **78**(11–12), 1314–1319 (1993).

Jäschke B., Jansen M. Synthese, Kristallstruktur und pektroskopische Charakterisierung der Phosphaniminato-Komplexe $[Cl_2BNPCl_2NPCl_3]_2$ und $[Br_2BNPCl_3]_2$, *Z. anorg. und allg. Chem.*, **628**(9–10), 2000–2004 (2002).

Jin Y.J., Tang X.H., Ke C., Yu S.Y., Zhang D.H. Bandgap engineering of InSb by N incorporation by metal-organic chemical vapor deposition, *J. Alloys Compd.*, **756**, 134–138 (2018).

Jin Y.J., Tang X.H., Ke C., Zhang D.H. $InSb_{1-x}N_x$ alloys on GaSb substrate by metal-organic chemical vapor deposition for long wavelength detection, *Thin Solid Films*, **616**, 624–627 (2016).

Jin Y.J., Tang X.H., Teng J.H., Zhang D.H. Optical properties and bonding behaviors of InSbN alloys grown by metal-organic chemical vapor deposition, *J. Cryst. Growth*, **416**, 12–16 (2015).

Johan Z., Slansky E., Povondra P. Vashegyite, a sheet aluminum phosphate: New data, *Canad. Mineralog.*, **21**(3), 489–498 (1983).

Kampf A.R., Adams P.M., Barwood H., Nash B.P. Fluorwavellite, $Al_3(PO_4)_2(OH)_2F{\cdot}5H_2O$, the fluorine analog of wavellite, *Am. Mineralog.*, **102**(4), 909–915 (2017a).

Kampf A.R., Adams P.M., Barwood H., Nash B.P. Fluorwavellite, IMA 2015-077. CNMNC Newsletter No. 28, *Mineralog. Mag.*, **79**(7), 1859–1864 (2015a).

Kampf A.R., Adams P.M., Housley R.M. Fluorowardite, IMA 2012-016. CNMNC Newsletter No. 13, *Mineralog. Mag.*, **76**(3), 807–817 (2012a).

Kampf A.R., Adams P.M., Housley R.M., Rossman G.R. Fluorowardite, $NaAl_3(PO_4)_2(OH)_2F_2{\cdot}2H_2O$, the fluorine analog of wardite from the Silver Coin mine, Valmy, Nevada, *Am. Mineralog.*, **99**(4), 804–810 (2014a).

Kampf A.R., Colombo F., Simmons W.B., Falster A.U., Nizamoff J.W. Galliskilte, $Ca_4Al_2(PO_4)_2F_8{\cdot}5H_2O$, a new mineral from the Gigante granitic pegmatite, Córdoba province, Argentina, *Am. Mineralog.*, **95**(2–3), 392–396 (2010).

Kampf A.R., Mills S.J., Housley R.M., Rossman G.R., Nash B.P, Dini M., Jenkins R.A. Joteite, $Ca_2CuAl[AsO_4][AsO_3(OH)]_2(OH)_2{\cdot}5H_2O$, a new arsenate with a sheet structure and unconnected acid arsenate groups, *Mineralog. Mag.*, **77**(6), 2811–2823 (2013a).

Kampf A.R., Mills S.J., Housley R.M., Rossman G.R., Nash B., Dini M., Jenkins R.A. Joteite, IMA 2012-091. CNMNC Newsletter No. 16, *Mineralog. Mag.*, **77**(6), 2695–2709 (2013b).

Kampf A.R., Mills S.J., Nash B., Dini M., Molina Donoso A.A. Currierite, IMA 2016-030. CNMNC Newsletter No. 32, *Mineralog. Mag.*, **80**(5), 915–922 (2016).

Kampf A.R., Mills S.J., Nash B.P., Dini M., Molina Donoso A.A. Currierite, $Na_4Ca_3MgAl_4(AsO_3OH)_{12}{\cdot}9H_2O$, a new acid arsenate with ferrinatrite-like heteropolyhedral chains from the Torrecillas mine, Iquique Province, Chile, *Mineralog. Mag.*, **81**(5), 1141–1149 (2017b).

Kampf A.R., Mills S.J., Nash B., Dini M., Molina Donoso A.A. Tapiaite, IMA 2014-024. CNMNC Newsletter No. 21, *Mineralog. Mag.*, **78**(4), 797–804 (2014b).

Kampf A.R., Mills S.J., Nash B.P., Dini M., Molina Donoso A.A. Tapiaite, $Ca_5Al_2(AsO_4)_4(OH)_4{\cdot}12H_2O$, a new mineral from the Jote mine, Tierra Amarilla, Chile, *Mineralog. Mag.*, **79**(2), 345–354 (2015b).

Kampf A.R., Mills S.J., Rossman G.R., Steele I.M., Pluth J.J., Favreau G. Afmite, $Al_3(OH)_4(H_2O)_3(PO_4)(PO_3OH){\cdot}H_2O$, a new mineral from Fumade, Tarn, France: Description and crystal structure, *Eur. J. Mineral.*, **23**(2), 269–277 (2011).

Kampf A.R., Mills S.J., Rumsey M.S., Sprat J., Favreau G. The crystal structure determination and redefinition of matulaite, $Fe^{3+}Al_7(PO_4)_4(PO_3OH)_2(OH)_8(H_2O)_8{\cdot}8H_2O$, *Mineralog. Mag.*, **76**(3), 517–534 (2012b).

Kampf A.R. Minyulite: Its atomic arrangement, *Am. Mineralog.*, **62**(3–4), 256–262 (1977).

Kampf A.R., Nash B., Dini M., Molina Donoso A.A. Juansilvinite, IMA 2015-080. CNMNC Newsletter No. 28, *Mineralog. Mag.*, **79**(7), 1859–1864 (2015c).

Kampf A.R., Nash B.P., Dini M., Molina Donoso A.A. Juansilvaite, $Na_5Al_3[AsO_3(OH)]_4[AsO_2(OH)_2]_2(SO_4)_2 \cdot 4H_2O$, a new arsenate-sulfate from the Torrecillas mine, Iquique Province, Chile, *Mineralog. Mag.*, **81**(3), 619–628 (2017c).

Kato T. Cell dimensions of the hydrated phosphate, kingite, *Am. Mineralog.*, **55**(3–4), Pt. 1, 515–517 (1970).

Kato T. Further refinement of the goyazite structure, *Mineralog. J.*, **13**(6), 390–396 (1987).

Kato T., Miúra Y. The crystal structure of jarosite and svanbergite, *Mineralog. J.*, **8**(8), 419–430 (1977).

Kazantsev S.S., Pushcharovsky D.Y., Pasero M., Merlino S., Zubkova N.V., Kabalov Y.K., Voloshin A.V. Crystal structure of holtite I, *Crystallogr. Rep.*, **50**(1), 42–47 (2005).

Kazantsev S.S., Zubkova N.V., Voloshin A.V. Refinement of composition and structure of holtite I, *Crystallogr. Rep.*, **51**(3), 412–413 (2006).

Khoury H.N., Sokol E.V., Kokh S.N., Seryotkin Y.V., Nigmatulina E.N., Goryainov S.V., Belogub E.V., Clark I.D. Tululite, IMA 2014-065. CNMNC Newsletter No. 23, *Mineralog. Mag.*, **79**(1), 51–58 (2015).

Kim J., Yamasue E., Okumura H., Ishihara K.N. Crystal structures and electronic band structures for hypothetic lithium boron nitride intercalation compounds, *J. Alloys Compd.*, **751**, 324–334 (2018).

Kim J., Yamasue E., Okumura H., Ishihara K.N., Michioka C. Structures of boron nitride intercalation compound with lithium synthesized by mechanical milling and heat treatment, *J. Alloys Compd.*, **685**, 135–141 (2016a).

Kim S.-C., Kwak H.-J., Yoo C.-Y., Yun H., Kim S.-J. Synthesis, crystal structure, and ionic conductivity of a new layered metal phosphate, $Li_2Sr_2Al(PO_4)_3$, *J. Solid State Chem.*, **243**, 12–17 (2016b).

King G.S.D., Sengier-Roberts L. Drugmanite, $Pb_2(Fe_{0.78}Al_{0.22})H(PO_4)_2(OH)_2$: Its crystal structure and place in the datolite group, *Bull. Minéral.*, **111**, 431–437 (1988).

Kniep R., Boy I., Engelhardt H. $RbFe[BP_2O_8(OH)]$: Ein neues Borophosphat mit offen-verzweigten Vierer-Einfach-Tetraedeifcetten, *Z. anorg. und allg. Chem.*, **625**(9), 1512–1516 (1999a).

Kniep R., Engelhardt H. $Na_{1.89}Ag_{0.11}[BP_2O_7(OH)]$ und $Na_2[BP_2O_7(OH)]$ – isotype Borophosphate mit Tetraeder-Schichtpaketen, *Z. anorg. und allg. Chem.*, **624**(8), 1291–1297 (1998).

Kniep R., Koch D., Borrmann H. Crystal structure of aluminum *catena*-[monohydrogenborate-dihydrogenborate-bis(monohydrogenphosphate)] monohydrate, $Al[B_2P_2O_7(OH)_5] \cdot H_2O$, *Z. Kristallogr., New Cryst. Str.*, **217**(2), 187–188 (2002a).

Kniep R., Koch D., Hartmann T. Crystal structure of potassium aluminum *catena*-(monohydrogenmonoborate)-bis(monophosphate), $KAl[BP_2O_8(OH)]$, *Z. Kristallogr., New Cryst. Str.*, **217**(1), 186 (2002b).

Kniep R., Schäfer G., Borrmann H. Crystal structure of ammonium [monozinco-monoboro-diphosphate], $NH_4[ZnBP_2O_8]$, *Z. Kristallogr., New Cryst. Str.*, **215**(3), 335–336 (2000).

Kniep R., Schäfer G., Engelhardt H., Boy I. $K[ZnBP_2O_8]$ and $A[ZnBP_2O_8]$ ($A = NH_4^+, Rb^+, Cs^+$): Zincoborophosphates as a new class of compounds with tetrahedral framework structures, *Angew. Chem. Int. Ed.*, **38**(24), 3641–3644 (1999b).

Kniep R., Schäfer G., Engelhardt H., Boy I. $K[ZnBP_2O_8]$ und $A[ZnBP_2O_8]$ ($A = NH_4^+, Rb^+, Cs^+$): Zinkoborophosphate als neue Klasse von Verbindungen mit Tetraeder-Gerüststrukturen, *Angew. Chem.*, **111**(24), 3857–3861 (1999c).

Kniep R., Will H.G., Boy I., Röhr C. 6_1-Helices aus Tetraederbändern $^1_\infty[BP_2O_8^{3-}]$: Isotype Borophosphate $M^IM^{II}(H_2O)_2[BP_2O_8] \cdot H_2O$ und ihre Dehydratisierung zu mikroporosen Phasen $M^IM^{II}(H_2O)[BP_2O_8]$, *Angew. Chem.*, **109**(9), 1052–1054 (1997a).

Kniep R., Will H.G., Boy I., Röhr C. 6_1 Helices from tetrahedral ribbons $^1_\infty[BP_2O_8^{3-}]$: Isostructural borophosphates $M^IM^{II}(H_2O)_2[BP_2O_8] \cdot H_2O$ (M^I = Na, K; M^{II} = Mg, Mn, Fe, Co, Ni, Zn) and their dehydration to microporous phases $M^IM^{II}(H_2O)[BP_2O_8]$, *Angew. Chem. Int. Ed. Eng.*, **36**(9), 1013–1014 (1997b).

Knoll S.M., Zhang S., Joyce T.B., Kappers M.J., Humphreys C.J., Moram M.A. Growth, microstructure and morphology of epitaxial ScGaN films, *Phys. status solidi (a)*, **209**(1), 33–40 (2012).

Knorring Von O., Fransolet A.-M. Gatumbaite, $CaAl_2(PO_4)_2(OH)_2{\cdot}H_2O$: A new species from Buranga pegmatite, Rwanda, *Neues Jahrb. Mineral., Monatsh.*, (12), 561–568 (1977).

Knorring Von O., Lehtinen M., Sahama T.G. Burangaite, a new phosphate mineral from Rwanda, *Bull. Geol. Soc. Finland*, **49**, 33–36 (1977).

Koch D., Kniep R. Crystal structure of sodium aluminum (monohydrogenmonophosphate-dihydrogenmonoborate-monophosphate), $NaAl[BP_2O_7(OH)_3]$, *Z. Kristallogr., New Cryst. Str.*, **214**(4), 441–442 (1999).

Kokkoros P. Über die Struktur des Durangit $NaAlF[AsO_4]$, *Z. Kristallogr.*, **99**(1), 38–49 (1938).

Kolitsch U., Giester G. The crystal structure of faustite and its copper analogue turquoise, *Mineralog. Mag.*, **64**(5), 905–913 (2000).

Kolitsch U. The crystal structure of wycheproofite, $NaAlZr(PO_4)_2(OH)_2{\cdot}H_2O$, *Eur. J. Mineral.*, **15**(6), 1029–1034 (2003).

Kolitsch U., Schwendtner K. Octahedral-tetrahedral framework structures of $InAsO_4{\cdot}H_2O$ and $PbIn(AsO_4)(AsO_3OH)$, *Acta Crystallogr.*, **C61**(9), i86–i89 (2005).

Kolitsch U., Tiekink E.R.T., Slade P.G., Taylor M.R., Pring P. Hinsdalite and plumbogummite, their atomic arrangements and disordered lead sites, *Eur. J. Mineral.*, **11**(6), 949–954 (1999).

Köllisch K., Höppe H.A., Huppertz H., Orth M., Schnick W. Neue Vertreter des $Er_6[Si_{11}N_{20}]O$-Strukturtyps. Hochtemperature-Synthesen und Kristallstrukturen von $Ln_{(6+x/3)}[Si_{(11-y)}Al_yN_{(1-x+y)}$ mit Ln = Nd, Er, Yb, Dy und $0 \le x \le 3$, $0 \le y \le 3$, *Z. anorg. allg. Chem.*, **627**(6), 1371–1376 (2001).

Kovtyukhova N.I., Wang Y., Lv R., Terrones M., Crespi V.H., Mallouk T.E. Reversible intercalation of hexagonal boron nitride with Brønsted acids, *J. Am. Chem. Soc.*, **135**(22), 8372–8381 (2013).

Kraus E.H., Seaman W.A., Slawson C.B. Seamanite, a new manganese phosphoborate from Iron County, Michigan, *Am. Mineralog.*, **15**(6), 220–225 (1930).

Krevel Van J.W.H., Hintzen H.T., Metselaar R. On the Ce^{3+} luminescence in the melilite-type oxide nitride compound $Y_2Si_{3-x}Al_xO_{3+x}N_{4-x}$, *Mater. Res. Bull.*, **35**(5), 747–754 (2000).

Kritikos M., Wikstad E., Walldén K. Hydrothermal synthesis, characterization and magnetic properties of three isostructural chain borophosphates; $NH_4M(III)[BP_2O_8(OH)]$ with M = V or Fe and $NH_4(Fe(III)_{0.53}V(III)_{0.47})[BP_2O_8(OH)]$, *Solid State Sci.*, **3**(6), 649–658 (2001).

Krol' O.F., Chernov V.I., Shipovalov Y.V., Khan G.A. Saryarkite: A new mineral [in Russian], *Zap. Vses. mineralog. obshch.*, **93**(2), 147–155 (1964).

Krutik V.M., Pushcharovskii D.Y., Pobedimskaya E.A., Belov N.V. Crystal structure of arrojadite, *Sov. Phys. Crystallogr.*, **24**(4), 425–429 (1979).

Kuang S.-F., Huang Z.-K., Sun W.-Y., Yen T.S. Phase relationships in the Li_2O–Si_3N_4–AlN system and the formation of lithium-α'-sialon, *J. Mater. Sci. Lett.*, **9**(1), 72–74 (1990a).

Kuang S.-F., Huang Z.-K., Sun W.-Y., Yen T.S. Phase relationships in the system MgO–Si_3N_4–AlN, *J. Mater. Sci. Lett.*, **9**(1), 69–71 (1990b).

Kubota T., Makino A., Inoue A. Low core loss of $Fe_{85}Si_2B_8P_4Cu_1$ nanocrystalline alloys with high B_s and B_{800}, *J. Alloys Compd.*, **509**(Suppl. 1), S416–S419 (2011).

Kumar V., Bhardwaj N., Tomar N., Thakral V., Uma S. Novel lithium-containing honeycomb structures, *Inorg. Chem.*, **51**(20), 10471–10473 (2012).

Kuo S.-Y., Jou Z.C., Virkar A.V., Rafaniello W. Fabrication, thermal treatment and microstructure development in SiC–AlN–Al_2OC ceramics, *J. Mater. Sci.*, **21**(9), 3019–3024 (1986).

Kurkutova E.N., Rau V.G., Rumanova I.M. Crystal structure of seamanite, $Mn_3[HO_4/BO_3]{\cdot}3H_2O$ = $Mn_3[PO_3OH][BO(OH)_3](OH)_2$ [in Russian], *Dokl. AN SSSR*, **197**(5), 1070–1073 (1971).

Kurushima T., Gundiah G., Shimomura Y., Mikami M., Kijima N., Cheetham A.K. Synthesis of Eu^{2+}-activated $MYSi_4N_7$ (M = Ca, Sr, Ba) and $SrYSi_{4-x}Al_xN_{7-x}O_x$ (x = 0–1) green phosphors by carbothermal reduction and nitridation, *J. Electrochem. Soc.*, **157**(3), J64–J68 (2010).

Lahti S.I., Pajunen A. New data on lacroixite, $NaAlFPO_4$. Part I. Occurrence, physical properties and chemical composition. Part II. Crystal structure, *Am. Mineralog.*, **70**(7–8), 849–855 (1985).

Landmann J., Sprenger J.A.P., Bertermann R., Ignat'ev N., Bernhardt-Pitchougina V., Bernhardt E., Willner H., Finze M. Convenient access to the tricyanoborate dianion $B(CN)_3^{2-}$ and selected reactions as a boron-centred nucleophile, *Chem. Commun.*, **51**(24), 4989–4992 (2015).

Larsen E.S. Overite and montgomeryite: Two new minerals from Fairfield, Utah, *Am. Mineralog.*, **25**(5), 315–326 (1940).

Lauterbach R., Irran E., Henry P.F., Weller M.T., Schnick W. High-temperature synthesis, single-crystal X-ray and neutron powder diffraction and materials properties of $Sr_3Ln_{10}Si_{18}Al_{12}O_{18}N_{36}$ (Ln = Ce, Pr, Nd) – novel sialons with an ordered distribution of Si, Al, O and N, *J. Mater. Chem.*, **10**(6), 1357–1364 (2000).

Lauterbach R., Schnick W. High-temperature synthesis and single-crystal X-ray structure determination of $Sr_{10}Sm_6Si_{30}Al_6O_7N_{54}$ – a layered sialon wth an ordered distribution of Si, Al, O and N, *Solid State Sci.*, **2**(4), 463–472 (2000a).

Lauterbach R., Schnick W. $Nd_3Si_5AlON_{10}$ – Synthese, Kristallstruktur und Eigenschaften eines Sialons im $La_3Si_6N_{11}$-Strukturtyp, *Z. anorg. allg. Chem.*, **626**(1), 56–61 (2000b).

Lauterbach R., Schnick W. $Sm_2Si_3O_3N_4$ und $Ln_2Si_{2.5}Al_{0.5}O_{3.5}N_{3.5}$ (Ln = Ce, Pr, Nd, Sm, Gd) – neuer synthetischer Zugang zu N-haltigen Melilith-Phasen und deren Einkristall-Röntgenstrukturanalyse, *Z. anorg. allg. Chem.*, **625**(3), 429–434 (1999).

Lauterbach R., Schnick W. Synthese, Kristallstrukture und Eigenschaften eines neuen Sialons – $SrSiAl_2O_3N_2$, *Z. anorg. allg. Chem.*, **624**(7), 1154–1158 (1998).

Leavens P.B., Rheingold A.L. Crystal structures of gordonite, $MgAl_2(PO_4)_2(OH)_2(H_2O)_6{\cdot}2H_2O$, and its Mn analog, *Neues Jahrb. Mineral., Monatsh.*, 265–270 (1988).

Leavens P.B., White, Jr. J.S., Robinson G.W., Nelen J.A. Mangangordonite, a new phosphate mineral from Kings Mountain, North Carolina and Newry, Maine, USA, *Neues Jahrb. Mineral., Monatsh.*, 169–176 (1991).

Le Bail A., Stephens P.W., Hubert F. A crystal structure for the souzalite/gormanite series from synchrotron powder diffraction data, *Eur. J. Mineral.*, **15**(4), 719–723 (2003).

Lee B.-J., Kim H.-D., Hong J.-H. Calculation of α/γ equilibria in SA508 grade 3 steels for intercritical heat treatment, *Metal. Mater. Trans.*, **A29**(5), 1441–1447 (1998).

Lee J.J., Kim J.D., Razeghi M. Growth and characterization of InSbBi for long wavelength infrared photodetectors, *Appl. Phys. Lett.*, **70**(24), 3266–3268 (1997).

Lefebvre J.-J., Gasparrini C. Florencite, an occurrence in the Zairian copperbelt, *Canad. Mineralog.*, **18**(3), 301–311 (1980).

Le Meins J.-M., Courbion G. Hydrothermai synthesis and crystal structure of $SrAl_2(PO_4)_2F_2$: A new three-dimensional framework with channels delimited by a helical anionic border, *Eur. J. Solid State Inorg. Chem.*, **35**(10–11), 639–653 (1998).

Lesage J., Guesdon A., Raveau B., Petricek V. The anionic 3D-framework $[Ga_2(PO_4)_3]_\infty$: A microporous host lattice for various species, *J. Solid State Chem.*, **177**(10), 3581–3589 (2004).

Lieb A., Kechele J.A., Kraut R., Schnick W. The sialons $MLn[Si_{4-x}Al_xO_xN_{7-x}]$ with *M* = Eu, Sr, Ba and *Ln* – Ho-Yb – twelve substitution variants with the MYb[Si_4N_7] structure type, *Z. anorg. allg. Chem.*, **633**(1), 166–171 (2007).

Lieb A., Schnick W. $BaSm_5[Si_9Al_3N_{20}]O$ – a nitridoaluminosilicate oxide with a new structure type composed of "star-shaped" $[N^{[4]}((Si,Al)N_3)_4]$ units as secondary building units, *Solid State Sci.*, **8**(2), 185–191 (2006).

Li H., Zhao Y., Pan S., Wu H., Yu H., Zhang F., Yang Z., Poeppelmeier K.R. Synthesis and structure of $KPbBP_2O_8$ – a congruent melting borophosphate with nonlinear optical properties, *Eur. J. Inorg. Chem.*, **2013**(18), 3185–3190 (2013).

Lii K.-H. $Rb_2[Ga_4(HPO_4)(PO_4)_4]{\cdot}0.5H_2O$: A new gallium phosphate containing four-, five, and six-coordinated gallium atoms, *Inorg. Chem.*, **35**(25), 7440–7442 (1996a).

Lii K.-H. $RbIn(OH)PO_4$: An indium(III) phosphate containing spirals of corner-sharing InO_6 octahedra, *J. Chem. Soc., Dalton Trans.*, (6), 815–818 (1996b).

Li M.-R., Liu W., Ge M.-H., Chen H.-H., Yang X.-X., Zhao J.-T. $NH_4[BPO_4F]$: A novel open-framework ammonium fluorinated borophosphate with a zeolite-like structure related to gismondine topology, *Chem. Commun.*, (11), 1272–1273 (2004).

Li M.-R., Mao S.-Y., Chen H.-H., Deng J.-F., Mi J.-X., Zhao J.-T. Crystal structure of diammonium indium (monophosphate-monohydrogen-monophosphate), $(NH_4)_2In[(PO_4)(HPO_4)]$, *Z. Kristallogr., New Cryst. Str.*, **217**(3), 309–310 (2002a).

Li M.-R., Mao S.-Y., Huang Y.-X., Mi J.-X., Wei Z.-B., Zhao J.-T., Kniep R. Crystal structure of ammonium gallium (monophosphate-hydrogenmonoborate-monophosphate), $(NH_4)Ga[BP_2O_8(OH)]$, *Z. Kristallogr., New Cryst. Str.*, **217**(1), 165–166 (2002b).

Lindberg M.L. Arrojadite, hühnerkobelite, and graftonite, *Am. Mineralog.*, **35**(1–2), 59–76 (1950).

Lindberg M.L., Christ C.L. Crystal structures of the isostructural minerals lazulite, scorzalite and barbosalite, *Acta Crystallogr.*, **12**(9), 695–697 (1959).

Lin Z.-S., Hoffmann S., Huang Y.-X., Prots Y., Zhao J.-T., Kniep R. Crystal structure of trilithium divanadium(III) borophosphate hydrogenphosphate, $Li_3V_2[BP_3O_{12}(OH)][HPO_4]$, *Z. Kristallogr., New Cryst. Str.*, **225**(1), 3–4 (2010).

Lin Z.-S., Huang Y.-X., Prots Y., Zhao J.-T., Kniep R. Crystal structure of potassium vanadium (monophosphatehydrogenmonoborate-monophosphate), $KV[BP_2O_8(OH)]$, *Z. Kristallogr., New. Cryst. Str.*, **223**(4), 323–324 (2008a).

Lin Z.-S., Huang Y.-X., Prots Y., Zhao J. T., Kniep R. $Na_5(NH_4)Mn_3[B_9P_6O_{33}(OH)_3]\cdot 1.5H_2O$, *Acta Crystallogr.*, **E64**(12), i82–i83 (2008b).

Lin Z.-S., Huang Y.-X., Wu Y.-H., Zhou Y. Lithium diaquamagnesium *catena*-borodiphosphate(V) monohydrate, $LiMg(H_2O)_2[BP_2O_8]\cdot H_2O$, at 173 K, *Acta Crystallogr.*, **E64**(6), i39–i40 (2008c).

Little M.E., Kordesch M.E. Band-gap engineering in sputter-deposited $Sc_xGa_{1-x}N$, *Appl. Phys. Lett.*, **78**(19), 2891–2892 (2001).

Liu F., Pang S., Li R., Zhang T. Ductile Fe–Mo–P–C–B–Si bulk metallic glasses with high saturation magnetization, *J. Alloys Compd.*, **483**(1–2), 613–615 (2009).

Liu W., Huang Y.-X., Cardoso R., Schnelle W., Kniep R. $Na_6Cu_3\{B_6P_6O_{27}(O_2BOH)_3\}\cdot 2H_2O$: A novel copper borophosphate with a tubelike borophosphate anion, *Z. anorg. und allg. Chem.*, **632**(12–13), 2143 (2006).

Liu W., Zhang L., Su G., Cao L.-X., Wang Y.-G. A novel polyoxochromium borophosphate with new 6-membered ring crown-shaped clusters, *J. Chem. Soc., Dalton Trans.*, **39**(31), 7262–7265 (2010).

Liu W., Zhao J. $(NH_4)[B_3PO_6(OH)_3]\cdot 0.5H_2O$, *Acta Crystallogr.*, **E63**(11), i185 (2007).

Li X., Sundaram S., El Gmili Y., Moudakir T., Genty F., Bouchoule S., Patriarche G., Dupuis R.D., Voss P.L., Salvestrini J.-P., Ougazzaden A. BAlN thin layers for deep UV applications, *Phys. status solidi (a)*, **212**(4), 745–750 (2015).

Li X., Wang S., Liu H., Ponce F.A., Detchprohm T., Dupuis R.D. 100-nm thick single-phase wurtzite BAlN films with boron contents over 10%, *Phys. status solidi (b)*, **254**(8), 1600699 (2017).

Li Y.Q., Hirosaki N., Xie R.-J., Mitomo M. Crystal, electronic and luminescence properties of Eu^{2+}-doped $Sr_2Al_{2-x}Si_{1+x}O_{7-x}N_x$, *Sci. Technol. Adv. Mater.*, **8**(7–8), 607–616 (2007).

Li Y.Q., Hirosaki N., Xie R.J., Takeda T., Mitomo M. Yellow-orange-emitting $CaAlSiN_3:Ce^{3+}$ phosphor: Structure, photoluminescence, and application in white LEDs, *Chem. Mater.*, **20**(21), 6704–6714 (2008).

Locock A.J., Ercit T.S., Kjellman J., Piilonen P.C. New mineral names, *Am. Mineralog.*, **91**(11–12), 1945–1954 (2006a).

Locock A.J., Kinman W.S., Burns P.C. The structure and composition of uranospathite, $Al_{1-x}\square_x[(UO_2)(PO_4)]_2(H_2O)_{20+3x}F_{1-3x}$, $0 < x < 0.33$, a non-centrosymmetric fluorine-bearing mineral of the autunite group, and of a related synthetic lower hydrate, $Al_{0.67}\square_{0.33}[(UO_2)(PO_4)]_2(H_2O)_{15.5}$, *Canad. Mineralog.*, **43**(3), 989–1003 (2005).

Locock A.J., Piilonen P.C., Ercit T.S., Rowe R. New mineral names, *Am. Mineralog.*, **91**(1), 216–224 (2006b).
Loehman R.E. Oxynitride glasses, *J. Non-Cryst. Solids*, **42**(1–3), 433–446 (1980).
Loehman R.E. Preparation and properties of yttrium–silicon–aluminum oxynitride glasses, *J. Am. Ceram. Soc.*, **62**(9–10), 4491–4494 (1979).
Loiseau T., Paulet C., Simon N., Munch V., Taulelle F., Férey G. Hydrothermal synthesis and structural characterization of $(NH_4)GaPO_4F$, KTP-type and $(NH_4)_2Ga_2(PO_4)(HPO_4)F_3$, pseudo-KTP-type materials, *Chem. Mater.*, **12**(5), 1393–1399 (2000).
Machatschki F. Synthese des Durangites *NaAlF*[AsO_4], *Z. Kristallogr.*, **103**(1), 221–227 (1941).
Magin, Jr. G.B., Jansen G.J., Levin B. Synthesis of sabugalite, *Am. Mineralog.*, **44**(3–4), 419–422 (1959).
Maisonneuve V., Evain M., Payen C., Cajipe V.B., Molinié P. Room-temperature crystal structure of the layered phase $Cu^{I}In^{III}P_2S_6$, *J. Alloys Compd.*, **218**(2), 157–164 (1995).
Makino A., Chang C., Kubota T., Inoue A. Soft magnetic Fe–Si–B–P–C bulk metallic glasses without any glass-forming metal elements, *J. Alloys Compd.*, **483**(1–2), 616–619 (2009).
Malinko S.V. The first finding of cahnite in the USSR [in Russian], *Dokl. AN SSSR*, **166**(3), 695–697 (1966).
Mandarino J.A., Sturman B.D., Corlett M.I. Penikisite, the magnesium analogue of kulanite, from Yukon Territory, *Canad. Mineralog.*, **15**(3), 393–395 (1977).
Mandarino J.A., Sturman B.D. Kulanite, a new barium iron aluminum phosphate from the Yukon Territory, Canada, *Canad. Mineralog.*, **14**(2), 127–131 (1976).
Mao S.-Y., Kang Y.-J., Liu W., Mi J.-X., Zhao J.-T. $K_5[W_4O_8(H_2BO_4)(HPO_4)_2(PO_4)_2]\cdot0.5H_2O$: A tungsten borophosphate with a novel sandwich-like layered structure, *Chem. Lett.*, **35**(6), 676–677 (2006).
Mao S.-Y., Li M.-R., Huang Y.-X., Mi J.-X., Chen H.-H., Wei Z.-B., Zhao J.-T. Hydrothermal synthesis and crystal structure of the first ammonium indium(III) phosphate $NH_4In(OH)PO_4$ with spiral chains of $InO_4(OH)_2$, *J. Solid State Chem.*, **165**(2), 209–213 (2002a).
Mao S.-Y., Li M.-R., Huang Y.-X., Mi J.-X., Wei Z.-B., Zhao J.-T., Kniep R. Crystal structure of potassium indium (monophosphate-hydrogenmonoborate-monophosphate), $KIn[BP_2O_8(OH)]$, *Z. Kristallogr., New Cryst. Str.*, **217**(1), 3–4 (2002b).
Mao S.-Y., Li M.-R., Mi J.-X., Chen H.-H., Deng J.-F., Zhao J.-T. Crystal structure of ammonium indium di(hydrogenmonophosphate), $(NH_4)In[PO_3(OH)]_2$, *Z. Kristallogr., New Cryst. Str.*, **217**(3), 311–312 (2002c).
Mao S.-Y., Li M.-R., Wei Z.-B., Mi J.-X., Zhao J.-T. Studies on the intergrowth structures in indium(In) borophosphates, *Chin. J. Inorg. Chem.*, **20**(1), 53–56 (2004a).
Mao S.-Y., Mi J.-X., Akselrud L.G., Huang Y.-X., Zhao J.-T., Kniep R. The layered gallium borophosphate $Ga[B_2P_2O_7(OH)_5]$ refined from X-ray powder data, *Acta Crystallogr.*, **E60**(12), i149–i150 (2004b).
Martini J. Sasaite, a new phosphate mineral from West Driefontein Cave, Transvaal, South Africa, *Mineralog. Mag.*, **42**(323), 401–404 (1978).
Marzoni Fecia di Cossato Y., Orlandi P., Vezzalini G. Rittmannite, a new mineral species of the whiteite group from the Mangualde granitic pegmatite, Portugal, *Canad. Mineralog.*, **27**(3), 447–449 (1989).
Marzouki R., Guesmia A., Driss A. The novel arsenate $Na_4Co_{7-x}A_{12/3}(AsO_4)_6$ (x = 1.37): Crystal structure, charge-distribution and bond-valence-sum investigations, *Acta Crystallogr.*, **C65**(10), i94–i98 (2009).
Ma Q., Zhao C., Xu H., Liu R., Ye H., Hu Y., Liu Y., Chen G. Effect of replacement of Al–O for Si–N on the properties of $Sr_6Si_{25.6}Al_{6.4}N_{41.6}O_{4.4}$: Eu^{2+} phosphors for white light-emitting diodes, *RSC Adv.*, **5**(16), 12323–12328 (2015).
Massiot D., Conanec R., Feldmann W., Marchand R., Laurent Y. NMR characterization of the $Na_3AlP_3O_9N$ and $Na_2Mg_2P_3O_9N$ nitridophosphates: Location of the (NaAl)/Mg_2 substitution, *Inorg. Chem.*, **35**(17), 4957–4960 (1996).

Matselko O., Burkhardt U., Gladyshevskii R., Grin Y. Crystal structure of $Ga_{0.62(3)}Sb_{0.38(3)}Pd_3$, *Z. Kristallogr., New Cryst. Str.*, **233**(1), 87–88 (2018a).

Matselko O., Burkhardt U., Grin Y., Gladyshevskii R. Crystal structure of $Ga_{0.47(1)}Sb_{0.53(1)}Pd_2$, *Z. Kristallogr., New Cryst. Str.*, **233**(1), 89–90 (2018b).

Matsumoto H., Urata A., Yamada Y., Inoue A. Novel $Fe_{(97-x-y)}P_xB_yNb_2Cr_1$ glassy alloys with high magnetization and low loss characteristics for inductor core materials, *IEEE Trans. Magn.*, **46**(2), 374–376 (2010).

McConnell D. Are vashegyite and kingite hydrous aluminum phyllophosphates with kaolinite-type structures?, *Mineralog. Mag.*, **39**(307), 802–806 (1974).

McConnell D. Clinobarrandite and the isodimorphous series, variscite-metavariscite, *Am. Mineralog.*, **25**(11), 719–725 (1940).

McConnel D. Griphite, a hydrophosphate garnetoid, *Am. Mineralog.*, **27**(6), 452–461 (1942).

McConnell D., Pondrom, Jr. W.L. X-ray cystallography of seamanite, *Am. Mineralog.*, **26**(7), 446–447 (1941).

McConnell D. Viséite, a zeolite with the analcime structure and containing linked SiO_4, PO_4, and H_xO_4 groups, *Am. Mineralog.*, **37**(7–8), 609–617 (1952).

McKie D. Goyazite and florencite from two African carbonatites, *Mineralog. Mag.*, **33**(259), 281–297 (1962).

Meagher E.P., Coates M.E., Aho A.E. Jagowerite: A new barium phosphate mineral from the Yukon Territory, *Canad. Mineralog.*, **12**(2), 135–136 (1973).

Meagher E.P., Gibbons C.S., Trotter J. The crystal structure of jagowerite: $BaAl_2P_2O_8(OH)_2$, *Am. Mineralog.*, **59**(3–4), 291–295 (1974).

Meer van der H. The crystal structure of a monoclinic form of aluminium metaphosphate, $Al(PO_3)_3$, *Acta Crystallogr.*, **B32**(8), 2423–2426 (1976).

Meisser N., Brugger J., Krivovichev S., Armbruster T., Favreau G. Description and crystal structure of maghrebite, $MgAl_2(AsO_4)_2(OH)_2{\cdot}8H_2O$, from Aghbar, Anti-Atlas, Morocco: First arsenate in the laueite mineral group, *Eur. J. Mineral.*, **24**(4), 717–726 (2012).

Menezes L., Chaves M.L.S.C., Cooper M.A., Ball N., Abdu Y., Sharp R., Hawthorne F.C, Day M. Brandãoite, IMA 2016-071a. CNMNC Newsletter No. 39, October 2017, page 934, *Eur. J. Mineral.*, **29**(5), 931–936 (2017a).

Menezes L., Chaves M.L.S.C., Cooper M.A., Ball N., Abdu Y., Sharp R., Hawthorne F.C., Day M. Brandãoite, IMA 2016-071a. CNMNC Newsletter No. 39, October 2017, page 1283, *Mineralog. Mag.*, **81**(5), 1279–1286 (2017b).

Menezes P.W., Hoffmann S., Prots Y., Kniep R. Crystal structure of barium cobalt(II) monophosphate-hydrogenmonoborate-monophosphate, $BaCo[BP_2O_8(OH)]$, *Z. Kristallogr., New. Cryst. Str.*, **223**(4), 339–340 (2008a).

Menezes P.W., Hoffmann S., Prots Y., Kniep R. Crystal structure of barium iron(II) monophosphate-hydrogenmonoborate-monophosphate, $BaFe[BP_2O_8(OH)]$, *Z. Kristallogr., New. Cryst. Str.*, **223**(4), 337–338 (2008b).

Menezes P.W., Hoffmann S., Prots Y., Kniep R. Crystal structure of calcium iron(II) hydrogenmonophosphate-dihydrogenmonoborate-monophosphate, $CaFe[BP_2O_7(OH)_3]$, *Z. Kristallogr., New. Cryst. Str.*, **223**(4), 335–336 (2008c).

Menezes P.W., Hoffmann S., Prots Y., Kniep R. Crystal structure of calcium nickel(II) hydrogenmonophosphate-dihydrogenmonoborate-monophosphate, $CaNi[BP_2O_7(OH)_3]$, *Z. Kristallogr., New. Cryst. Str.*, **221**(4), 429–430 (2006a).

Menezes P.W., Hoffmann S., Prots Y., Kniep R. Crystal structure of dicaesium diaquatricobalt(II) (phosphate-borate-hydrogenphosphate), $Cs_2Co_3(H_2O)_2[B_4P_6O_{24}(OH)_2]$, *Z. Kristallogr., New Cryst. Str.*, **224**(1), 1–2 (2009a).

Menezes P.W., Hoffmann S., Prots Y., Kniep R. Crystal structure of hemicalcium diaquairon(II) *catena*-(monoborodiphosphate) monohydrate, $Ca_{0.5}Fe(H_2O)_2[BP_2O_8]{\cdot}H_2O$, *Z. Kristallogr., New. Cryst. Str.*, **223**(1), 9–10 (2008d).

Menezes P.W., Hoffmann S., Prots Y., Kniep R. Crystal structure of hemicalcium diaquanickel(II) catena-(monoborodiphosphate) monohydrate, $Ca_{0.5}Ni(H_2O)_2[BP_2O_8]{\cdot}H_2O$, *Z. Kristallogr., New Cryst. Str.*, **222**(1), 1–2 (2007a).

Menezes P.W., Hoffmann S., Prots Y., Kniep R. Crystal structure of lithium diaquacobalt(II) catena-monoborodophosphate monohydrate, $LiCo(H_2O)_2[BP_2O_8]{\cdot}H_2O$, *Z. Kristallogr., New. Cryst. Str.*, **223**(4), 333–334 (2008e).

Menezes P.W., Hoffmann S., Prots Y., Kniep R. Crystal structure of potassium scandium (monophosphate-hydrogenmonoborate-monophosphate), $KSc[BP_2O_8(OH)]$, *Z. Kristallogr., New. Cryst. Str.*, **221**(3), 251–252 (2006b).

Menezes P.W., Hoffmann S., Prots Y., Kniep R. Crystal structure of rubidium scandium (monophosphate-hydrogenmonoborate-monophosphate), $RbSc[BP_2O_8(OH)]$, *Z. Kristallogr., New. Cryst. Str.*, **221**(3), 253–254 (2006c).

Menezes P.W., Hoffmann S., Prots Y., Kniep R. $CsSc[B_2P_3O_{11}(OH)_3]$: A new borophosphate oligomer containing boron in CN = 3 and 4, *Z. anorg. und allg. Chem.*, **632**(12–13), 2131 (2006d).

Menezes P.W., Hoffmann S., Prots Y., Kniep R. $CsSc[B_2P_3O_{11}(OH)_3]$: A new borophosphate oligomer containing boron in three- and fourfold coordination, *Inorg. Chem.*, **46**(18), 7503–7508 (2007b).

Menezes P.W., Hoffmann S., Prots Y., Kniep R. $NaSc[BP_2O_6(OH)_3][(HO)PO_3]$: Synthesis and crystal structure of an alkali-metal scandium borophosphate hydrogenphosphate, *Z. anorg. und allg. Chem.*, **636**(1), 19–22 (2010a).

Menezes P.W., Hoffmann S., Prots Y., Kniep R. Synthesis and crystal structure of $CaCo(H_2O)[BP_2O_8(OH)]{\cdot}H_2O$, *Z. anorg. und allg. Chem.*, **635**(4–5), 614–617 (2009b).

Menezes P.W., Hoffmann S., Prots Y., Schnelle W., Kniep R. Preparation and characterization of the layered borophosphates $M^{II}(H_2O)_2[B_2P_2O_8(OH)_2]{\cdot}H_2O$ (M^{II} = Fe, Co, Ni), *Inorg. Chim. Acta*, **363**(15), 4299–4306 (2010b).

Menezes P.W., Hoffmann S., Prots Y., Schnelle W., Kniep R. Synthesis and crystal structure of $SrFe[BP_2O_8(OH)_2]$, *Z. anorg. und allg. Chem.*, **635**(8), 1153–1156 (2009c).

Merlino S., Mellini M., Zanazzi P.F. Structure of arrojadite, $KNa_4CaMn_4Fe_{10}Al(PO_4)_{12}(OH,F)_2$, *Acta Crystallog.*, **B37**(9), 1733–1736 (1981).

Metselaar R. Terminology for compounds in the Si–Al–O–N system, *J. Eur. Ceram. Soc.*, **18**(3), 183–184 (1998).

Mertz J.L., Ding N., Kanatzidis M.G. Three-dimensional frameworks of cubic $(NH_4)_5Ga_4SbS_{10}$, $(NH_4)_4Ga_4SbS_9(OH){\cdot}H_2O$, and $(NH_4)_3Ga_4SbS_9(OH_2){\cdot}2H_2O$, *Inorg. Chem.*, **48**(23), 10898–10900 (2009).

Meyer H.-J. Ca_3Cl_2CBN, eine Verbindung mit dem neuen Anion CBN^{4-}, *Z. anorg. und allg. Chem.*, **594**(1), 113–118 (1991).

Meyer L.M., Haushalter R.C. The first octahedral-trigonal bipyramidal-tetrahedral framework oxide: Hydrothermal synthesis and structure of $K[Ni(H_2O)Al_2(PO_4)_3]$, *Chem. Mater.*, **6**(4), 349–350 (1994).

Miao H., Chang C., Li Y., Wang Y., Jia X., Zhang W. Fabrication and properties of soft magnetic Fe–Co–Ni–P–C–B bulk metallic glasses with high glass-forming ability, *J. Non-Cryst. Solids*, **421**, 24–29 (2015).

Michiue Y., Shioi K., Hirosaki N., Takeda T., Xie R.-J., Sato A., Onoda M., Matsushita Y. $Eu_3Si_{15-x}Al_{1+x}O_xN_{23-x}$ ($x \approx 5/3$) as a commensurate composite crystal, *Acta Crystallogr.*, **B65**(5), 567–575 (2009).

Michor H., Hilscher G., Krendelsberger R., Rogl P., Bourée F. Effect of Ni-site substitutions in superconducting $La_3Ni_2B_2N_{3-\delta}$, *Phys. Rev. B*, **58**(22), 15045–15052 (1998).

Mi J.-X., Borrmann H., Huang Y.-X., Mao S.-Y., Zhao J.-T., Kniep R. Crystal structure of caesium gallium(III) catena-[monohydrogen-monoborate-bis(monophosphate)], $CsGa[BP_2O_8(OH)]$, *Z. Kristallogr., New Cryst. Str.*, **218**(2), 171–172 (2003a).

Mi J.-X., Borrmann H., Huang Y.-X., Zhao J.-T., Kniep R. Crystal structure of the 1M-modification of caesium gallium(III) monohydrogen triphosphate, 1M-$CsGaHP_3O_{10}$, *Z. Kristallogr., New Cryst. Str.*, **218**(2), 169–170 (2003b).

Mi J.-X., Borrmann H., Huang Y.-X., Zhao J.-T., Kniep R. Crystal structure of the α-modification of caesium gallium(III) monohydrogen triphosphate, α-$CsGaHP_3O_{10}$, *Z. Kristallogr., New Cryst. Str.*, **218**(2), 167–168 (2003c).

Mi J.-X., Borrmann H., Mao S.-Y., Huang Y.-X., Zhang H., Zhao J.-T., Kniep R. Crystal structure of rubidium gallium *catena*-[monohydrogen-monoborate-bis(monophosphate)] $RbGa[BP_2O_8(OH)]$, *from a twinned crystal, Z. Kristallogr., New Cryst. Str.*, **218**(1), 17–18 (2003d).

Mi J.-X., Deng J.-F., Mao S.-Y., Huang Y.-X., Borrmann H., Zhao J.-T., Kniep R. Crystal structure of dilithium indium (monophosphate-monohydrogen-monophosphate), $Li_2In[(PO_4)(HPO_4)]$, *Z. Kristallogr., New Cryst. Str.*, **217**(3), 307–308 (2002a).

Mi J.-X., Huang Y.-X., Deng J.-F., Borrmann H., Zhao J.-T., Kniep R. Crystal structure of ammonium aluminum *catena*-[monohydrogen-monoborate-bis(monophosphate)], $(NH_4)Al[BP_2O_8(OH)]$, *Z. Kristallogr., New Cryst. Str.*, **217**(3), 305–306 (2002b).

Mi J.-X., Huang Y.-X., Deng J.-F., Borrmann H., Zhao J.-T., Kniep R. Crystal structure of caesium aluminum *catena*-[monohydrogen-monoborate-bis(monophosphate)], $CsAl[BP_2O_8(OH)]$, *Z. Kristallogr., New Cryst. Str.*, **217**(1), 169–170 (2002c).

Mi J.-X., Huang Y.-X., Mao S.-Y., Borrmann H., Zhao J.-T., Kniep R. Crystal structure of potassium galium (monophosphate-hydrogenmonoborate-monophosphate), $KGa[BP_2O_7(OH)_3]$, *Z. Kristallogr., New Cryst. Str.*, **217**(1), 167–168 (2002d).

Mi J.-X., Huang Y.-X., Mao S.-Y., Huang X.-D., Wei Z.-B., Huang Z.-L., Zhao J.-T. Hydrothermal synthesis and crystal structure of $Na_2In_2[PO_3(OH)]_4\ H_2O$ with a new structure type, *J. Solid State Chem.*, **157**(1), 213–219 (2001).

Mi J.-X., Li M.-R., Mao S.-Y., Huang Y.-X., Wei Z.-B., Zhao J.-T., Kniep R. Crystal structure of ammonium indium (monophosphate-hydrogenmonoborate-monophosphate), $(NH_4)In[BP_2O_8(OH)]$, *Z. Kristallogr., New Cryst. Str.*, **217**(1), 5–6 (2002e).

Mi J.-X., Zhao J.-T., Huang Y.-X., Deng J.-F., Borrmann H., Kniep R. Crystal structure of rubidium aluminum *catena*-[monohydrogen-monoborate-bis(monophosphate)], $RbAl[BP_2O_8(OH)]$, *Z. Kristallogr., New Cryst. Str.*, **217**(1), 171–172 (2002f).

Mills S.J., Christy A.G., Favreau G. The crystal structure of ceruleite, $CuAl_4[AsO_4]_2(OH)_8(H_2O)_4$, from Cap Garonne, France, *Mineralog. Mag.*, **82**(1), 181–187 (2018).

Mills S.J., Christy A.G., Kampf A.R., Housley R.M., Favreau G., Boulliard J.-C., Bourgoin V. Zincalstibite-9R: The first nine-layer polytype with the layered double hydroxide structure-type, *Mineralog. Mag.*, **76**(5), 1337–1345 (2012a).

Mills S.J., Christy A.G., Schnyder C., Favreau G., Price J.R. The crystal structure of camerolaite and structural variation in the cyanotrichite family of merotypes, *Mineralog. Mag.*, **78**(7), 1527–1552 (2014).

Mills S.J., Grey I.E., Kampf A.R., Birch W.D., Macrae C.M., Smith J.B., Keck E. Kayrobertsonite, $MnAl_2(PO_4)_2(OH)_2{\cdot}6H_2O$, a new phosphate mineral related to nordgauite, *Eur. J. Mineral.*, **28**(3), 649–654 (2016).

Mills S.J., Grey I.E., Kampf A.R., MacRae C.M., Smith J.B., Davidson C.J., Glenn A.M. Ferraioloite, IMA 2015-066. CNMNC Newsletter No. 28, December 2015, page 1862, *Mineralog. Mag.*, **79**(7), 1859–1864 (2015).

Mills S.J., Grey I.E., Kampf A.R., Macrae C.M., Smith J.B., Davidson C.J., Glenn A.M. Ferraioloite, $MgMn^{2+}{}_4(Fe^{2+}{}_{0.5}Al^{3+}{}_{0.5})_4Zn_4(PO_4)_8(OH)_4(H_2O)_{20}$, a new secondary phosphate mineral from the Foote mine, USA, *Eur. J. Mineral.*, **28**(3), 655–661 (2016b).

Mills S.J., Groat L.A., Wilson S.A., Birch W.D., Whitfield P.S., Raudsepp M. Angastonite, $CaMgAl_2(PO_4)_2(OH)_4{\cdot}7H_2O$: A new phosphate mineral from Angaston, South Australia, *Mineralog. Mag.*, **72**(5), 1011–1020 (2008).

Mills S.J., Kampf A.R., McDonald A.M., Favreau G., Chiappero P.-J. Forêtite, a new secondary arsenate mineral from the Cap Garonne mine, France, *Mineralog. Mag.*, **76**(3), 769–775 (2012b).

Mills S.J., Kampf A.R., McDonald A.M., Favreau G., Chiappero P.-J. Forêtite, IMA 2011-100. CNMNC Newsletter No. 13, June 2012, page 809, *Mineralog. Mag.*, **76**(3), 807–817 (2012c).

Mills S.J., Kampf A.R., Raudsepp M., Birch W.D. The crystal structure of waylandite from Wheal Remfry, Cornwall, United Kingdom, *Miner. Petrol.*, **100**(3–4), 249–253 (2010a).

Mills S., Kampf A.R., Raudsepp M., Christy A.G. The crystal structure of Ga-rich plumbogummite from Tsumeb, Namibia, *Mineralog. Mag.*, **73**(5), 837–845 (2009).

Mills S.J., Kampf A.R., Sejkora J., Adams P.M., Birch W.D., Plášil J. Iangreyite: A new secondary phosphate mineral closely related to perhamite, *Mineralog. Mag.*, **75**(2), 327–336 (2011a).

Mills S.J., Kartashov P.M., Kampf A.R., Raudsepp M. Arsenoflorencite-(La), a new mineral from the Komi Republic, Russian Federation: Description and crystal structure, *Eur. J. Mineral.*, **22**(4), 613–621 (2010b).

Mills S.J., Ma C., Birch W.D. A contribution to understanding the complex nature of peisleyite, *Mineralog. Mag.*, **75**(6), 2733–2737 (2011b).

Mills S., Mumme G., Grey I., Bordet P. The crystal structure of perhamite, *Mineralog. Mag.*, **70**(2), 201–209 (2006).

Mills S.J., Rumsey M.S., Favreau G., Spratt J., Raudsepp M., Dini M. Bariopharmacoalumite, a new mineral species from Cap Garonne, France and Mina Grande, Chile, *Mineralog. Mag.*, **75**(1), 135–144 (2011c).

Mills S.J., Sejkora J., Kampf A.R., Grey I.E., Bastow T.J., Ball N.A., Adams P.M., Raudsepp M., Cooper M.A. Krásnoite, IMA 2011-040. CNMNC Newsletter No. 10, October 2011, page 2557, *Mineralog. Mag.*, **75**(5), 2549–2561 (2011d).

Mills S.J., Sejkora J., Kampf A.R., Grey I.E., Bastow T.J., Ball N.A., Adams P.M., Raudsepp M., Cooper M.A. Krásnoite, the fluorophosphate analogue of perhamite, from the Huber open pit, Czech Republic and the Silver Coin mine, Nevada, USA, *Mineralog. Mag.*, **76**(3), 625–634 (2012d).

Moore P.B., Araki T. $Ba(Mn,Fe)^{2+}{}_2Al_2(OH)_3[PO_4]_3$: Its atomic arrangement, *Am. Mineralog.*, **59**(5–6), 567–572 (1974a).

Moore P.B., Araki T. Hematolite: A complex dense-packed sheet structure, *Am. Mineralog.*, **63**(1), 150–159 (1978).

Moore P.B., Araki T., Merlino S., Mellini M., Zanazzi P.F. The arrojadite-dickinsonite series, $KNa_4Ca(Fe,Mn)^{2+}{}_{14}Al(OH)_2(PO_4)_{12}$: Crystal structure and crystal chemistry, *Am. Mineralog.*, **66**(9–10), 1034–1049 (1981).

Moore P.B., Araki T. Montgomeryite, $Ca_4Mg(H_2O)_{12}[Al_4(OH)_4(PO_4)_6]$: Its crystal structure and relation to vauxite, $Fe^{2+}{}_2(H_2O)_4[Al_4(OH)_4(H_2O)_4(PO_4)_4]\cdot 4H_2O$, *Am. Mineralog.*, **59**(7–8), 843–850 (1974b).

Moore P.B., Araki T. Overite, segelerite, and jahnsite: A study in combinatorial polymorphism, *Am. Mineralog.*, **62**(7–8), 692–702 (1977a).

Moore P.B., Araki T. Palermoite, $SrLi_2[Al_4(OH)_4(PO_4)_4]$: Its atomic arrangement and relationship to carminite, $Pb_2[Fe_4(OH)_4(AsO_4)_4]$, *Am. Mineralog.*, **60**(5–6), Pt. 1, 460–465 (1975a).

Moore P.B., Araki T. Samuelsonite: Its crystal structure and relation to apatite and octacalcium phosphate, *Am. Mineralog.*, **62**(3–4), 229–245 (1977b).

Moore P.B. Cell data of orientite and its relation to ardennite and zoisite, *Canad. Mineralog.*, **8**(1), 262–265 (1965).

Moore P.B. Derivative structures based on the alunite octahedral sheet: Mitridatite and englishite, *Mineralog. Mag.*, **40**(316), 863–866 (1976).

Moore P.B., Ghose S. A novel face-sharing octahedral trimer in the crystal structure of seamanite, *Am. Mineralog.*, **56**(9–10), 1527–1538 (1971).

Moore P.B., Ito J. I. Jahnsite, segelerite, and robertsite, three new transition metal phosphate species. II. Redefinition of overite, an isotype of segelerite. III. Isotypy of robertsite, mitridatite, and arseniosiderite, *Am. Mineralog.*, **59**(1–2), 48–59 (1974).

Moore P.B., Ito J. Kraisslite, a new platy arsenosilicate from Sterling Hill, New Jersey, *Am. Mineralog.*, **63**(9–10), 938–940 (1978a).

Moore P.B., Ito J.I. Whiteite, a new species, and a proposed nomenclature for the jahnsite–whiteite complex series. II. New data on xanthoxenite. III. Salmonsite discredited, *Mineralog. Mag.*, **42**(323), 309–323 (1978b).

Moore P.B., Ito J. Wyllieite, $Na_2Fe^{2+}{}_2Al(PO_4)_3$: A new species, *Mineralog. Rec.*, **4**(3), 131–136 (1973).

Moore P.B., Irving A.J., Kampf A.R. Foggite, $CaAl(OH)_2(H_2O)[PO_4]$; goedkenite, $(Sr,Ca)_2Al(OH)[PO_4]_2$; and samuelsonite, $(Ca,Ba)Fe^{2+}{}_2Mn^{2+}{}_2Ca_8Al_2(OH)_2[PO_4]_{10}$: Three new species from the Palermo No. I pegmatite, North Groton, New Hampshire, *Am. Mineralog.*, **60**(11–12), 957–964 (1975b).

Moore P.B., Kampf A.R., Araki T. Foggite, $Ca(H_2O)_2[CaAl_2(OH)_4(PO_4)_2]$: Its atomic arrangement and relationship to calcium Tschermak's pyroxene, *Am. Mineralog.*, **60**(11–12), 965–971 (1975c).

Moore P.B., Lund D.H., Keester K.L. Bjarebyite, $(Ba,Sr)(Mn,Fe,Mg)_2Al_2(OH)_3(PO_4)_3$, a new species, *Mineralog. Rec.*, **4**(6), 282–285 (1973).

Moore P.B., Molin-Case J. Contribution to pegmatite phosphate giant crystal paragenesis: II. The crystal chemistry of wyllieite, $Na_2Fe^{2+}{}_2Al[PO_4]_3$, a primary phase, *Am. Mineralog.*, **59**(3–4), 280–290 (1974).

Moore P.B., Shen J. An X-ray structural study of cacoxenite, a mineral phosphate, *Nature*, **306**(5941), 356–358 (1983).

Moram M.A., Zhang Y., Joyce T.B., Holec D., Chalker P.R., Mayrhofer P.H., Kappers M.J., Humphreys C.J. Structural properties of wurtzitelike ScGaN films grown by NH_3-molecular beam epitaxy, *J. Appl. Phys.*, **106**(11), 113533_1–113533_6 (2009).

Moreno-Armenta M.G., Mancera L., Takeuchi N. First principles total energy calculations of the structural and electronic properties of $Sc_xGa_{1-x}N$, *Phys. status solidi (b)*, **238**(1), 127–135 (2003).

Morgan K.R., Gainsford G.J., Milestone N.B. A new type of layered aluminium phosphate $[NH_4]_3[Co(NH_3)_6]_3[Al_2(PO_4)_4]_2$ assembled about a cobalt(III) hexammine complex, *Chem. Commun.*, (1), 61–62 (1997).

Moss A.A., Fejer E.E., Embrey P.G. On the X-ray identification of amblygonite and montebrasite, *Mineralog. Mag.*, **37**(287), 414–422 (1969).

Mrose M.E. Palermoite and goyazite, two strontium minerals from the Palermo Mine, North Groton, New Hampshire, *Am. Mineralog.*, **38**(3–4), 354 (1953).

Murashova E.V., Chudinova N.N. Condensed cesium indium phosphates, *Inorg. Mater.*, **37**(12), 1521–1524 (2001).

Nagashima M., Armbruster T. Ardennite, tiragalloite and medaite: Structural control of $(As^{5+},V^{5+},Si^{4+})O_4$ tetrahedra in silicates, *Mineralog. Mag.*, **74**(1), 55–71 (2010).

Naik I.K., Tien T.Y. Subsolidus phase relations in part of the system Si,Al,Y/N,O, *J. Am. Ceram. Soc.*, **62**(9–10), 642–643 (1979).

Nakajima A., Furukawa Y., Yokoya H., Yonezu H. Growth of $B_xAl_{1-x}N$ layers using decaborane on SiC substrates, *J. Cryst. Growth*, **278**(1–4), 437–442 (2005).

Nickel E.H., Temperly J.E. Arsenoflorencite-(Ce): A new arsenate mineral from Australia, *Mineralog. Mag.*, **51**(362), 605–609 (1987).

Noreika A.J., Takei W.J., Francombe M.H., Wood C.E.C. Indium antimonide-bismuth compositions grown by molecular beam epitaxy, *J. Appl. Phys.*, **53**(7), 4932–4937 (1982).

Norrish K., Rogers L.E.R., Shapter R.E. Kingite, a new hydrated aluminum phosphate mineral from Robertstown, South Australia, *Mineralog. Mag.*, **31**(236), 351–357 (1957).

Novikova E.M., Vasil'ev M.G., Evseev V.A., Yershova S.A. Phase diagram of the Sn–GaAs–ZnSe quasiternary system from the Sn-rich side [in Russian], *Deposited in VINITI*, № 3039–74Dep (1974).

Oeckler O., Kechele J.A., Koss H., Schmidt P.J., Schnick W. $Sr_5Al_{5+x}Si_{21-x}N_{35-x}O_{2+x}:Eu^{2+}$ ($x \approx 0$) – a novel green phosphor for white-light pcLEDs with disordered intergrowth structure, *Chem. Eur. J.*, **15**(21), 5311–5319 (2009).

Oe K. Characteristics of semiconductor alloy $GaAs_{1-x}Bi_x$, *Jpn. J. Appl. Phys.*, **41**(5A), Pt. 1, 2801–2806 (2002).

Okamoto H., Oe K. Growth of metastable alloy InAsBi by low-pressure MOVPE, *Jpn. J. Appl. Phys.*, **37**(3B), Pt. 1, 1608–1613 (1998).

Okamoto H., Oe K. Structural and energy-gap characterization of metalorganic-vapor-phase-epitaxy-grown InAsBi, *Jpn. J. Appl. Phys.*, **38**(2B), Pt. 1, 1022–1025 (1999).

Onabe K. Unstable regions in III-V quaternary solid solutions composition plane calculated with strictly regular solution approximation, *Jpn. J. Appl. Phys.*, **21**(6), Pt. 2, L323–L325 (1982).

Onac B.P., Ettinger K., Kearns J., Balasz I.I. A modern, guano-related occurrence of foggite, $CaAl(PO_4)(OH)_2{\cdot}H_2O$ and churchite-(Y), $YPO_4{\cdot}2H_2O$ in Cioclovina Cave, Romania, *Mineral. Petrol.*, **85**(3–4), 291–302 (2005).

Orlandi P., Biagioni C., Pasero M., Mellini M. Lavoisierite, IMA 2012-009. CNMNC Newsletter No. 13, June 2012, page 815, *Mineralog. Mag.*, **76**(3), 807–817 (2012).

Orlandi P., Biagioni C., Pasero M., Mellini M. Lavoisierite, $Mn^{2+}{}_8[Al_{10}(Mn^{3+}Mg)][Si_{11}P]O_{44}(OH)_{12}$, a new mineral from Piedmont, Italy: The link between "ardennite" and sursassite, *Phys. Chem. Minerals*, **40**(3), 239–249 (2013).

Orsal G., Maloufi N., Gautier S., Alnot M., Sirenko A.A., Bouchaour M., Ougazzaden A. Effect of boron incorporation on growth behaviour of BGaN/GaN by MOVPE, *J. Cryst. Growth*, **310**(23), 5058–5062 (2008).

Orth M., Hoffmann R.-D., Pöttgen R., Schnick W. Orthonitridoborate ions $[BN_3]^{6-}$ in oxonitridosilicate cages: Synthesis, crystal structure, and magnetic properties of $Ba_4Pr_7[Si_{12}N_{23}O][BN_3]$, $Ba_4Nd_7[Si_{12}N_{23}O][BN_3]$, and $Ba_4Sm_7[Si_{12}N_{23}O][BN_3]$, *Chem. Eur. J.*, **7**(13), 2791–2797 (2001).

Ouerfelli N., Guesmi A., Mazza D., Madani A., Zid M.F., Driss A. Synthesis, crystal structure and mono-dimensional thallium ion conduction of $TlFe_{0.22}Al_{0.78}As_2O_7$, *J. Solid State Chem.*, **180**(4), 1224–1229 (2007).

Ougazzaden A., Gautier S., Moudakir T., Djebbour Z., Lochner Z., Choi S., Kim H.J., Ryou J.-H., Dupuis R.D., Sirenko A.A. Bandgap bowing in BGaN thin films, *Appl. Phys. Lett.*, **93**(8), 083118_1–083118_3 (2008).

Ougazzaden A., Gautier S., Sartel C., Maloufi N., Martin J., Jomar F. BGaN materials on GaN/sapphire substrate by MOVPE using N_2 carrier gas, *J. Cryst. Growth*, **298**, 316–319 (2007).

Overweg A.R., Haan de J.W., Magusin P.C.M.M., Santen van R.A., Sankar G., Thomas J.M. Synthesis and characterization of some heterometal-substituted ammonium gallophosphates, *Chem. Mater.*, **11**(7), 1680–1686 (1999).

Owens J.P., Altschuler Z.S., Berman R. Millisite in phosphorite from Homeland, Florida, *Am. Mineralog.*, **45**(5–6), 547–561 (1960).

Pabst A. Schoderite, a new locality and a redescription, *Am. Mineralog.*, **64**(7–8), 713–720 (1979).

Pabst A. Some computations on svanbergite, woodhouseite and alunite, *Am. Mineralog.*, **32**(1–2), 16–30 (1947).

Pajunen A., Lahti S.I. The crystal structure of viitaniemiite, *Am. Mineralog.*, **69**(9–10), 961–966 (1984).

Palache C., Bauer L.H. Cahnite, a new boro-arsenate of calcium from Franklin, New Jersey, *Am. Mineralog.*, **12**(4), 149–153 (1927).

Palache C., Richmond W.E., Wolfe C.W. On amblygonite, *Am. Mineralog.*, **28**(1), 39–53 (1943).

Pamplin B.R., Hasoon F.S. The system InAs–$ZnGeAs_2$–$CuInSe_2$–Cu_2GeSe_3, *Progr. Cryst. Growth Charact.*, **10**, 213–215 (1985).

Pamplin B.R., Shah J.S. Quinary adamantine semiconductors, *Nature*, **207**(4993), 180–181 (1965).

Pan W., Wu L., Jiang Y., Huang Z. Solid solution and phase relations in SiC–AlN–Pr_2O_3 system, *J. Phase Equilib. Dif.*, **38**(5), 676–683 (2017).

Panz C., Polborn K., Behrens P. LMU-3: A new cobalt aluminophosphate with exclusively five-coordinated aluminium and octahedrally coordinated cobalt, *Inorg. Chim. Acta*, **269**(1), 73–82 (1998).

Pâques-Ledent M.T., Tarte P. Spectre infa-rouge et structure des composés $GaPO_4{\cdot}2H_2O$ et $GaAsO_4{\cdot}2H_2O$, *Spectrochim. Acta, Part A*, **25**(6), 1115–1125 (1969).

Parise J.B. Preparation and structure of the aluminium ammonium phosphate dihydrate $Al_2[NH_4](OH)(PO_4){\cdot}2.2H_2O$: A tunnel structure with ammonium ions in the channels, *Acta Crystallogr.*, **C40**(10), 1641–1643 (1984).

Park W.B. The $SrLiAl_3N_4{:}Eu^{2+}$ phosphor synthesized by the raw material model obtained by DFT calculations, *J. Kor. Ceram. Soc.*, **54**(3), 217–221 (2017).

Peacor D.R., Dunn P.J., Roberts W.L., Campbell T.J., Simmons W.B. Sinkankasite, a new phosphate from the Barker pegmatite, South Dakota, *Am. Mineralog.*, **69**(3–4), 380–382 (1984a).

Peacor D.R., Dunn P.J., Simmons W.B. Paulkerrite, a new titanium phosphate from Arizona, *Mineral. Record*, **15**(5), 303–306 (1984b).

Pekov I.V., Kleimenov D.A., Chukanov N.V., Yakubovich O.V., Massa W., Belakovskiy D.I., Pautov L.A. Bushmakinite $Pb_2Al(PO_4)(VO_4)(OH)$, a new mineral of the brackebuschite group from the oxidized zone of the Berezovskoye gold deposit, the middle Urals [in Russian], *Zap. Vseros. mineralog. obshch.*, **131**(2), 62–71 (2002).

Pekov I.V., Zubkova N.V., Belakovskiy D.I., Yapaskurt V.O., Vigasina M.F., Sidorov E.G., Pushcharovsky D.Y. New arsenate minerals from the Arsenatnaya fumarole, Tolbachik volcano, Kamchatka, Russia. IV. Shchurovskyite, $K_2CaCu_6O_2(AsO_4)_4$ and dmisokolovite, $K_3Cu_5AlO_2(AsO_4)_4$, *Mineralog. Mag.*, **79**(7), 1737–1753 (2015).

Pekov I.V., Zubkova N.V., Belakovskiy D.I., Yapaskurt V.O., Vigasina M.F., Sidorov E.G., Pushcharovsky D.Y. Dmisokolovite, IMA 2013-079. CNMNC Newsletter No. 18, December 2013, page 3253, *Mineralog. Mag.*, **77**(8), 3249–3258 (2013).

Pekov I.V., Zubkova N.V., Koshlyakova N.N., Belakovskiy D.I., Vigasina M.F., Agakhanov A.A., Britvin S.N., Turchkova A.G., Sidorov E.G., Pushcharovsky D.Y. Achyrophanite, IMA 2018-011. CNMNC Newsletter No 43, June 2018, page 650, *Eur. J. Mineral.*, **30**(3), 647–652 (2018a).

Pekov I.V., Zubkova N.V., Koshlyakova N.N., Belakovskiy D.I., Vigasina M.F., Agakhanov A.A., Britvin S.N., Turchkova A.G., Sidoro, E.G., Pushcharovsky D.Y. Achyrophanite, IMA 2018-011. CNMNC Newsletter No 43, June 2018, page 783, *Mineralog. Mag.*, **82**(3), 779–785 (2018b).

Peng L., Li J., Yu J., Li G., Fang Q., Xu R. Solvothermal synthesis, structure and Mössbauer spectroscopy of a new mixed-valence iron aluminophosphate $[Fe^{II}(H_2O)_2Fe_{0.8}{}^{III}Al_{1.2}(PO_4)_3]$. H_3O, *C. R. Chimie*, **8**(3–4), 541–547 (2005).

Pfeiff R., Kniep R. Darstellung von quaternären Selenodiphosphaten(IV) aus Halogenidschmelzen: Die Kristallstruktur des $CuAl[P_2Se_6]$, *Z. Naturforsch.*, **48B**(9), 1270–1274 (1993).

Piao X., Machida K.-I., Horikawa T., Hanzawa H., Shimomura Y., Kijima N. Preparation of $CaAlSiN_3{:}Eu^{2+}$ phosphors by the self-propagating high-temperature synthesis and their luminescent properties, *Chem. Mater.*, **19**(18), 4592–4599 (2007).

Piilonen P.C., Rowe R., Poirier G., Tait K.T. New mineral names, *Am. Mineralog.*, **96**(10), 1654–1661 (2011).

Pilkington E.S., Segnit E. R., Watts J., Francis G. Kleemanite, a new zinc aluminium phosphate, *Mineralog. Mag.*, **43**(325), 93–95 (1979).

Pilkington E.S., Segnit E.R., Watts J.A. Peisleyite, a new sodium aluminium sulphate phosphate, *Mineralog. Mag.*, **46**(341), 449–452 (1982).

Piret P., Declercq J.-P., Wauters-Stoop D. Structure of threadgoldite, *Acta Crystallogr.*, **B35**(12), 3017–3020 (1979a).

Piret P., Declercq J.-P. Structure cristalline de l'upalite $Al[(UO_2)_3O(OH)(PO_4)_2]{\cdot}7H_2O$. Un exemple de macle mimétique, *Bull. Minéral.*, **106**, 383–389 (1983).

Piret P., Deliens M. Les phosphates d'uranyle et d'aluminium de Kobokobo. IX. L'althupite $AlTh(UO_2)[(UO_2)_3O(OH)(PO_4)_2]_2(OH)_3{\cdot}15H_2O$, nouveau minéral; propriétés et structure cristalline, *Bull. Minéral.*, **110**(1), 65–72 (1987).

Piret P., Piret-Meunier J., Declercq J.-P. Structure of phuralumite, *Acta Crystallogr.*, **B35**(8), 1880–1882 (1979b).

Pluth J.J., Smith J.V., Bennett J.M., Cohen J.P. Structure of $NH_4Al_2(OH)(H_2O)(PO_4)_2{\cdot}H_2O$, the ammonium-aluminum analog of $GaPO_4{\cdot}2H_2O$ and leucophosphite, *Acta Crystallogr.*, **C40**(12), 2008–2011 (1984).

Pogrebnjak A., Rogoz V., Ivashchenko V., Bondar O., Shevchenko V., Jurga S., Coy E. Nanocomposite Nb–Al–N coatings: Experimental and theoretical principles of phase transformations, *J. Alloys Compd.*, **718**, 260–269 (2017).

Poirier G., Tait K.T., Rowe R. New mineral names, *Am. Mineralog.*, **96**(5–6), 936–945 (2011).

Portnova N.L., Kononova G.N., Soklakov A.I. $2MgO{\cdot}P_2O_5{\cdot}B_2O_3{\cdot}7H_2O$, a new magnesium borate phosphate [in Russian], *Zhurn. neorgan. khimii*, **14**(2), 344–347 (1969).

Potoriy M.V., Motrya S.F., Pritz I.P., Milyan P.M. Physical-chemical interaction in the system $AgInS_2$–"P_2S_4" and growing crystals $AgInP_2S_6$ [in Ukrainian], *Nauk. visnyk Volyns'k. nats. un-tu im. Lesi Ukrayinky*, (24), 20–22 (2009).

Potoriy M.V., Pritz I.P., Motrya S.F., Milyan P.M., Mikaylo O.A. Formation pattern and growing of $CuInP_2S_6$ single crystals, *Nauk. visnyk Volyns'k. nats. un-tu im. Lesi Ukrayinky*, (30), 51–55 (2010).

Pough F.H., Henderson E.P. Brazilianite, a new phosphate mineral, *Am. Mineralog.*, **30**(9–10), 572–582 (1945).

Pouliot G., Hofmann H.J. Florencite: A first occurrence in Canada, *Canad. Mineralog.*, **19**(4), 535–540 (1981).

Prewitt C.T., Buerger M.J. The crystal structure of cahnite, $Ca_2BAsO_4(OH)_4$, *Am. Mineralog.*, **46**(9–10), 1077–1085 (1961).

Pryce M.W. Holtite: A new mineral allied to dumortierite, *Mineralog. Mag.*, **38**(293), 21–25 (1971).

Pust P., Weiler V., Hecht C., Tücks A., Wochnik A.S., Henß A.-K., Wiechert D., Scheu C., Schmidt P.J., Schnick W. Narrow-band red-emitting $Sr[LiAl_3N_4]{:}Eu^{2+}$ as a next-generation LED-phosphor material, *Nature Mater.*, **13**(9), 891–896 (2014).

Raade G., Grice J.D., Rowe R. Ferrivauxite, a new phosphate mineral from Llallagua, Bolivia, *Mineralog. Mag.*, **80**(2), 311–324 (2016).

Raade G., Grice J., Rowe R. Ferrivauxite, IMA 2014-003. CNMNC Newsletter No. 20, June 2014, page 555, *Mineralog. Mag.*, **78**(3), 549–558 (2014).

Raghavan V. Al–C–Cr–Fe–Mn–Mo–N–Ni–Si–V, *J. Phase Equilibr. Dif.*, **28**(3), 297 (2007).

Ramik R.A., Sturman B.D., Dunn P.J., Povarennykh A.S. Tancoite, a new lithium sodium aluminum phosphate from the Tanco Pegmatite, Bernic Lake, Manitoba, *Canad. Mineralog.*, **18**(2), 185–190 (1980).

Ramik R.A., Sturman B.D., Roberts A.C., Dunn P.J. Viitaniemiite from the Francon Quarry, Montreal, Quebec, *Canad. Mineralog.*, **21**(3), 499–502 (1983).

Reckeweg O., Schulz A., DiSalvo F.J. One more compound containing the CBN^{4-} ion – synthesis, single-crystal structure and vibrational spectra of $Ca_{15}(CBN)_6(C_2)_2F_2$, *Z. Naturforsch.*, **65B**(12), 1409–1415 (2010).

Repina S.A., Popova V.I., Churin E.I., Belogub E.V., Khiller V.V. Florencite-(Sm) – (Sm,Nd) $Al_3(PO_4)_2(OH)_6$: A new mineral species of the alunite-jarosite group from the Subpolar Urals, *Geol. Ore Deposits*, **53**(7), 564–574 (2011).

Richardson J.M., Roberts A.C., Grice J.D., Ramik R.A. Mcauslanite, a supergene hydrated iron aluminum fluorophosphate from the East Kemptville tin mine, Yarmouth County, Nova Scotia, *Canad. Mineralog.*, **26**(4), 917–921 (1988).

Roberts A.C., Cooper M.A., Hawthorne F.C., Gault R.A., Jensen M.C., Foord E.E. Goldquarryite, a new Cd-bearing phosphate mineral from the Gold Quarry mine, Eureka County, Nevada, *Mineralog. Rec.*, **34**(3), 237–240 (2003).

Rochère de la M., Kahn A., d'Yvoire F., Bretey E. Crystal structure and cation transport properties of the ortho-diphosphates $Na_7(MP_2O_7)_4PO_4$ (M = Al, Cr, Fe), *Mater. Res. Bull.*, **20**(1), 27–34 (1985).

Rocherulle J., Guyader J., Verdier P., Laurent Y. Li–Si–Al–O–N and Li–Si–O–N oxynitride glasses study and characterization, *J. Mater. Sci.*, **24**(12), 4525–4530 (1989a).

Rocherulle J., Verdier P., Laurent Y. Preparation and properties of gadolinium oxide and oxynitride glasses, *Mater. Sci. Eng. B*, **2**(4), 265–268 (1989b).

Rodgers K.A., Greatrex R., Hyland M., Simmons S.F., Browne P.R.L. A modern, evaporitic occurrence of teruggite, $Ca_4MgB_{12}As_2O_{28}{\cdot}18H_2O$, and nobleite, $CaB_6O_{10}{\cdot}4H_2O$, from the El Tatio geothermal field, Antofagasta Province, Chile, *Mineralog. Mag.*, **66**(2), 253–259 (2002).

Rondeau B., Devouard B., Jacob D., Roussel P., Stéphant N., Boulet C., Mollé V., Corre M., Fritsch E., Ferraris C., Parodi G.C. Lasnierite, IMA 2017-084. CNMNC Newsletter No. 42, April 2018, page 405, *Eur. J. Mineral.*, **30**(2), 403–408 (2018a).

Rondeau B., Devouard B., Jacob D., Roussel P., Stéphant N., Boulet C., Mollé V., Corre M., Fritsch E., Ferraris C., Parodi G.C. Lasnierite, IMA 2017-084. CNMNC Newsletter No. 42, April 2018, page 447, *Mineralog. Mag.*, **82**(2), 445–451 (2018b).

Ruan J., Xie R.-J., Hirosaki N., Takeda T. Nitrogen gas pressure synthesis and photoluminescent properties of orange-red $SrAlSi_4N_7{:}Eu^{2+}$ phosphors for white light-emitting diodes, *J. Am. Ceram. Soc.*, **94**(2), 536–542 (2011).

Rumsey M.S., Mills S.J., Spratt J. Natropharmacoalumite, $NaAl_4[(OH)_4(AsO_4)_3]{\cdot}4H_2O$, a new mineral of the pharmacosiderite supergroup and the renaming of aluminopharmacosiderite to pharmacoalumite, *Mineralog. Mag.*, **74**(5), 929–936 (2010).

Russel A. On rashleighite, a new mineral from Cornwall, intermediate between turquoise and chalcosiderite, *Mineralog. Mag.*, **28**(202), 353–358 (1948).

Sabelli C. The crystal structure of chalcophyllite, *Z. Kristallogr.*, **151**(1–2), 129–140 (1980).

Sakae T., Sudo T. Taranakite from the Onino-Iwaya Limestone Cave at Hiroshima Prefecture, Japan: A new occurrence, *Am. Mineralog.*, **60**(3–4), 331–334 (1975).

Sarp H., Bertrand J., Deferne J. Présence de goudeyite, $Cu_6Al(AsO_4)_2(OH)_6{\cdot}3H_2O$, dans des échantillons provenant de l'ancienne mine de Chessy (Rhône, France), *Schweiz. Mineral. Petrog. Mitt.*, **61**, 173–176 (1981).

Sarp H., Cerny R. Barrotite, IMA 2011-061a. CNMNC Newsletter No. 16, August 2013, page 2698, *Mineralog. Mag.*, **77**(6), 2695–2709 (2013).

Sarp H., Cerny R., Pushcharovsky D.Y., Schouwink P., Teyssier J., Williams P.A., Babalik H., Mari G. La barrotite, $Cu_9Al(HSiO_4)_2[(SO_4)(HAsO_4)_{0.5}](OH)_{12}{\cdot}8H_2O$, un nouveau minéral de la mine de Roua (Alpes-Maritimes, France), *Riviéra Sci.*, **98**, 3–22 (2014).

Schäfer G., Borrmann H., Kniep R. Synthesis and crystal structure of $NH_4[(Zn_{1-x}Co_x)BP_2O_8]$ ($0 \leq x \leq 0.14$), a metallo-borophosphate analogue of the zeolite gismondine, *Micropor. Mesapor. Mater.*, **41**(1–3), 161–167 (2000).

Schäfer G., Carrillo-Cabrera W., Schnelle W., Borrmann H., Kniep R. Synthesis and crystal structure of $\{(NH_4)_xCo_{((3-x)/2)}\}(H_2O)_2[BP_2O_8]{\cdot}(1-x)H_2O$ ($x \approx 0.5$), *Z. anorg. und allg. Chem.*, **628**(1), 289–294 (2002).

Schaller W.T. Crystallized turquoise from Virginia, *Am. J. Sci.*, **33**(193), 35–40 (1912).

Schlüter J., Klaska K.-H., Friese K., Adiwidjaja G. Kastningite, $(Mn,Fe,Mg)Al_2(PO_4)(OH)_2{\cdot}8H_2O$, a new phosphate mineral from Waidhaus, Bavaria, Germany, *Neues Jahrb. Mineral., Monatsh.*, 40–48 (1999).

Schlüter J., Malcherek T. Galloplumbogummite, IMA 2010-088. CNMNC Newsletter No. 9, August 2011, page 2537, *Mineralog. Mag.*, **75**(4), 2535–2540 (2011).

Schmid-Beurmann P., Morteani G., Cemič L. Experimental determination of the upper stability of scorzalite, $FeAl_2[OH/PO_4]_2$, and the occurrence of minerals with a composition intermediate between scorzalite and lazulite(ss) up to the conditions of the amphibolite facies, *Mineral. Petrol.*, **61**(1–4), 211–222 (1997).

Schnick W., Huppertz H., Lauterbach R. High temperature syntheses of novel nitride- and oxonotrido-silicates and sialons using rf furnaces, *J. Mater. Chem.*, **9**(1), 289–296 (1999).

Schölch J., Dierkes T., Enseling D., Ströbele M., Jüstel M., Meyer H.-J. Synthesis and photoluminescence properties of the red-emitting phosphor $Mg_3(BN_2)N$ doped with Eu^{2+}, *Z. anorg. allg. Chem.*, **641**(5), 803–808 (2015).

Schumer B.N., Yang H., Downs R.T. Natropalermoite, IMA 2013-118. CNMNC Newsletter No. 19, February 2014, page 170, *Mineralog. Mag.*, **78**(1), 165–170 (2014).
Schumer B.N., Yang H., Downs R.T. Natropalermoite, $Na_2SrAl_4(PO_4)_4(OH)_4$, a new mineral isostructural with palermoite, from the Palermo No. 1 mine, Groton, New Hampshire, USA, *Mineralog. Mag.*, **81**(4), 833–840 (2017).
Schwab R.G., Herold H., Götz C., Pinto de Oliveira N. Compounds of the crandallite type: Synthesis and properties of pure goyazite, gorceixite and plumbogummite, *Neues Jahrb. Mineral., Monatsh.*, (3), 113–126 (1990).
Schwendtner K., Kolitsch U. $CsAl(H_2AsO_4)_2(HAsO_4)$: A new monoclinic protonated arsenate with decorated kröhnkite-like chains, *Acta Crystallogr.*, **C63**(3), i17–i20 (2007).
Schwendtner K., Kolitsch U. $CsGa(H_{1.5}AsO_4)_2(H_2AsO_4)$ and isotypic $CsCr(H_{1.5}AsO_4)_2(H_2AsO_4)$: Decorated kröhnkite-like chains in two unusual hydrogen arsenates, *Acta Crystallogr.*, **C61**(9), i90–i93 (2005).
Schwendtner K. $TlInAs_2O_7$, $RbInAs_2O_7$, and $(NH_4)InAs_2O_7$: Synthesis and crystal structures of three isotypic microporous diarsenates – representatives of a novel structure type, *J. Alloys Compd.*, **421**(1–2), 57–63 (2006).
Sejkora J., Grey I.E., Kampf A.R., Price J.R., Čejka J. Tvrdýite, $Fe^{2+}Fe^{3+}{}_2Al_3(PO_4)_4(OH)_5(OH_2)_4 \cdot 2H_2O$, a new phosphate mineral from Krásno near Horní Slavkov, Czech Republic, *Mineralog. Mag.*, **80**(6), 1077–1088 (2016).
Sejkora J., Grey I.E., Kampf A.R., Price J.R. Tvrdýite, IMA 2014-082. CNMNC Newsletter No. 23, February 2015, page 57, *Mineralog. Mag.*, **79**(1), 51–58 (2015).
Selway J.B., Cooper M.A., Hawthorne F.C. Refinement of the crystal structure of burangaite, *Canad. Mineralog.*, **35**(6), 1515–1522 (1997).
Shablinskii A.P., Vergasova L.P., Filatov S.K., Avdontseva E.Y., Moskaleva S.V. Ozerovaite, IMA 2016-019. CNMNC Newsletter No. 32, August 2016, page 917, *Mineralog. Mag.*, **80**(5), 915–922 (2016).
Shen B., Akiba M., Inoue A. Effect of Cr addition on the glass-forming ability, magnetic properties, and corrosion resistance in FeMoGaPCBSi bulk glassy alloys, *J. Appl. Phys.*, **100**(4), 043523_1–043523_5 (2006).
Shen B., Inoue A. Bulk glassy Fe–Ga–P–C–B–Si alloys with high glass-forming ability, high saturation magnetization and good soft magnetic properties, *Mater. Trans.*, **43**(5), 1235–1239 (2002).
Shen B., Kimura H., Inoue A., Mizushima T. Bulk glassy Fe–Co–Ga–P–C–B alloys with high glass-forming ability, high saturation magnetization and good soft magnetic properties, *Mater. Trans., JIM*, **41**(12), 1675–1678 (2000a).
Shen B., Kimura H., Inoue A., Mizushima T. Bulk glassy $Fe_{78-x}Co_xGa_2P_{12}C_4B_4$ alloys with high saturation magnetization and good soft magnetic properties, *Mater. Trans.*, **42**(6), 1052–1055 (2001).
Shen B., Koshiba H., Kimura H., Inoue A. High strength and good soft magnetic properties of bulk glassy Fe–Mo–Ga–P–C–B alloys with high glassforming ability, *Mater. Trans., JIM*, **41**(11) 1478–1481 (2000b).
Shen B.-L., Koshiba H., Mizushima T., Inoue A. Bulk amorphous Fe-Ga-P-B-C alloys with a large supercooled liquid region, *Mater. Trans., JIM*, **41**(7), 873–876 (2000c).
Shen C., Mayorga S.G., Biagioni R., Piskoti C., Ishigami M., Zettl A., Bartlett N. Intercalation of hexagonal boron nitride by strong oxidizers and evidence for the metallic nature of the products, *J. Solid State Chem.*, **147**(1), 74–81 (1999a).
Shen J., Peng Z. The crystal structure of furongite, *Acta Crystallogr*, **A37**(Suppl.), C186 (1981).
Shen T.D., Schwarz R.B. Bulk ferromagnetic glasses prepared by flux melting and water quenching, *Appl. Phys. Lett.*, **75**(1), 49–51 (1999).
Shen Z., Grins J., Esmaeilzadeh S., Ehrenberg H. Preparation and crystal structure of a new Sr containing sialon phase $Sr_2Al_xSi_{12-x}N_{16-x}O_{2+x}$ ($x \approx 2$), *J. Mater. Chem.*, **9**(4), 1019–1022 (1999b).

Shen Z., Nygren M., Wang P., Feng J. Eu-doped α-sialon and related phases, *J. Mater. Sci. Lett.*, **17**(20), 1703–1706 (1998).

Shi H., Feng Y., Huang Q., Qiu D., Li M., Liu K. Ti[$BP_2O_7(OH)_3$]: The first titanium borophosphate containing novel anionic partial structure, *CrystEngComm*, **13**(24), 7185–7188 (2011).

Shi H., Li M., Tangbo H., Kong A., Chen B., Shan Y. A novel open-framework copper borophosphate: $Cu(H_2O)_2[B_2P_2O_8(OH)_2]$, *Inorg. Chem.*, **44**(23), 8179–8181 (2005).

Shi H.-Z., Chang J.-Z., Tang-Bo H.-J., Ding H.-M., Shan Y.-K. Synthesis, structure, thermal and magnetic properties of a new open-framework borophosphate: $NH_4Mn(H_2O)_2BP_2O_8{\cdot}H_2O$, *Chin. J. Chem.*, **24**(9), 1255–1258 (2006).

Shi H.-Z., Dan Y.-K., Dai L.-Y., Liu Z.-Y., Weng L.-H. Crystal structure of borophosphate with 62 screw axis helices, *Chin. J. Struct. Chem.*, **22**(4), 391–394 (2003a).

Shi H.-Z., Shan Y.-K., He M.-Y., Liu Y.-Y. Crystal structure of diaquamagnesium (dihydrogen-monoboratemonophosphate), $Mg[BPO_4(OH)_2](H_2O)_2$, *containing isolated six-membered rings of tetrahedra, Z. Kristallogr., New Cryst. Str.*, **218**(1), 21–22 (2003b).

Shi H.-Z., Shan Y.-K., He M.-Y., Liu Y.-Y., Weng L.-H. Synthesis and structure of the first zincoborophosphate: $(H_3O)Zn(H_2O)_2BP_2O_8{\cdot}H_2O$, *Chin. J. Chem.*, **21**(9), 1170–1173 (2003c).

Shioi K., Michiue Y., Hirosaki N., Xie R.-J., Takeda T., Matsushita Y., Tanaka M., Li Y.Q. Synthesis and photoluminescence of a novel Sr-SiAlON:Eu^{2+} blue-green phosphor ($Sr_{14}Si_{68-s}Al_{6+s}O_s N_{106-s}:Eu^{2+}$ ($s \approx 7$)), *J. Alloys Compd.*, **509**(2), 332–337 (2011).

Shi Y., Pan S., Dong X., Wang Y., Zhang M., Zhang F., Zhou Z. $Na_3Cd_3B(PO_4)_4$: A new noncentrosymmetric borophosphate with zero-dimensional anion units, *Inorg. Chem.*, **51**(20), 10870–10875 (2012).

Sleight A.W., Bouchard R.J. A new cubic $KSbO_3$ derivative structure with interpenetrating networks. Crystal structure of $Bi_3GaSb_2O_{11}$, *Inorg. Chem.*, **12**(10), 2314–2316 (1973).

Smith J.P., Brown W.E. *X*-ray studies of aluminum and iron phosphates containing potassium or ammonium, *Am. Mineralog.*, **44**(1–2), 138–142 (1959).

Smith R.L., Simons F.S., Vlisidis A.C. Hidalgoite, a new mineral, *Am. Mineralog.*, **38**(11–12), 1218–1224 (1953).

Somer M., Carrillo-Cabrera W., Peters K., Schnering Von H.G. Crystal structure of trisodium tetraphosphidoaluminate, $Na_3(Sr_{2.88}Na_{0.12})[(Al_{0.94}Nb_{0.06})P_4]$, *Z. Kristallogr., New Cryst. Str.*, **213**(1), 8 (1998).

Spencer L.J., Bannister F.A., Hey M.H., Bennett H. Minyulite (hydrous K-Al fluophosphate) from South Australia. *Mineralog. Mag.*, **26**(181), 309–314 (1943).

Spielvogel B.F., Ahmed F.U., Das M.K., McPhail A.T. Synthesis of sodium and tetra-*n*-butylammonium dicyanodihydroborates, *Inorg. Chem.*, **23**(20), 3263–3265 (1984).

Sprenger J.A.P., Landmann J., Drisch M., Ignat'ev N., Finze M. Syntheses of tricyanofluoroborates $M[BF(CN)_3]$ (M = Na, K): $(CH_3)_3SiCl$ catalysis, countercation effect, and reaction intermediates, *Inorg. Chem.*, **54**(7), 3403–3412 (2015).

Stanley C.R. Hinsdalite and other products of oxidation at the Daisy Creek stratabound copper–silver prospect, northwestern Montana, *Canad. Mineralog.*, **25**(2), 213–220 (1987).

Stoica M., Eckert J., Roth S., Yavari A.R., Schultz L. $Fe_{65.5}Cr_4Mo_4Ga_4P_{12}C_5B_{5.5}$ BMGs: Sample preparation, thermal stability and mechanical properties, *J. Alloys Compd.*, **434–435**, 171–175 (2007).

Sturman B.D., Mandarino J.A., Mrose M.E., Dunn P.J. Gormanite, $Fe^{2+}{}_3Al_4(PO_4)_4(OH)_6{\cdot}2H_2O$, the ferrous analogue of souzalite, and new data for souzalite, *Canad. Mineralog.*, **19**(3), 381–387 (1981).

Stutz D., Greil P., Petzow G. Two-dimensional solid-solution formation of Y-containing α-Si_3N_4, *J. Mater. Sci. Lett.*, **5**(3), 335–336 (1986).

Sumiyoshi A., Hyodo H., Kimura K. Li-intercalation into hexagonal boron nitride, *J. Phys. Chem. Solids*, **71**(4), 569–571 (2010).

Sumiyoshi A., Hyodo H., Kimura K. Structural analysis of Li-intercalated hexagonal boron nitride, *J. Solid State Chem.*, **187**, 208–210 (2012).

Sumiyoshi A., Hyodo H., Sato Y., Terauchi M., Kimura K. Good reproductive preparation method of Li-intercalated hexagonal boron nitride and transmission electron microscopy – Electron energy loss spectroscopy analysis, *Solid State Sci.*, **47**, 68–72 (2015).

Sun W., Huang Z., Cao G., Yan D. Phase relationships of O′-β′ in Y–Si–Al–O–N system, *Sci. China. Ser. A.*, **31**(6), 742–747 (1988).

Sun W.-Y., Tien T.-Y., Yen T.-S. Solubility limits of α′-SiAlON solid solutions in the system Si,Al,Y/N,O, *J. Am. Ceram. Soc.*, **74**(10), 2547–2550 (1991a).

Sun W.-Y., Tien T.-Y., Yen T.-S. Subsolidus phase relationships in part of the system Si,Al,Y/N,O: The system Si_3N_4–AlN–YN–Al_2O_3–Y_2O_3, *J. Am. Ceram. Soc.*, **74**(11), 2753–2758 (1991b).

Sun W.Y., Yan D.S., Gao L., Mandal H., Liddel K., Thompson D.P. Subsolidus phase relationships in the systems Ln_2O_3–Si_3N_4–AlN–A_2O_3 (Ln = Nd, Sm), *J. Eur. Ceram. Soc.*, **15**(4), 349–355 (1995).

Sun W.Y., Yan D.S., Gao L., Mandal H., Thompson D.P. Subsolidus phase relationships in the system Dy_2O_3–Si_3N_4–AlN–A_2O_3, *J. Eur. Ceram. Soc.*, **16**(11), 1277–1282 (1996).

Sun W.-Y., Yan D.-S., Yen T.-S., Tian Z.-Y., Tien T.-Y. Graphical representation of the sub-solidus phase relationships in the reciprocal salt system Si,Al,Y/N,O, *Sci. China. Ser. A*, **35**(7), 877–888 (1992).

Su T., Xing H., Xu J., Yu J., Xu R. Ionothermal syntheses and characterizations of new open-framework metal borophosphates, *Inorg. Chem.*, **50**(3), 1073–1078 (2011).

Tait K.T., Ercit T.S., Abdu Y.A., Černý P., Hawthorn F.C. The crystal structure and crystal chemistry of manitobaite, ideally $(Na_{16}\square)Mn^{2+}{}_{25}Al_8(PO_4)_{30}$, from Cross Lake, Manitoba, *Canad. Mineralog.*, **49**(5), 1221–1242 (2011).

Tait K.T., Hawthorne F.C., Černý P., Galliski M.A. Bobfergusonite from the Nancy pegmatite, San Luis Range, Argentina: Crystal-structure refinement and chemical composition, *Canad. Mineralog.*, **42**(3), 705–716 (2004).

Takano T., Kurimoto M., Yamamoto J., Kawanishi H. Epitaxial growth of high quality BAlGaN quaternary lattice matched to AlN on 6H–SiC substrate by LP-MOVPE for deep-UV emission, *J. Cryst. Growth*, **237–239**, Pt. 2, 972–977 (2002).

Tang X.-J., Gentiletti M.J., Lachgar A. Synthesis and crystal structure of indium arsenate and phosphate dihydrates with variscite and metavariscite structure types, *J. Chem. Thermodyn.*, **31**(1), 45–50 (2001).

Tang X., Jones A., Lachgar A., Gross B.J., Yarger J.L. Synthesis, crystal structure, NMR studies, and thermal stability of mixed iron–indium phosphates with quasi-one-dimensional frameworks, *Inorg. Chem.*, **38**(26), 6032–6038 (1999).

Tang X., Lachgar A. Hydrothermal synthesis and crystal structure of $CaIn_2(PO_4)_2(HPO_4)$: An octahedral-tetrahedral framework ternary calcium indium(III) phosphate, *Z. anorg. allg. Chem.*, **622**(3), 513–517 (1996).

Tanh Jeazet H.B., Menezes P.W., Hoffmann S., Prots Y., Kniep R. Crystal structures of lead(II) cobalt(II) (monophosphate-hydrogenmonoborate-monophosphate), $PbCo[BP_2O_8(OH)]$, and lead(II) zinc(II) (monophosphate-hydrogenmonoborate-monophosphate), $PbZn[BP_2O_8(OH)]$, *Z. Kristallogr., New. Cryst. Str.*, **221**(4), 431–433 (2006).

Tassel Van R., Fransolet A.-M., Abraham K. Drugmanite, $Pb_2(Fe^{3+},Al)(PO_4)_2(OH)\cdot H_2O$, a new mineral from Richelle, Belgium, *Mineralog. Mag.*, **43**(328), 463–467 (1979).

Ten Kate O.M., Xie R.-J., Wang C.-Y., Funahashi S., Hirosaki N. Eu^{2+}-doped $Sr_2B_{2-2x}Si_{2+3x}Al_{2-x}N_{8+x}$: A boron-containing orange-emitting nitridosilicate with interesting composition-dependent photoluminescence properties, *Inorg. Mater.*, **55**(21), 11331–11336 (2016).

Tran Qui D., Hamdoune S. Structure of the orthorhombie phase of $Li_{1+x}Ti_{2-x}In_xP_3O_{12}$, $x = 1.08$, *Acta Crystallogr.*, **C44**(8), 1360–1362 (1988).

Tu H.Y., Sun W.Y., Wang P.L., Yan D.S. Glass-forming region in the Sm–Si–Al–O–N system, *J. Mater. Sci. Lett.*, **14**(16), 1118–1122 (1995).

Turkevich V., Taniguchi T., Andreev A., Itsenko P. Kinetics and mechanism of cubic boron nitride formation in the AlN–BN system at 6 GPa, *Diamond Relat. Mater.*, **13**(1), 64–68 (2004).

Tyagi P., Ramesh C., Kushvaha S.S., Mishra M., Gupta G., Yadav B.S., Kumar M.S. Dependence of Al incorporation on growth temperature during laser molecular beam epitaxy of $Al_xGa_{1-x}N$ epitaxial layers on sapphire (0001), *J. Alloys Compd.*, **739**, 122–128 (2018).

Urushihara D., Kaga M., Asaka T., Nakano H., Fukuda K. Synthesis and structural characterization of Al7C3N3-homeotypic aluminium oxycarbonitride $(Al_{7-x}Si_x)(O_yC_zN_{6-y-z})$ ($x \sim 1.2$, $y \sim 1.0$, and $z \sim 3.5$), *J. Solid State Chem.*, **184**(8), 2278–2284 (2011).

Uvarova Y., Gagne O.C., Belakovskiy D.I. New mineral names, *Am. Mineralog.*, **100**(1), 334–339 (2015).

Van Wambeke L. Eylettersite, un nouveau phosphate de thorium appartenant à la série de la crandallite, *Bull. Soc. fr. Minéral.*, **95**, 98–105 (1972).

Vasil'yev M.G., Vigdorovich V.N., Selin A.A., Khanin V.A. Phase equilibria in the Sn–In–P, Sn–Ga–In–As and Sn–Ga–In–As–P systems [in Russian], *Legir. Poluprovodn. Materialy*, Moscow, Nauka Publish., 61–65 (1985).

Vergasova L.P., Krivovichev S.V., Britvin S.N., Burns P.C., Ananiev V.V. Filatovite, $K[(Al,Zn)_2(As,Si)_2O_8]$, a new mineral species from the Tolbachik volcano, Kamchatka Peninsula, Russia, *Eur. J. Mineral.*, **16**(3), 533–536 (2004).

Vignola P., Hatert F., Baijot M., Dal Bo F., Andò S., Bersani D., Risplendente A., Vanini F. Arrojadite-(BaNa), IMA 2014-071. CNMNC Newsletter No. 23, February 2015, page 55, *Mineralog. Mag.*, **79**(1), 51–58 (2015).

Vochten R., Pelsmaekers J. Synthesis, solubility, electrokinetic properties and refined crystallographic data of sabugalite, *Phys. Chem. Minerals*, **9**(1), 23–29 (1983).

Voloshin A.V., Pakhomovskiiy Y.A., Tyusheva F.N. Lun'okite, a new phosphate, the manganese analogue of overite, from granitic pegmatites of the Kola Peninsula [in Russian], *Zap. Vses. mineralog. obshch.*, **112**(2), 232–237 (1983).

Wade R.C., Sullivan E.A., Berschied, Jr. J.R., Purcell K.F. Synthesis of sodium cyanotrihydroborate and sodium isocyanotrihydroborate, *J. Am. Chem. Soc.*, **9**(9), 2146–2150 (1970).

Wakabayashi R.H., Abruña H.D., DiSalvo F.J. Synthesis and electrochemical characterization of $Ti_xTa_yAl_zN_{1-\delta}O_\gamma$ for fuel cell catalyst supports, *J. Solid State Chem.*, **246**, 293–301 (2017).

Walenta K. Cualstibit, ein neues Sekundärmineral aus der Grube Clara im mittleren Schwarzwald (BRD), *Chem. Erde*, **43**, 255–260 (1984).

Walenta K., Dun P.J. Arsenogoyazit, ein neues Mineral der Crandallitgruppe aus dem Schwarzwald, *Schweiz. Mineral. Petrog. Mitt.*, **64**(1–2), 11–19 (1984).

Walenta K. Mineralien der Beudantit-Crandallitgruppe aus dem Schwarzwald: Arsenocrandallit und sulfatfreier Weilerit, *Schweiz. Mineral. Petrog. Mitt.*, **61**, 23–35 (1981).

Walenta K. Uranospathite and arsenuranospathite, *Mineralog. Mag.*, **42**(321), 117–128 (1978).

Wang G., Mudring A.-V. A new open-framework iron borophosphate from ionic liquids: $KFe[BP_2O_8(OH)]$, *Crystals*, **1**(1), 22–27 (2011).

Wang G., Valldor M., Lorbeer C., Mudring A.-V. Ionothermal synthesis of the first luminescent open-framework manganese borophosphate with switchable magnetic properties, *Eur. J. Inorg. Chem.*, **2012**(18), 3032–3038 (2012a).

Wang J., Cao W., Wang L., Zhu S., Guan S., Huang L., Li R., Zhang T. Fe–Al–P–C–B bulk metallic glass with good mechanical and soft magnetic properties, *J. Alloys Compd.*, **637**, 5–9 (2015).

Wang M.-Z. Crystal structure of potassium diaquamanganese(II) borophosphate monohydrate, $K[Mn(H_2O)_2(BP_2O_8)]\cdot H_2O$, *Z. Kristallogr., New Cryst. Str.*, **227**(1), 3–4 (2012).

Wang P.L., Tu H.Y., Sun W.Y., Yan D.S., Nygren M., Ekström T. Study on the solid solubility of Al in the melilite systems $R_2Si_{3-x}Al_xO_{3+x}N_{4-x}$ with R = Nd, Sm, Gd, Dy and Y, *J. Eur. Ceram. Soc.*, **15**(7), 689–695 (1995).

Wang P.-L., Werner P.-E., Gao L., Harris R.K., Thompson D.P. Ordering of nitrogen and oxygen in nitrogen-containing melilites $Y_2Si_3O_3N_4$ and $Nd_2Si_{2.5}Al_{0.5}O_{3.5}N_{3.5}$, *J. Mater. Chem.*, **7**(10), 2127–2130 (1997).

Wang P.L., Werner P.-E. Study on the structures of N-containing melilite $Y_2Si_3O_3N_4$ and $Nd_2Si_{2.5}Al_{0.5}O_{3.5}N_{3.5}$, *J. Mater. Sci.*, **32**(7), 1925–1929 (1997).

Wang S., Alekseev E.V., Depmeier W., Albrecht-Schmitt T.E. Further insights into intermediate- and mixed-valency in neptunium oxoanion compounds: Structure and absorption spectroscopy of $K_2[(NpO_2)_3B_{10}O_{16}(OH)_2(NO_3)_2]$, *Chem. Commun.*, **46**(22), 3955–3957 (2010).

Wang S., Li X., Fischer A.M., Detchprohm T., Dupuis R.D., Ponce F.A. Crystal structure and composition of BAlN thin films: Effect of boron concentration in the gas flow, *J. Cryst. Growth*, **475**, 334–340 (2017).

Wang Y., Pan S., Shi Y. Further examples of the P–O–P connection in borophosphates: Synthesis and characterization of $Li_2Cs_2B_2P_4O_{15}$, $LiK_2BP_2O_8$, and $Li_3M_2BP_4O_{14}$ (M = K, Rb), *Chem. Eur. J.*, **18**(38), 12046–12051 (2012b).

Wang Y., Pan S., Zhang M., Han S., Su X., Dong L. Synthesis, crystal growth and characterization of a new noncentrosymmetric borophosphate: $RbPbBP_2O_8$, *CrystEngComm*, **15**(24), 4956–4962 (2013).

Watanabe H, Kijima N. Crystal structure and luminescence properties of $Sr_xCa_{1-x}AlSiN_3:Eu^{2+}$ mixed nitride phosphors, *J. Alloys Compd.*, **475**(1–2), 434–439 (2009).

Watanabe H., Wada H., Seki K., Itou M., Kijima N. Synthetic method and luminescence properties of $Sr_xCa_{1-x}AlSiN_3:Eu^{2+}$ mixed nitride phosphors, *J. Electrochem. Soc.*, **155**(3), F31–F36 (2008a).

Watanabe H., Yamane H., Kijima N. Crystal structure and luminescence of $Sr_{0.99}Eu_{0.01}AlSiN_3$, *J. Solid State Chem.*, **181**(8), 1848–1852 (2008b).

Wei C.H., Edgar J.H. Thermodynamic analysis of $Ga_xB_{1-x}N$ growth by MOVPE, *J. Cryst. Growth*, **217**(1–2), 109–114 (2000).

Wei C.H., Xie Z.Y., Edgar J.H., Zeng K.C., Lin J.Y., Jiang H.X., Chaudhuri J., Ignatiev C., Braski D.N. MOCVD growth of GaBN on 6H-SiC (0001) substrates, *J. Electron. Mater.*, **29**(4), 452–456 (2000).

Wei C.H., Xie Z.Y., Edgar J.H, Zeng K.C., Lin J.Y., Jiang H.X., Ignatiev C., Chaudhuri J., Braski D.N. Growth and characterization of $B_xGa_{1-x}N$ on 6H-SiC (0001) by MOVPE, *MRS Internet J. Nitride Semicond.*, **4**(S1), 429–434 (1999).

Welch M.D., Smith D.G.W., Camara F., Gatta G.D. New mineral names, *Am. Mineralog.*, **98**(1), 279–282 (2013).

Williams S.A. Curetonite – a new phosphate from Nevada, *Mineralog. Rec.*, **10**(4), 219–221 (1979).

Williams S.A. Luetheite, $Cu_2Al_2(AsO_4)_2(OH)_4{\cdot}H_2O$, a new mineral from Arizona, compared with chenevixite, *Mineralog. Mag.*, **41**(317), 27–32 (1977).

Williams S.A. Zapatalite, a new mineral from Sonora, Mexico, *Mineralog. Mag.*, **38**(297), 541–544 (1972).

Williams S. Mendozavilite and paramendozavilite, two new minerals from Cumobabi, Sonora, *Boletín de Mineralogía*, **2**(1), 13–19 (1986).

Wilson M.J., Bain D.C. Spheniscidite, a new phosphate mineral from Elephant Island, British Antarctic Territory, *Mineralog. Mag.*, **50**(356), 291–293 (1986).

Winkler B., Hytha M., Hantsch U., Milman V. Theoretical study of the structures and properties of $SrSiAl_2O_3N_2$ and $Ce_4[Si_4O_4N_6]O$, *Chem. Phys. Let.*, **343**(5–6), 622–626 (2001).

Wise W.S., Loh S.E. Equilibria and origin of minerals in the system Al_2O_3–$AlPO_4$–H_2O, *Am. Mineralog.*, **61**(5–6), 409–413 (1976).

Wise W.S. Parnauite and goudeyite, two new copper arsenate minerals from the Majuba Hill Mine, Pershing County, Nevada, *Am. Mineralog.*, **63**(7–8), 704–708 (1978).

Wittig G., Raff P. Darstellung von Lithium-monocyano-borhydrid, *Z. Naturforsch.*, **6B**(4), 225–226 (1951).

Wolfe C.W. The unit cell of dickinsonite, *Am. Mineralog.*, **26**(5), 338–342 (1941).

Womelsdorf H., Meyer H.-J. Über die Synthese von Erdalkalimetalldihalogeniden und die Strukturen von Ca_3Br_2CBN und Sr_3Cl_2CBN, *Z. anorg. allg. Chem.*, **620**(2), 258–261 (1994).

Wörle M., Muhr H.-J., Meyer zu Altenschildesche H., Nesper R. $Ca_{15}(CBN)_6(C_2)_2O$ – a new cubic structure type containing CBN4– anions together with acetylide and oxo anions, *J. Alloys Compd.*, **260**(1–2), 80–87 (1997).

Wu S., Wang S., Diwu J., Depmeier W., Malcherek T., Alekseev E.V., Albrecht-Schmidt T.E. Complex clover cross-sectioned nanotubules exist in the structure of the first uranium borate phosphate, *Chem. Commun.*, **48**(29), 3479–3481 (2012).

Xing H., Li Y., Su T., Xu J., Yang W., Zhu E., Yu J., Xu R. Spontaneous crystallization of a new chiral open-framework borophosphate in the ionothermal system, *Dalton Trans.*, **39**(7), 1713–1715 (2010).

Yakhontova L.K. New mineral smolianinovite [in Russian], *Dokl. AN SSSR*, **109**(4), 849–850 (1956).

Yakovenchuk V.N., Keck E., Krivovichev S.V., Pakhomovsky Y.A., Selivanova E.A., Mikhailova J.A., Chernyatieva A.P., Ivanyuk G.Y. Whiteite-(CaMnMn), $CaMnMn_2Al_2[PO_4]_4(OH)_2 \cdot 8H_2O$, a new mineral from the Hagendorf-Süd granitic pegmatite, Germany, *Mineralog. Mag.*, **76**(7), 2761–2771 (2012).

Yakubovich O.V., Kotelnikov A.R., Suk N.I., Chukanov N.V. The novel arsenate $Na_3Al_5O_2(OH)_2(AsO_4)_4$: Crystal structure and topological relations with minerals of the palermoite and carminite–attakolite families, *Eur. J. Mineral.*, **28**(1), 83–91 (2016).

Yakubovich O.V., Steele I.M., Chernyshev V.V., Zayakina N.V., Gamyanin G.N., Karimova O.V. The crystal structure of arangasite, $Al_2F(PO_4)(SO_4) \cdot 9H_2O$ determined using low-temperature synchrotron data, *Mineralog. Mag.*, **78**(4), 889–903 (2014a).

Yakubovich O.V., Steele I.M., Dimitrova O.V. $Na(H_2O)[Mn(H_2O)_2(BP_2O_8)]$: Crystal structure refinement, *Crystallogr. Rep.*, **54**(1), 13–18 (2009a).

Yakubovich O., Steele I., Karimova O. Crystal structure interconnections in a family of hydrated phosphate-sulfates, *Acta Crystallogr.*, **A70**(Suppl.), C1091 (2014b).

Yakubovich O.V., Steele I.M., Massa W., Dimitrova O.V. 12-Membered borophosphate rings in $KNi_5[P_6B_6O_{23}(OH)_{13}]$, *Acta Crystallogr.*, **C65**(12), i94–i98 (2009b).

Yakubovich O.V., Steele I.M., Dimitrova O.V. Polymorphism of the borophosphate anion in $K(Fe,Al)[BP_2O_8(OH)]$ and $Rb(Al,Fe)[BP_2O_8(OH)]$ crystal structures, *Crystallogr. Rep.*, **55**(5), 760–768 (2010).

Yamane H., Shimooka S., Uheda K. Synthesis, crystal structure and photoluminescence of a newEu-doped Sr containing sialon $(Sr_{0.94}Eu_{0.06})(Al_{0.3}Si_{0.7})_4(N_{0.8}O_{0.2})_6$, *J. Solid State Chem.*, **190**, 264–270 (2012).

Yamane H., Yoshimura F. Synthesis, crystal structure and photoluminescence of $(Ba_{0.99}Eu_{0.01})Al_3Si_4N_9$, *J. Solid State Chem.*, **228**, 258–265 (2015).

Yan C., Zhao M., Gao J., Yao J. Crystal structure of yttrium gallium antimonide, $Y_5Ga_{1.24}Sb_{2.77}$, *Z. Kristallogr., New Cryst. Str.*, **232**(2), 331–332 (2017).

Yang H., Downs R.T., Evans S.H., Morrison S.M., Schumer B.N. Lefontite, IMA 2014-075. CNMNC Newsletter No. 23, February 2015, page 55, *Mineralog. Mag.*, **79**(1), 51–58 (2015).

Yang M., Chen P., Welz-Biermann U. Synthesis, crystal structure, and characterization of a new metal borophosphate: $Pb^{II}_4\{Co_2[B(OH)_2P_2O_8](PO_4)_2\}Cl$, *Z. anorg. und allg. Chem.*, **636**(8), 1454–1460 (2010a).

Yang M., Li X., Yu J., Zhu J., Liu X., Chen G., Yan Y. $LiCu_2[BP_2O_8(OH)_2]$: A chiral open-framework copper borophosphate via spontaneous asymmetrical crystallization, *Dalton. Trans.*, **42**(18), 6298–6301 (2013).

Yang M., Xu F., Liu Q., Yan P., Liu X., Wang C., Welz-Biermann U. Chelated orthoborate ionic liquid as a reactant for the synthesis of a new cobalt borophosphate containing extra-large 16-ring channels, *Dalton Trans.*, **39**(44), 10571–10573 (2010b).

Yang M., Yan P., Xu F., Ma J., Welz-Biermann U. Role of boron-containing ionic liquid in the synthesis of manganese borophosphate with extra-large 16-ring pore openings, *Micropor. Mesopor. Mater.*, **147**(1), 73–78 (2012).

Yang M., Yu J., Chen P., Li J., Fang Q., Xu R. Synthesis and characterization of metalloborophosphates with zeotype ANA framework by the boric acid 'flux' method, *Micropor. Mesopor. Mater.*, **87**(2), 124–132 (2005).

Yang M., Yu J., Di J., Li J., Chen P., Fang Q., Chen Y., Xu R. Syntheses, structures, ionic conductivities, and magnetic properties of three new transition-metal borophosphates $Na_5(H_3O)\{M^{II}_3[B_3O_3(OH)]_3(PO_4)_6\}\cdot 2H_2O$ (M^{II} = Mn, Co, Ni), *Inorg. Chem.*, **45**(9), 3588–3593 (2006a).

Yang M., Yu J., Shi L., Chen P., Li G., Chen Y., Xu R. Synthesis, structure, and magnetic property of a new open-framework manganese borophosphate, $[NH_4]_4[Mn_9B_2(OH)_2(HPO_4)_4(PO_4)_6]$, *Chem. Mater.*, **18**(2), 476–481 (2006b).

Yang T., Li G., Ju J., Liao F., Xiong M., Lin J. A series of borate-rich metalloborophosphates $Na_2[M^{II}B_3P_2O_{11}(OH)]\cdot 0.67H_2O$ (M^{II} = Mg, Mn, Fe, Co, Ni, Cu, Zn): Synthesis, structure and magnetic susceptibility, *J. Solid State Chem.*, **179**(8), 2534–2540 (2006c).

Yang T., Ju J., Liao F., Sasaki J., Toyota N., Lin J. Chirality and ferromagnetism in $NiBPO_4(OH)_2$ containing helix edge-sharing NiO_6 chains, *J. Solid State Chem.*, **181**(6), 1110–1115 (2008a).

Yang T., Sun J., Li G., Eriksson L., Zou X., Liao F., Lin J. $Na_8[Cr_4B_{12}P_8O_{44}(OH)_4][P_2O_7]\cdot nH_2O$: A 3D borophosphate framework with spherical cages, *Chem. Eur. J.*, **14**(24), 7212–7217 (2008b).

Yang W., Li J., Na T., Xu J., Wang L., Yu J., Xu R. $(NH_4)_6[Mn_3B_6P_9O_{36}(OH)_3]\cdot 4H_2O$: A new open-framework manganese borophosphate synthesized by using boric acid flux method, *Dalton Trans.*, **40**(11), 2549–2554 (2011a).

Yang W., Li J., Pan Q., Jin Z., Yu J., Xu R. $Na_2[VB_3P_2O_{12}(OH)]\cdot 2.92H_2O$: A new open-framework vanadium borophosphate containing extra-large 16-ring pore openings and $12^8 16^6$ super cavities synthesized by using the boric acid flux method, *Chem. Mater.*, **20**(15), 4900–4905 (2008c).

Yang Y., Ni Y., Guo Y., Zhang Y., Liu J. First discovery of kulanite in China [in Chinese], *Yanshi Kuangwuxue Zashi*, **5**, 119–127 (1986).

Yang Y., Wu J., Wang Y., Zhu J., Liu R., Meng C. Synthesis, crystal structure and characterization of a new protonated magnesium borophosphate: $(H_3O)Mg(H_2O)_2[BP_2O_8]\cdot H_2O$, *Z. anorg. und allg. Chem.*, **637**(1), 137–141 (2011b).

Yilmaz A., Yildirim L.T., Bu X., Kizilyalli M., Stucky G.D. New zeotype borophosphates with chiral tetrahedral topology: $(H)_{0.5}M_{1.25}(H_2O)_{1.5}[BP_2O_8]\cdot H_2O$ (M = Co(II) and Mn(II)), *Cryst. Res. Technol.*, **40**(6), 579–585 (2005).

Yilmaz A., Bu X., Kizilyalli M., Stucky G.D. $Fe(H_2O)_2BP_2O_8\cdot H_2O$, a first zeotype ferriborophosphate with chiral tetrahedral framework topology, *Chem. Mater.*, **12**(11), 3243–3245 (2000).

Yong T., Yang H., Downs R.T. Ferrobobfergusonite, IMA 2017-006. CNMNC Newsletter No. 37, June 2017, page 739, *Mineralog. Mag.*, **81**(3), 737–742 (2017).

Yoshimoto M., Murata S., Chayahara A., Horino Y., Saraie J., Oe K. Metastable GaAsBi alloy grown by molecular beam epitaxy, *Jpn. J. Appl. Phys.*, **42**(10B), Pt. 2, L1235–L1237 (2003).

Yoshimura F., Yamane H., Nagasako M. Synthesis and crystal structure of a new aluminum-silicon-nitride phosphor containing boron, $Ba_5B_2Al_4Si_{32}N_{52}$:Eu, *J. Solid State Chem.*, **251**, 43–49 (2017).

Yoshimura F., Yamane H., Nagasako M. Synthesis, crystal structure, and luminescence properties of a new nitride polymorph, β-$Sr_{0.98}Eu_{0.02}AlSi_4N_7$, *J. Solid State Chem.*, **258**, 664–673 (2018).

Young E.J., Weeks A.D., Meyrowitz R. Coconinoite, a new uranium mineral from Utah and Arizona, *Am. Mineralog.*, **51**(5–6), 651–663 (1966).

Yue L., Chen X., Zhang Y., Zhang F., Wang L., Shao J., Wang S. Molecular beam epitaxy growth and optical properties of high bismuth content $GaSb_{1-x}Bi_x$ thin films, *J. Alloys Compd.*, **742**, 780–789 (2018).

Yuh S.J., Rockett T.J. The system CaO–Al_2O_3–P_2O_5–H_2O at 200°C, *J. Am. Ceram. Soc.*, **64**(8), 491–493 (1981).

Zabukovec Logar N., Mrak M., Kaučič V., Golobič A. Syntheses and structures of two ammonium zinc gallophosphates: Analcime and paracelsian analogs, *J. Solid State Chem.*, **156**(2), 480–486 (2001).

Zhang A.-Y., Zhang L.-N., Zheng J. Crystal structure of a lithium zinc borophosphate, *J. Henan Polytech. Univ. (Natural Sci.)*, **30**(1), 100–103 (2011).

Zhang H., Chen Z., Weng L., Zhou Y., Zhao D. Hydrothermal synthesis of new berylloborophosphates M^IBeBPO ($M^I = K^+$, Na^+ and NH_4^+) with zeolite ANA framework topology, *Micropor. Mesopor. Mater.*, **57**(3), 309–316 (2003).

Zhang L.-R., Zhang H., Borrmann H., Kniep R. Crystal structure of sodium vanadium(III) (monohydrogen-monophosphate-dihydrogenmonoborate-monophosphate), $NaV[BP_2O_7(OH)_3]$. *Z. Kristallogr., New Cryst. Str.*, **217**(4), 477–478 (2002).

Zhang Q., Chung I., Jang J.I., Ketterson J.B., Kanatzidis M.G. Chalcogenide chemistry in ionic liquids: Nonlinear optical wave-mixing properties of the double-cubane compound $[Sb_7S_8Br_2](AlCl_4)_3$, *J. Am. Chem. Soc.*, **131**(29), 9896–9897 (2009).

Zhang S., Holec D., Fu W.Y., Humphreys C.J., Moram M.A. Tunable optoelectronic and ferroelectric properties in Sc-based III-nitrides, *J. Appl. Phys.*, **114**(13), 133510_1–133510_11 (2013).

Zhang T., Liu F., Pang S., Li R. Ductile Fe-based bulk metallic glass with good soft-magnetic properties, *Mater. Trans.*, **48**(5), 1157–1160 (2007).

Zhang W., Cheng W., Zhang H., Geng L., Li Y., Lin C., He Z. Syntheses and characterizations of $Cs_2Cr_3(BP_4O_{14})(P_4O_{13})$ and $CsFe(BP_3O_{11})$ compounds with novel borophosphate anionic partial structures, *Inorg. Chem.*, **49**(5), 2550–2556 (2010).

Zhang W.-L., He Z.-Z., Xia T.-L., Luo Z.-Z., Zhang H., Lin C.-S., Cheng W.-D. Syntheses and magnetic properties study of isostructural $BiM_2BP_2O_{10}$ (M = Co, Ni) containing a quasi-1D linear chain structure, *Inorg. Chem.*, **51**(16), 8842–8847 (2012).

Zhang W., Li C., Du Z. A thermodynamic database of the Al–Ga–In–P–As–Sb–C–H System and its application in the design of an epitaxy process for III-V semiconductors, *J. Phase Equilibr.*, **22**(4), 475–481 (2001).

Zhang W., Miao H., Li Y., Chang C., Xie G., Jia X. Glass-forming ability and thermoplastic formability of ferromagnetic $(Fe,Co,Ni)_{75}P_{10}C_{10}B_5$ metallic glasses, *J. Alloys Compd.*, **707**, 57–62 (2017).

Zhang Y., Franke P., Seifert H.J. CALPHAD modeling of metastable phases and ternary compounds in Ti–Al–N system, *Calphad*, **59**, 142–153 (2017).

Zhao C., Ma Q., Xu H., Chen K., Hu Y., Ye H., Liu R., Liu Y., Zhou X. Photoluminescence and thermal quenching properties of tunable green-emitting oxynitride $(Ca_xSr_{1-x})_6Si_{25.6}Al_{6.4}N_{41.6}O_{4.4}{:}Eu^{2+}$ phosphors, *J. Rare Earths*, **34**(5), 470–474 (2016).

Zhao D., Cheng W.-D., Zhang H., Huang S.-P., Xie Z., Zhang W.-L., Yang S.-L. $KMBP_2O_8$ (M = Sr, Ba): A new kind of noncentrosymmetry borophosphate with thr three-dimensional diamond-like framework, *Inorg. Chem.*, **48**(14), 6623–6629 (2009).

Zheng J., Zhang A. An open-framework borophosphate, $LiCu_2BP_2O_8(OH)_2$, *Acta Crystallogr.*, **E65**(5), i40 (2009a).

Zheng J., Zhang A. Lithium diaquanickel(II) *catena*-borodiphosphate(V) monohydrate, *Acta Crystallogr.*, **E65**(6), i42 (2009b).

Zhesheng M., Nicheng S., Zhizhong P. Crystal structure of a new phosphatic mineral – qingheiite, *Sci. Sinica, Sér. B*, **26**(8), 876–884 (1983).

Zhou J., Zhong J., Chen L., Zhang L., Du Y., Liu Z.-K., Mayrhofer P.H. Phase equilibria, thermodynamics and microstructure simulation of metastable spinodal decomposition in c–$Ti_{1-x}Al_xN$ coatings, *Calphad*, **56**, 92–101 (2017).

Zhou L., Li H., Zeng Y. Effects of annealing treatment on the microstructure and mechanical properties of the AlN–SiC–TiB_2 ceramic composites prepared by SHS-HIP, *J. Alloys Compd.*, **552**, 499–503 (2013).

Zhou Y., Hoffmann S., Huang Y.-X., Prots Y., Schnelle W., Menezes P.W., Carrillo-Cabrera W., Sichelschmidt J., Mi J.-X., Kniep R. $K_3Ln[OB(OH)_2]_2[HOPO_3]_2$ (*Ln* = Yb, Lu): Layered rare-earth dihydrogen borate monohydrogen phosphates, *J. Solid State Chem.*, **184**(6), 1517–1522 (2011a).

Zhou Y., Hoffmann S., Menezes P.W., Carrillo-Cabrera W., Huang Y.-X., Vasylechko L., Schmidt M., Prots Yu., Deng J.-F., Mi J.-X., Kniep R. Nanoporous titanium borophosphates with rigid gainesite-type framework structure, *Chem. Commun.*, **47**(42), 11695–11696 (2011b).

Zhuang R.-C., Chen X.-Y., Mi J.-X. Lithium manganese(II) diaquaborophosphate monohydrate, *Acta Crystallogr.*, **E64**(8), i46 (2008).

Zouihri H., Saadi M., Jaber B., El Ammari L. Silver(I) diaqucobalt(II) *catena*-borodiphosphate(V) hydrate, $(Ag_{0.79}Co_{0.11})Co(H_2O)_2[BP_2O_8]{\cdot}0.67H_2O$, *Acta Crystallogr.*, **E68**(1), i3 (2012).

Zouihri H., Saadi M., Jaber B., El Ammari L. Silver(I) diaquamagnesium catena-borodiphosphate(V) monohydrate, $AgMg(H_2O)_2[BP_2O_8]{\cdot}H_2O$, *Acta Crystallogr.*, **E67**(6), i39 (2011a).

Zouihri H., Saadi M., Jaber B., El Ammari L. Silver(I) diaquanickel(II) catena-borodiphosphate(V) hydrate, $(Ag_{0.57}Ni_{0.22})Ni(H_2O)_2[BP_2O_8]{\cdot}0.67H_2O$, *Acta Crystallogr.*, **E67**(8), i44 (2011b).

Zubkova N.V., Pushcharovskiĭ D.Y., Kabalov Y.K., Kazantsev S.S., Voloshin A.V. Crystal structure of holtite II, *Crystallogr. Rep.*, **51**(1), 16–22 (2006).

Index

A

B

C

L

M

P

R

T

Y

Z